A
COMPLETE GCSE
MATHEMATICS

GENERAL COURSE

Second Edition

A COMPLETE GCSE MATHEMATICS

GENERAL COURSE

A. Greer

Formerly Senior Lecturer,
Gloucestershire College of Arts and Technology

SECOND EDITION

Stanley Thornes (Publishers) Ltd

© A. Greer 1986, 1989
© Diagrams ST(P) Ltd. 1986, 1989
Chapter 38 © C.H. Hopkins, 1989

First published in 1986 by
Stanley Thornes (Publishers) Ltd
Old Station Drive
Leckhampton
CHELTENHAM GL53 0DN

Reprinted 1987 (three times)
Reprinted 1988
Second Edition 1989

British Library Cataloguing in Publication Data

Greer, A. (Alex)
    A complete GCSE mathematics: a general
    course—2nd edition
    1. Mathematics
    510

    ISBN 0-7487-0075-7 (paperback)
    ISBN 0-7487-0102-8 (casebound)

Typeset by Tech-Set, Gateshead, Tyne & Wear in 10/12pt Century.
Printed and bound in Great Britain at The Bath Press, Avon.

Contents

Preface

This book is one of a series of three books intended to cover all the topics prescribed by the National Criteria in Mathematics at three levels. It is suitable for those students who expect to obtain a grade C, D or E in the General Certificate of Secondary education (GCSE), the target grade being D. Also I have tried to cover the additional topics introduced by the five regional Examining Groups.

Because this is a revision book, the sections on Arithmetic, Algebra, Geometry, Graphical Work and Statistics have been dealt with separately. The emphasis is on a simple approach and a teacher and the class may work, if desired, through the book chapter by chapter.

A large number of examples and exercises has been included which supplement and amplify the text. Many practical examples of mathematics in the real world have been used.

At the end of most chapters there is a 'miscellaneous exercise' which has been divided into two parts, Section A and Section B. This is because most Examining Groups have devised overlapping examination papers. The basic-level students will take papers 1 and 2, the general-level students will take papers 2 and 3 and the higher-level students will take papers 3 and 4. In the miscellaneous exercises, Section B in the Basic book appears as Section A in the General book, while Section B in the General book appears as Section A in the Higher book.

Also at the end of many chapters are sets of multi-choice questions and mental tests. It is hoped that these, together with the miscellaneous exercises, will give the students confidence in answering examination questions.

I would like to thank my son David for working out the answers to the exercises.

A. Greer   Gloucester 1988

Coursework

We are aware of the demands made upon GCSE candidates and it is not suggested that they should attempt all the tasks which are offered in the chapter on coursework. The investigations and other studies are provided as a guide for those students who are not sure what is expected of them in this part of the examination.

The readers may like to choose a few of the suggestions for practice. We hope the topics will provide some inspiration when candidates attempt their own particular coursework projects.

I was pleased to accept the author's invitation to provide a section dealing with coursework and it is a source of personal satisfaction to be associated with this very successful series of books.

C.H. Hopkins   Gloucester 1989

Acknowledgements

The author and publishers are grateful to the following who provided material for inclusion:

Cheltenham and Gloucester Omnibus Co. Ltd. for the timetable (1980), page 112.

British Rail for the timetable, page 107.

1 Operations in Arithmetic

Types of Numbers

Counting numbers or **natural numbers** are the numbers 1, 2, 3, 4, 5, etc.

Whole numbers are the numbers 0, 1, 2, 3, 4, 5, etc.

Positive numbers are numbers which are greater than zero. They either have no sign in front of them or a + sign. Some examples of positive numbers are 5, +9, 8, +21, +293 and 564.

Negative numbers have a value less than zero. They are always written with a − sign in front of them. Some examples of negative numbers are −4, −15, −168 and −783.

Directed numbers or **integers** are whole numbers but they include negative numbers. Some examples of integers are −28, −15, 0, 2, +9, 24, +371 and −892.

Directed numbers are used on thermometers to show Celsius temperatures above

Fig. 1.1

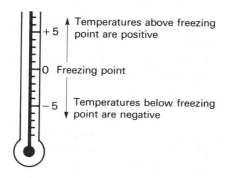

and below freezing point (Fig. 1.1). They are also used on graphs (Fig. 1.2).

Fig. 1.2

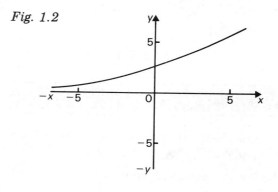

It is also possible to have negative time, for example −12 seconds to blast off.

An overdraft at the bank can be regarded as negative money, for instance −£200 would represent an overdraft of £200 (i.e. the customer owes the bank £200).

The Number Line

Directed numbers may be represented on a number line (Fig. 1.3). Numbers to the right of zero are positive whilst those to the left of zero are negative numbers.

Fig. 1.3

Negative numbers become smaller in value the further we move to the left of zero. For instance −8 is smaller in value than −3 and −56 is smaller in value than −27.

Positive numbers, however, become greater in value the further we move to the right of zero. For instance 12 has a greater value than 7 and +89 has a greater value than +34.

Example 1

Rearrange in order of size, smallest first, the numbers

$$-3, 5, -7, -2, 8, -4 \text{ and } 3$$

The order is:

$$-7, -4, -3, -2, 3, 5 \text{ and } 8$$

Exercise 1.1

1. On a number line represent the numbers −6, 3, +7, −8, −2, 0, 5 and −4.

2. Which has the greater value, −6 or +3?

3. Write down all the positive numbers from the following: −2, 7, +5, −6, 3, +2 and −3.

4. Write down all the whole numbers from the following set: 7, −4, 3, −6, +2, 0, −9, 5 and −2.

5. Rearrange in order of size, smallest first, the numbers −3, −5, −1, −9 and −4.

6. Rearrange in order of size, largest first, the numbers 3, −8, −15, +25, 18, −6, −2 and 12.

7. From the set of numbers −6, 8, +5, −10, 0, 12, −8 and +7, pick out (a) the largest number, (b) the smallest number, (c) all the natural numbers.

8. Which number is the smaller, −5 or −3?

9. Which of the following numbers are positive integers: +5, −3, 0, −1, 7, +11 and −15?

10. On a Celsius thermometer a temperature of −8 degrees is recorded. Is the temperature below the freezing point of water?

Rounding Off

Calculators save a lot of time and effort when used for addition, subtraction, multiplication and division. We do, however, need to be able to estimate the size of the answer we expect as a check on our use of the calculator.

One way of doing this is to round off the numbers involved. It is generally good enough to round off

> a number between 10 and 100 to the nearest number of tens
>
> a number between 100 and 1000 to the nearest number of hundreds
>
> a number between 1000 and 10 000 to the nearest number of thousands

and so on.

Example 2

(a) Round off the number 67 to the nearest number of tens.

> 67 is roughly 7 tens so 67 is rounded up to 70

(b) Round off 362 to the nearest number of hundreds.

> 362 is roughly 4 hundreds so 362 is rounded up to 400

(c) Round off 8317 to the nearest number of thousands.

> 8317 is roughly 8000 so 8317 is rounded down to 8000

(d) Round off 75 to the nearest number of tens.

> 75 lies exactly half way between 70 and 80. In such cases we always round up so that 75 is rounded up to 80

Round off the following numbers to the nearest number of tens:

1. 42   2. 63   3. 85
4. 92   5. 76

Round off the following numbers to the nearest number of hundreds:

6. 698   7. 703   8. 886
9. 350   10. 885

Round off the following numbers to the nearest number of thousands:

11. 2631   12. 5378   13. 8179
14. 8500   15. 5765

Adding Numbers

$+$ is the addition sign, meaning plus. Thus $5 + 3 = 8$, which is read as 'five plus three equals eight'. The result obtained by adding numbers is called their **sum**. Thus the sum of 5, 6 and 8 is $5 + 6 + 8 = 19$. *The order in which numbers are added does not affect their sum.*

$$4 + 6 + 8 = 6 + 4 + 8$$
$$= 8 + 6 + 4$$
$$= 18$$

This rule, which is known as the **commutative law,** is very useful when adding numbers mentally:

$$15 + 8 + 5 = (15 + 5) + 8$$
$$= 20 + 8$$
$$= 28$$

It is also useful for checking the result of an addition:

$$
\begin{array}{r}
315 \\
17 \\
692 + \\
\hline
1024
\end{array}
\qquad
\begin{array}{r}
315 \\
17 \\
692 + \\
\hline
1024
\end{array}
$$

Add upwards   Check downwards

More often than not a calculator will be used to add numbers greater than 10.

Example 3

Using a calculator find the sum of 278, 89 and 3924.

For a rough estimate we will round off the numbers to

$$300 + 90 + 4000 = 4390$$

Input	Display
278	278.
+	278.
89	89.
+	367.
3924	3924.
=	4291.

The rough estimate shows that the answer, it is 4291, is sensible because the rough estimate and the answer are of the same size.

The whole object of doing arithmetic is to produce correct answers. It is pointless to produce wrong answers. Therefore all calculations must be checked. When using a calculator we make use of the commutative law and input the numbers in reverse order.

Input	Display
3924	3924.
+	3924.
89	89.
+	4013.
278	278.
=	4291.

Since the same answer has been produced twice we can be confident that

$$278 + 89 + 3924 = 4291$$

Exercise 1.3

Questions 1 to 5 inclusive should be done mentally.

1. Find the sum of 3, 7 and 6.

2. Add 6, 8 and 4.

3. Find the total of 12, 7 and 8.

4. Add 3, 6, 7 and 4.

5. What is the sum of 2, 5, 8 and 15?

Work out the following and check the answers:

6. $96 + 348 + 18$

7. $219 + 69 + 3487 + 935$

8. $35\,078 + 21\,009 + 807 + 1288 + 43$

9. $23\,589 + 7\,986\,232 + 245\,899 + 7892 + 335\,681$

10. $17\,438 + 1344 + 2636 + 107\,924 + 11\,358$

Adding Integers

To add integers we make use of the number line. Thus in Fig. 1.4 to find the value of $+3 + 4$, which is usually written $3 + 4$, we measure 3 units to the right of 0 to represent $+3$ and from this point we measure 4 units to the right to represent $+4$. We see from the diagram that

$$3 + 4 = 7$$

Fig. 1.4

To find the value of $-3 + (-2)$, which is usually written $-3 - 2$, we measure 3 units to the left of 0 (Fig. 1.5) to represent -3.

Then from this points we measure 2 units to the left to represent -2. We see from the diagram that

$$-3 - 2 = -5$$

Fig. 1.5

We see that the rules for the addition of integers having the same sign are as follows:

(1) If all the numbers to be added are positive then add them together and place a + sign (or no sign) in front of the sum.

(2) If all the numbers to be added are negative then add the numbers together and place a − sign in front of the sum.

Example 4

(a) $3 + 5 + 6 + 2 = 16$

(b) $-4 - 6 - 9 - 3 = -22$

To find $4 - 6$ we proceed as shown in Fig. 1.6 and we see that

$$4 - 6 = -2$$

Fig. 1.6

When the signs are different:

(1) For two numbers find the difference between them and place the sign of the larger number in front of the difference.

(2) If there are more than two numbers add the positive numbers together. Next add the negative numbers together. The set of numbers is thereby reduced to two numbers, one positive and the other negative. We then proceed as for two numbers.

Example 5

(a) $8 - 5 = 3$

(b) $-5 + 8 = 3$

(c) $2 - 6 = -4$

(d) $-9 + 3 = -6$

Example 6

$-5 + 8 - 7 + 3 + 6 - 9 + 4 + 1 - 2$

$$= (8 + 3 + 6 + 4 + 1)$$
$$\quad - (5 + 7 + 9 + 2)$$
$$= 22 - 23$$
$$= -1$$

When using a calculator we can input positive and negative numbers as they occur. There is no need to add the positive and negative numbers separately.

Example 7

Use a calculator to work out $72 - 43 - 59 + 34 + 112 - 89$.

Rough estimate, $70 - 40 - 60 + 30 + 110 - 90 = 20$

Input	Display
72	72.
$-$	72.
43	43.
$-$	29.
59	59.
$+$	$-30.$
34	34.
$+$	4.
112	112.
$-$	116.
89	89.
$=$	27.

The rough estimate shows that the answer, 27, is sensible.

The answer produced by the machine can be checked by making use of the commutative law and inputting the numbers in reverse order.

Input	Display
$-$	0.
89	89.
$+$	$-89.$
112	112.
$+$	23.
34	34.
$-$	57.
59	59.
$-$	$-2.$
43	43.
$+$	$-45.$
72	72.
$=$	27.

Since the same answer has been produced by two different methods we can be confident that

$$72 - 43 - 59 + 34 + 112 - 89 = 27$$

Exercise 1.4

Questions 1 to 10 inclusive should be done mentally. Find the value of each of the following:

1. $8 + 5$

2. $-6 - 3$

3. $-14 - 18$

4. $-7 - 6 - 3$

5. $8 - 12$

6. $9 - 17$

7. $8 - 7 - 15$

8. $-6 + 10$

9. $-3 + 5 + 7 - 4 - 2$

10. $6 + 4 - 3 - 5 - 7 + 2$

Use a calculator to work out the following, in each case making a rough estimate of the answer:

11. $-40 - 23 + 72 + 15 - 18$

12. $324 - 78 - 169 - 89 + 54 + 178 - 26$

13. $-63 - 98 + 37 - 75 + 19 - 28 + 68 - 87$

14. $152 - 78 + 43 - 81 - 128 + 57 - 63$

15. $278 + 452 + 96 + 73 - 156 - 238 - 417 - 82 + 175$

Subtracting Numbers

$-$ is the subtraction sign meaning minus. Thus $8 - 3 = 5$, which is read as eight minus three equals five.

The difference between two numbers is the larger number minus the smaller number.

Example 8

(a) Subtract 7 from 11.
$$11 - 7 = 4$$

(b) Take 2 from 8.
$$8 - 2 = 6$$

(c) Find the difference between 5 and 12.
$$\text{Difference} = 12 - 5 = 7$$

Example 9

Subtract 87 from 395.

Rough estimate, $400 - 90 = 310$

$$\begin{array}{r} 395 \\ 87 \, - \\ \hline 308 \end{array}$$

The rough estimate shows that the answer is sensible.

The calculation can be checked by adding the bottom two numbers. If this sum equals the top number then the calculation is correct.

$$\begin{array}{r} 87 \\ 308 \, + \\ \hline 395 \end{array}$$

Most subtractions will be done on a calculator as shown in Example 10.

Example 10

Take 2987 from 12 613.

Rough estimate,
$13\,000 - 3000 = 10\,000$

Input	Display
12613	12613.
$-$	12613.
2987	2987.
$=$	9626.

Hence $12\,613 - 2987 = 9626$.

The rough estimate shows that the answer is sensible.

To check the answer we add 2987 and 9626 to see if this sum equals 12 613.

Input	Display
2987	2987.
$+$	2987.
9626	9626.
$=$	12613.

Subtracting Integers

Suppose we want to find the value of $-(+5) - 4$. To represent $+5$ we measure 5 units to the right of 0. Therefore to represent $-(+5)$ we must reverse direction and measure 5 units to the left of 0 (Fig. 1.7).

Hence $-(+5)$ is the same as -5.

Therefore
$$-(+5) - 4 = -5 - 4$$
$$= -9$$

Fig. 1.7

To subtract a directed number change its sign and add the resulting number.

Example 11

(a) $5 - (-3) = 5 + 3 = 8$
(b) $-7 - (+4) = -7 - 4 = -11$
(c) $8 - (+5) = 8 - 5 = 3$

Questions 1 to 5 inclusive should be done mentally.

1. Subtract 5 from 9.
2. Take 3 from 8.
3. Find the difference between 7 and 4.
4. Work out $8 - 2$.
5. How many less is 4 than 9?

Use a calculator to work out each of the following, in each case making a rough estimate of the answer:

6. $247 - 89$
7. $1253 - 357$
8. $2788 - 1359$
9. $11\,913 - 9876$
10. $234\,582 - 98\,687$

Questions 11 to 15 inclusive should be done mentally. Find the value of each of the following:

11. $7 - (+6)$
12. $-5 - (-6)$
13. $5 - (-3)$
14. $-3 - (-4)$
15. $-10 - (-4)$

Multiplication

$\times$ is the multiplication sign, meaning multiplied by, or times. Thus $6 \times 8 = 48$ which is read as six times eight equals forty-eight.

Multiplication is a quick way of adding equal numbers.

$$7 + 7 + 7 + 7 + 7 = 35$$

$$5 \times 7 = 35 \quad \text{(i.e. five sevens equal thirty-five)}$$

35 is called the **product** of 5 and 7.

The product of 4, 5 and 8 is $4 \times 5 \times 8 = 160$.

The order in which the numbers are multiplied does not affect the product. Thus

$$3 \times 4 \times 6 = 6 \times 4 \times 3$$
$$= 4 \times 3 \times 6$$
$$= 72$$

This rule is called the **commutative law for multiplication**. It may be used:

(1) To simplify multiplication when done mentally:

$$4 \times 9 \times 25 = (4 \times 25) \times 9$$
$$= 100 \times 9$$
$$= 900$$

(2) To check a product by interchanging numbers:

$$28 \times 73 = 2044$$
$$73 \times 28 = 2044$$

More often than not a calculator will be used for multiplication because of the time taken when using pen and paper.

Example 12

Use a calculator to find the product of 356 and 98.

Rough estimate, $350 \times 100 = 35\,000$

Input	Display
356	356.
$\times$	356.
98	98.
$=$	34888.

Hence $356 \times 98 = 34\,888$.

The rough estimate shows that the answer is sensible.

To check the product produced by the machine multiply in the reverse order, i.e. 98×356.

Example 13

Work out $27 \times 15 \times 54 \times 29$.

Rough estimate, $30 \times 20 \times 50 \times 30 = 30 \times 30 \times 1000 = 900\,000$

Input	Display
27	27.
×	27.
15	15.
×	405.
54	54.
×	21870.
29	29.
=	634230.

Although there is a big difference between the answer and the rough estimate the rough estimate shows the answer is 634 230 and not 63 423 or 6 342 300.

Hence $27 \times 15 \times 54 \times 29 = 634\,230$.

This product may be checked by multiplying in the reverse order, i.e. $29 \times 54 \times 15 \times 27$.

Multiplication of Directed Numbers

$$4 + 4 + 4 = 12$$

$$3 \times 4 = 12$$

Hence two positive numbers multiplied together give a positive product.

$$-4 - 4 - 4 = -12$$

$$3 \times (-4) = -12$$

Hence a positive number multiplied by a negative number gives a negative product.

Suppose we wish to find the value of $(-3) \times (-4)$:

$$(-3) \times (4 - 4) = (-3) \times 0$$

$$= 0$$

$$(-3) \times 4 + (-3) \times (-4) = 0$$

$$-12 + (-3) \times (-4) = 0$$

$$\therefore \qquad (-3) \times (-4) = 12$$

Hence a negative number multiplied by a negative number gives a positive product.

Therefore the rule for multiplication is:

(1) Numbers with like signs give a positive product.

(2) Numbers with unlike signs give a negative product.

Example 14

(a) $2 \times 3 = 6$

(b) $(-2) \times 3 = -6$

(c) $2 \times (-3) = -6$

(d) $(-2) \times (-3) = 6$

(e) $(-3) \times (-5) \times (-8) = 15 \times (-8)$
$$= -120$$

(f) $(-2) \times 3 \times (-4) = (-6) \times (-4) = 24$

Exercise 1.6

Questions 1 to 5 inclusive should be done mentally.

1. Find values for
   (a) $2 \times 3 \times 5$
   (b) $4 \times 7 \times 5$

2. Multiply 5 by 8.

3. Find the products of
   (a) 3, 4 and 5
   (b) 5, 7 and 8

4. Work out $2 \times 3 \times 5 \times 7$.

5. What is the product of 2, 25, 5 and 4?

Using a calculator work out the following:

6. 42×75

7. 235×39

8. $18 \times 27 \times 35$

9. $8 \times 12 \times 15 \times 22$

10. $9 \times 11 \times 14 \times 16 \times 5$

Questions 11 to 18 inclusive should be done mentally.

Find values for each of the following:

11. $8 \times (-3)$ **12.** $(-5) \times 2$

13. $6 \times (-3)$ **14.** $(-2) \times (-4)$

15. $(-2) \times (-5) \times (-6)$

16. $4 \times (-3) \times (-2)$

17. $2 \times (-5) \times 4$

18. $(-3) \times (-4) \times 5$

Division

$\div$ is the division sign, meaning divided by.

There are several ways of writing division problems. 6 divided by 3 may be written:

(1) $6 \div 3$, which reads six divided by three.

(2) $\frac{6}{3}$, which reads six over three.

(3) $3\overline{)6}$, which reads three into six.

Division will generally be done by using a calculator.

Example 15

Divide 1554 by 42.

Rough estimate, $1600 \div 40 = 40$

Input	Display
1554	1554.
$\div$	1554.
42	42.
=	37.

Hence $1554 \div 42 = 37$.

The rough estimate shows that the answer is sensible.

The result of a division can always be checked by multiplying back. Thus if

$$1554 \div 42 = 37$$

then $42 \times 37 = 1554$

We can check that this is so by using a calculator in the usual way.

Remainders

Consider $17 \div 3$.

3 will not divide into 17 exactly and the answer is 5 remainder 2.

Note that $3 \times 5 + 2 = 15 + 2 = 17$.

For more difficult numbers the remainder may be found by using a calculator as shown in Example 16.

Example 16

Find the remainder when 10 798 is divided by 29.

Input	Display	
10 798	10 798.	
$\div$	10 798.	
29	29.	
$-$	372.34 482	(this is the answer to the division)
372	372	
$\times$	0.34 482	(this is the remainder in decimal form)
29	29.	
=	9.999 78	(this is the required remainder, see note below)

Hence $10\,798 \div 29 = 372$ remainder 10.

The remainder must be a whole number but because the calculator is limited to 8 digits calculations of this kind are never exactly right, and we expect small errors to occur. 9.999 78 is so near to 10 that this must be the remainder. Note that

$$372 \times 29 + 10 = 10\,798$$

Division of Directed Numbers

The rules for division must be similar to the rules for multiplication. Since

$$3 \times (-4) = -12$$

$$\frac{-12}{3} = -4$$

and $\frac{-12}{-4} = 3$

Therefore the rule is:

(1) Numbers with like signs, when divided, give a positive answer.

(2) Numbers with unlike signs, when divided, give a negative answer.

Example 17

(a) $\dfrac{8}{4} = 2$ (b) $\dfrac{8}{-4} = -2$

(c) $\dfrac{-8}{4} = -2$ (d) $\dfrac{-8}{-4} = 2$

Combined Multiplication and Division

Example 18

Work out $\dfrac{722 \times 420}{19 \times 168}$

Rough estimate, $\dfrac{700 \times 400}{20 \times 200} = 70$

Input	Display
722	722.
×	722.
420	420.
÷	303240.
19	19.
÷	15960.
168	168.
=	95.

Hence $\dfrac{722 \times 420}{19 \times 168} = 95.$

The rough estimate shows that the answer is sensible.

Again we can check the calculation by multiplying back.

If the correct answer is 95, then

$$722 \times 420 = 19 \times 168 \times 95$$

which can be verified by using a calculator.

Example 19

Work out $\dfrac{4 \times (-8) \times 3}{(-2) \times 6 \times (-4)}$

$$\dfrac{4 \times (-8) \times 3}{(-2) \times 6 \times (-4)} = \dfrac{-96}{48}$$

$$= -2$$

Exercise 1.7

Using a calculator work out the following. In each case make a rough estimate of the answer:

1. $1064 \div 8$ 2. $255 \div 17$

3. $567 \div 21$ 4. $32\overline{)512}$

5. $71\overline{)3053}$ 6. $61\overline{)3599}$

7. $\dfrac{378}{21}$ 8. $\dfrac{16\,308}{453}$

9. $\dfrac{49\,182}{2342}$

10. $2\,233\,700 \div 6382$

Questions 11 to 18 inclusive should be done mentally.

11. $8 \div (-2)$ 12. $(-9) \div 3$

13. $(-8) \div (-2)$ 14. $(-14) \div (-7)$

15. $10 \div (-5)$ 16. $(-12) \div (-4)$

17. $18 \div (-6)$ 18. $(-24) \div (-8)$

Work out each of the following:

19. $\dfrac{-16}{(-2) \times (-4)}$

20. $\dfrac{2 \times (-8) \times 6}{(-3) \times (-4) \times (-2)}$

21. $\dfrac{15 \times (-3) \times 2}{(-5) \times (-6)}$

22. $\dfrac{(-9) \times (-5) \times (-4)}{3 \times 15 \times (-2)}$

23. $\dfrac{648 \times 252}{108 \times 28}$ **24.** $\dfrac{651 \times 252}{27 \times 217}$

25. $\dfrac{79\,380 \times 30\,429}{49 \times 27}$

26. $\dfrac{90\,343 \times 16\,695}{43 \times 45}$

Work out each of the following and state the remainder:

27. $15 \div 2$ **28.** $23 \div 3$ **29.** $19 \div 4$

30. $11 \div 6$ **31.** $33 \div 5$ **32.** $47 \div 8$

33. $302 \div 11$ **34.** $601 \div 19$

35. $5631 \div 21$

Sequence of Arithmetical Operations

In working out problems such as

$$5 + (3 + 6)$$

or $\qquad 15 - (7 - 2)$

or $\qquad 2 \times 4 + 2$

or $\qquad 7 + 6 \div 2$

a certain sequence of operations must be observed.

(1) Work out the contents of any brackets first:

$$4 \times (3 + 5) = 4 \times 8$$
$$= 32$$
$$15 - (7 - 2) = 15 - 5$$
$$= 10$$

We get the same answer and sometimes make the work easier if we expand the bracket as shown below:

$$8 \times (14 + 29) = 8 \times 14 + 8 \times 29$$
$$= 112 + 232$$
$$= 344$$

Sometimes the multiplication sign is omitted.

$$5(8 - 4) \text{ means } 5 \times (8 - 4)$$
$$= 5 \times 4$$
$$= 20$$

(2) Multiply and divide before adding and subtracting:

$$3 \times 6 + 5 = 18 + 5$$
$$= 23$$
$$5 - 6 \div 2 = 5 - 3$$
$$= 2$$

Example 20

Find the value of $11 - 12 \div 4 + 3 \times (6 - 2)$.

$$11 - 12 \div 4 + 3 \times (6 - 2)$$
$$= 11 - 12 \div 4 + 3 \times 4$$

(by working out the contents of the bracket)

$$= 11 - 3 + 12$$

(by multiplying and dividing)

$$= 20$$

(by adding and subtracting)

Exercise 1.8

Work out the value of each of the following:

1. $7 + 4 \times 3$

2. $2 \times 6 - 3$

3. $5 \times 4 - 3 \times 6 + 5$

4. $8 \times 5 - 15 \div 5 + 7$

5. $3 + 6 \times (3 + 2)$

6. $9 - 4 \times (3 - 2)$

7. $10 - 12 \div 6 + 3(8 - 3)$

8. $15 \div (4 + 1) - 9 \times 3 + 7(4 + 3)$

9. $35 \div (25 - 20)$

10. $24 \div 4 - 24 \div 6$

11. $3(2 - 5) + 4$

12. $25 \div (-5) - 7$

13. $3 - (-12) \times 4 - 15 \div (-3)$

14. $36 - (-4) + 36 \div (-6)$

15. $4(2 - 8) + (-12) \div (-3)$

Some Important Facts

(1) When zero is added to any number the sum is the number.

$$873 + 0 = 873 \quad \text{and} \quad 0 + 74 = 74$$

(2) When any number is multiplied by 1 the product is the number.

$$9483 \times 1 = 9483 \quad \text{and} \quad 1 \times 534 = 534$$

(3) When a number is multiplied by zero the product is zero.

$$178 \times 0 = 0 \quad \text{and} \quad 0 \times 2398 = 0$$

(4) It is impossible to divide a number by zero. Thus, for instance, $8 \div 0$ is meaningless.

(5) When zero is divided by any number the result is zero.

$$0 \div 79 = 0 \quad \text{and} \quad 0 \div 7 = 0$$

Exercise 1.9

This exercise should be done mentally.

Work out the value of each of the following:

1. $768 + 0$ 2. 256×1

3. 0×50 4. $0 \times 4 \times 5$

5. $3 \times 6 \times 1$ 6. $7 \times 0 \times 6$

7. $16 - 8 \times 0 - 3$

8. $12 - 24 \div 6 + 0 \times (7 - 3)$

9. $0 \div 32$

10. $(5 - 5) \times 3$

11. $7 \times (3 - 9) + 0 \div 7$

12. $24 \times 0 + 3(7 - 7) - 5$

Sequences

A set of numbers connected by some definite law is called a **sequence of numbers**. Each of the numbers in the sequence is called a **term** in the sequence.

Example 21

(a) Write down the next two terms of the sequence $4, 8, 12, 16, \ldots$

Each term of the sequence is formed by adding 4 to the previous term. Therefore

$$5\text{th term} = 16 + 4$$
$$= 20$$
$$6\text{th term} = 20 + 4$$
$$= 24$$

(b) Find the next two terms of the sequence $5, 2, -1, -4, \ldots$

Each term of the sequence is 3 less than the previous term. Therefore

$$5\text{th term} = -4 - 3$$
$$= -7$$
$$6\text{th term} = -7 - 3$$
$$= -10$$

(c) In the sequence $1, 3, 9, ?, 81, ?$ write down the terms denoted by a question mark.

Each term is formed by multiplying the previous term by 3. Therefore

$$4\text{th term} = 9 \times 3$$
$$= 27$$
$$6\text{th term} = 81 \times 3$$
$$= 243$$

(d) Find the next two terms of the sequence 1, 2, 3, 5, 8, ...

Each term is formed by adding the two previous terms. Therefore

$$6\text{th term} = 5 + 8$$
$$= 13$$
$$7\text{th term} = 8 + 13$$
$$= 21$$

(e) Write down the next two terms of the sequence 2, 7, 17, 37, ...

Each term is formed by multiplying the previous term by 2 and adding 3 to the product. Therefore

$$5\text{th term} = 37 \times 2 + 3$$
$$= 74 + 3$$
$$= 77$$
$$6\text{th term} = 77 \times 2 + 3$$
$$= 154 + 3$$
$$= 157$$

Exercise 1.10

Write down the next two terms for each of the following sequences:

1. 2, 10, 50, ...
2. 1, 5, 9, 13, ...
3. 3, 9, 15, 21, ...
4. 176, 88, 44, ...
5. 1, 3, 6, 10, ...
6. 1, 1, 2, 3, 5, 8, 13, ...
7. 3, 8, 18, 38, ...
8. 5, 9, 21, 57, ...
9. 2, 0, −2, −4, ...
10. −5, −3, −1, 1, ...

Write down the terms denoted by the question marks in the following sequences:

11. 0, 2, 4, ?, ?, 10, 12, 14
12. 3, 5, 8, 13, ?, 34, 55, ?
13. 2, 4, 10, 28, ?, ?, 730
14. 5, 11, 29, 83, ?, 731, ?
15. 243, 81, 27, ?, 3, ?

Miscellaneous Exercise 1

Section A

1. What is the difference between the two numbers 45 and 72?

2. Work out the value of $8 + 5 \times (4 - 2)$.

3. Calculate the value of $23 \times 0 + 5(3 - 3)$.

4. What is the remainder when 33 is divided by 5?

5. Write down the terms denoted by ? in the following sequence of numbers: 0, 2, 4, ?, ?, 10, 12, 14, ?

6. The instructions for cooking a turkey are 'cook for 20 minutes per pound plus an additional 25 minutes'. How long will it take to cook a turkey weighing 8 pounds?

7. Write down the next three terms of the sequence 21, 17, 13 ...

8. If the temperature rises from −5°C to 3°C, what is the increase in temperature?

Section B

1. Write down the next three terms of the following sequences:
   (a) 21, 17, 13, ...
   (b) 40, 32, 24, ...

2. Subtract one hundred thousand from one million.

3. $53 + 29 = 37 + ?$ What is the missing number?

4. A car travels 9 kilometres on 1 litre of petrol. How many litres will be needed for a journey of 324 kilometres?

5. Write down the next two numbers of the sequence 1, 4, 9, 16, 25, ...

6. Find the value of
   (a) $(-5) \times (-8)$ (b) $-5 - 8$
   (c) $8 \div (-2)$ (d) $30 \times (-10)$

7. Write down in order of size, smallest first, $-3, 0, -5$ and 3.

8. A wall is 342 bricks long. It has 23 courses of bricks. How many bricks have been used in building the wall?

9. Which is the larger and by how much: $(5 + 2) \times (7 - 4)$ or $5 + 2(7 - 4)$?

10. Fill in the blanks in the following sequence: ?, ?, 12, 24, 48, ?

11. Calculate the answers to the following:
    (a) $(7 - 3) \times (7 + 4)$
    (b) $7 - 3(7 + 4)$

12. In each of the following insert either a $+$ sign or a $-$ sign to make the statement correct:
    (a) $27 + 5 - 12 = 27 + (5 \ldots 12)$
    (b) $13 - 4 + 9 = 13 \ldots (4 - 9)$

Multi-Choice Questions 1

1. The next number in the sequence 1, 3, 6, 10, ... is
   A 11   B 15   C 16   D 20

2. The missing number in the sequence 55, 45, 36, ?, 21, 15, 10 is
   A 27   B 28   C 29   D 30

3. The value of $\dfrac{(-1) - (-1)}{1}$ is
   A -2 B -1 C 0   D 2

4. If the temperature rises from $-5\,^\circ$C to $15\,^\circ$C, what is the increase in temperature?
   A 52° B 20° C 10° D -10°

5. What is the difference between the two numbers 45 and 72?
   A 23   B 27   C 33   D 117

6. The value of $(+4) \times (-5) \times (-2)$ is
   A 40   B 18   C 11   D -11

7. The product of 9 and 3 is
   A 3   B 6   C 12   D 27

8. The next two numbers in the sequence 5, 4, 6, 5, 7, 6, 8, are
   A 9 and 7   B 7 and 6
   C 9 and 6   D 7 and 9

9. $7 \times 6 - 12 \div 3 + 1$ is equal to
   A 40   B 39   C 21   D -44

10. The sum of 3, 5 and 7 is
    A 15   B 21   C 35   D 105

Mental Test 1

Try to answer the following questions without writing anything down except the answer.

1. Add 3, 6, 7 and 4.

2. Find the sum of 2, 4, 6 and 8.

3. What is the difference between 17 and 28?

4. Take 14 away from 23.

5. Multiply 14 by 5.

6. Find the product of 15 and 6.

7. Divide 72 by 8.

8. Find the value of $2 \times 3 + 5$.

9. Work out $17 - 2(6 - 4)$.

10. Work out $3 \times 5 - 12 \div (3 + 1)$.

11. Find the value of $-6 - 3$.

12. Work out $-8 - 7 - 15$.

13. Subtract $+6$ from 4.

14. Find the value of $-12 - (-15)$.

15. Find the product of $-3, -4$ and -5.

16. Work out $(-16) \div (-4)$.

17. Find the value of $18 \div (2 - 5)$.

18. Find the next term of the series $2, 6, 18, \ldots$

19. Find the next term of the following pattern of numbers:

$$32, -16, 8, -4, \ldots$$

20. Find the next term of the sequence $6, 4, 2, 0, \ldots$

Factors and Multiples

Odd and Even Numbers

Counting in twos, starting with 2, the sequence is 2, 4, 6, 8, 10, 12, 14, ... Continuing the sequence, we discover that each number in it ends in either 0, 2, 4, 6 or 8. Such numbers are called **even numbers**. Thus 90, 282, 654, 8736 and 29 348 are all even numbers. If we now start at 1 and count in twos the sequence is 1, 3, 5, 7, 9, 11, 13, 15, 17, 19, ... Carrying on the sequence, we discover that each number in it ends in either 1, 3, 5, 7 or 9. Such numbers are called **odd numbers**. Thus 91, 103, 825, 9867 and 23 659 are all odd numbers. Note that any number which when divided by 2 has a remainder of 1 is an odd number.

Powers of Numbers

The quantity $5 \times 5 \times 5$ is usually written as 5^3, and 5^3 is called the third **power** of 5. The number 3 which indicates the number of fives to be multiplied together is called the **index** (plural: indices).

$$2^5 = 2 \times 2 \times 2 \times 2 \times 2$$
$$= 32$$
$$9^2 = 9 \times 9$$
$$= 81$$

When a number is multiplied by itself it is called the **square** of the number. Thus the square of a number is the number raised to the power of 2.

$$\text{Square of 3} = 3^2$$
$$= 3 \times 3$$
$$= 9$$
$$\text{Square of 15} = 15^2$$
$$= 15 \times 15$$
$$= 225$$

The cube of a number is the number raised to the power of three.

$$\text{Cube of 7 is } 7^3 = 7 \times 7 \times 7$$
$$= 343$$
$$\text{Cube of 85 is } 85^3 = 85 \times 85 \times 85$$
$$= 614\,125$$

Powers of numbers can easily be calculated using a calculator.

Example 1

Using a calculator find the value of 4^5.

Input	Display
4	4.
× ×	4.
=	16.
=	64.
=	256.
=	1024.

Hence $4^5 = 1024$.

Square Roots of Numbers

The square root of a number is the number whose square equals the given number. Since $5^2 = 25$, the square root of 25 is 5. The sign $\sqrt{}$ is used to denote a positive square root and we write $\sqrt{25} = 5$.

Similarly, since $7^2 = 49$, $\sqrt{49} = 7$.

Most calculators possess a square-root key which is used in the way shown in Example 2.

Example 2

Using a calculator find $\sqrt{529}$.

Input	Display
529	529.
$\sqrt{}$	23.

Hence $\sqrt{529} = 23$.

To check, note that $23^2 = 529$.

Square Root of a Product

The square root of a product is the product of the square roots of the individual numbers. Thus

$$\sqrt{4 \times 9} = \sqrt{4} \times \sqrt{9}$$
$$= 2 \times 3$$
$$= 6$$

Example 3

Find the square root of $16 \times 25 \times 49$.

$$\sqrt{16 \times 25 \times 49} = \sqrt{16} \times \sqrt{25} \times \sqrt{49}$$
$$= 4 \times 5 \times 7$$
$$= 140$$

Exercise 2.1

Questions 1 to 20 inclusive should be done mentally. State whether the numbers given below are even or odd:

1. 95
2. 936
3. 852
4. 5729
5. 8887
6. 63 752
7. 456 310
8. 9367
9. From the set of numbers 17, 36, 59, 98, 121, 136 and 259 write down (a) all the odd numbers, (b) all the even numbers.
10. Consider the numbers 16, 19, 35, 68, 89, 137 and 342. Write down all the odd numbers.

Find values for each of the following:

11. The square of 6
12. The cube of 5
13. (a) 7^2 (b) 4^3
14. The square root of 64
15. 9 squared
16. (a) $\sqrt{16}$ (b) $\sqrt{144}$
17. The square root of 9×36
18. $\sqrt{64 \times 64}$
19. $\sqrt{9 \times 25 \times 36}$
20. $\sqrt{64 \times 36 \times 49}$
21. Find the value of each of the following: (a) 3^4 (b) 2^5 (c) 4^6
22. Find the sum of 2^6 and 6^2.
23. Find the product of 2^4 and 3^3.
24. Find the difference between 2^5 and 5^2.
25. Find the square root of the following numbers:
    (a) 169   (b) 625
    (c) 1024   (d) 7569
    (e) 15 625

The Digit Sum of a Number

The digits of a number are the single figures which make up the number. For instance the digits of the number 312 are 3, 1 and 2.

The digit sum of a number is found by adding the digits of the number together.

Thus the digit sum of 312 is $3 + 1 + 2 = 6$.

Sometimes the digit sum contains more than one digit. In such cases the digits of the sum are added together until a single figure is obtained.

Example 4

Find the digit sum of the number 9587.

Adding the digits of 9587:
$$9 + 5 + 8 + 7 = 29$$

Adding the digits of 29:
$$2 + 9 = 11$$

Adding the digits of 11:
$$1 + 1 = 2$$

Tests for Divisibility

The following tests for divisibility are useful for work on factors, which follow later in this chapter.

If a number is even, then 2 will divide into it without leaving a remainder, i.e. the number is divisible by 2. Thus 38, 40, 56, 84 and 930 are all divisible by 2.

Numbers which are divisible by 2 are called multiples of 2. Thus 20, 42, 54, 136 and 978 are all multiples of 2.

A number is divisible by 3 if it divides exactly into the digit sum of the number without leaving a remainder.

Example 5

Is the number 4317 divisible by 3?

Adding the digits of 4317:
$$4 + 3 + 1 + 7 = 15$$

Adding the digits of 15:
$$1 + 5 = 6$$

Therefore the digit sum of 4317 is 6. Since 3 divides into 6 without leaving a remainder, the number 4317 is divisible by 3.

Numbers which are divisible by 3 are called multiples of 3. Thus 6, 9, 15, 24 and 36 are all multiples of 3.

A number is divisible by 5 if the number ends in 0 or 5. Hence 20, 35, 95, 8000 and 16 340 are all divisible by 5. The numbers 132, 984, 7906 and 8173 are not divisible by 5 because they do not end in 0 or 5.

Numbers which are divisible by 5 are called multiples of 5. Thus 15, 20, 30, 50 and 85 are all multiples of 5.

Numbers which end in 0 are divisible by 10. Thus 60, 9730, 27 000 and 16 876 320 are all divisible by 10.

Exercise 2.2

Find the digit sum of the following numbers:

1. 305	2. 171	3. 324
4. 976	5. 6283	6. 9457
7. 9207	8. 9875	

Which of the following numbers are divisible by 2?

9. 12	10. 39	11. 82
12. 1986	13. 6032	14. 26 735
15. 29 008		

Which of the following numbers are multiples of 2?

16. 92   **17.** 870   **18.** 1368

19. 7357   **20.** 9651

Which of the following numbers are divisible by 3?

21. 417   **22.** 484   **23.** 2740

24. 9354   **25.** 18 567

State which of the following numbers are multiples of 3:

26. 144   **27.** 286   **28.** 264

29. 5874   **30.** 29 739

Which of the following numbers are divisible by 5?

31. 105   **32.** 1237   **33.** 9180

34. 3415   **35.** 28 735

State which of the following numbers are multiples of 5:

36. 81   **37.** 180   **38.** 965

39. 4329   **40.** 2695

Which of the following numbers are divisible by 10?

41. 992   **42.** 1500   **43.** 4983

44. 9830   **45.** 26 325 410

Factors

A number is a factor of another number if it divides into that number without leaving a remainder.

Since 63 is divisible by 7, 7 is a factor of 63.

63 has other factors, namely 1, 3, 9, 21 and 63 since each of these numbers divides into 63 without leaving a remainder.

Exercise 2.3

Questions 1 to 5 should be done mentally.

1. Is 6 a factor of
    (a) 12   (b) 31   (c) 42
    (d) 66   (e) 86

2. Is 5 a factor of
    (a) 8   (b) 12   (c) 25
    (d) 32   (e) 50

3. Is 9 a factor of
    (a) 30   (b) 45   (c) 72
    (d) 128   (e) 135

4. Which of the following numbers are factors of 42?
    (a) 2   (b) 3   (c) 4
    (d) 5   (e) 6   (f) 7
    (g) 8   (h) 9

5. Which if the following numbers are factors of 72?
    (a) 2   (b) 3   (c) 4
    (d) 5   (e) 6   (f) 7
    (g) 8   (h) 9   (i) 11
    (j) 12   (k) 13   (l) 14

6. Consider the numbers 15, 31, 35, 49, 62, 84 and 109. Three of these numbers have a common factor. What is it?

Prime Numbers

Every number has itself and 1 as factors. If a number has no other factors it is said to be a **prime number**. It is useful to learn all of the prime numbers up to 100. They are

2, 3, 5, 7, 11, 13, 17, 19, 23, 29, 31, 37, 41, 43, 47, 53, 59, 61, 67, 71, 73, 79, 83, 89, 97

Notice that, with the exception of 2, all the prime numbers are odd numbers. Also note that although 1 is a factor of all other numbers it is not regarded as being a prime number.

Prime Factors

A factor which is a prime number is called a **prime factor.** In the statement $63 = 7 \times 9$, 7 is a prime factor of 63 but 9 is not a prime factor of 63.

Example 6

Find the prime factors of 63.

$$63 = 3 \times 3 \times 7$$

Since $3 \times 3 = 3^2$ we can write

$$63 = 3^2 \times 7$$

and 3^2 and 7 are the prime factors of 63.

The prime factors of any number can be found by successive division as shown in Example 7. Note that we start by dividing by the smallest of the prime factors and end by dividing by the largest of them.

Example 7

Find the prime factors of 420.

2	420
2	210
3	105
5	35
7	7
	1

Hence

$$420 = 2 \times 2 \times 3 \times 5 \times 7$$
$$= 2^2 \times 3 \times 5 \times 7$$

Example 8

Find all the factors of 60.

The factors are:

1, 2, 3, 4, 5, 6, 10, 12, 15, 20, 30, 60,

If we pair off the factors from each end as shown above and multiply them together we get

$$1 \times 60 = 60 \qquad 2 \times 30 = 60$$
$$3 \times 20 = 60 \qquad 4 \times 15 = 60$$
$$5 \times 12 = 60 \qquad 6 \times 10 = 60$$

We see that each pair gives a product of 60. We can use this method of pairing to check that all the factors have been found.

Lowest Common Multiple (LCM)

The multiples of 3 are 3, 6, 9, 12, 15, 18, 21, 24, 27, 30, 33, 36, 39, 42, etc.

The multiples of 4 are 4, 8, 12, 16, 20, 24, 28, 32, 36, 40, 44, etc.

We see that 12, 24 and 36 are multiples which are common to both 3 and 4. We say that 12 is the LCM of 3 and 4.

Note that 12 is the smallest number into which 3 and 4 will divide exactly.

Example 9

Find the LCM of 4, 10 and 12.

$$4 = 2 \times 2$$
$$= 2^2$$
$$10 = 2 \times 5$$
$$12 = 2 \times 2 \times 3$$
$$= 2^2 \times 3$$

The LCM is the product of the highest power of each prime factor. Hence

$$LCM = 2^2 \times 3 \times 5$$
$$= 60$$

Note that 60 is the smallest number into which 4, 10 and 12 will divide exactly.

Highest Common Factor (HCF)

The factors of 30 are 1, 2, 3, 5, 6, 10, 15 and 30.

The factors of 40 are 1, 2, 4, 5, 8, 10, 20 and 40.

We see that 1, 2, 5 and 10 are factors which are common to both 30 and 40. The highest of these common factors is 10 and we say that the HCF of 30 and 40 is 10. Note that 10 is the highest number which will divide exactly into 30 and 40.

Example 10

Find the HCF of 42, 98 and 112.

$$42 = 2 \times 3 \times 7$$
$$98 = 2 \times 7 \times 7$$
$$= 2 \times 7^2$$
$$112 = 2 \times 2 \times 2 \times 2 \times 7$$
$$= 2^4 \times 7$$

We see that only 2 and 7 are factors of all three numbers. Hence

$$HCF \text{ of } 42, 98 \text{ and } 112 = 2 \times 7$$
$$= 14$$

Note that 14 is the highest number which will divide exactly into 42, 98 and 112.

Exercise 2.4

1. Write down all the factors of
   (a) 15   (b) 64   (c) 48

2. Write down the next two prime numbers larger than 19.

3. From the set of numbers 24, 33, 45, 61, 49 and 27 write down (a) an even number, (b) an odd number, (c) a prime number, (d) a multiple of 7.

4. Consider the numbers 11, 21, 31, 77 and 112.

   (a) Two of these numbers are prime. What is the number which lies exactly half-way between them?

   (b) Three of these numbers have a common factor. What is it?

5. Express each of the following numbers as a product of its prime factors:
   (a) 24   (b) 72   (c) 45

6. Find the LCM of the following sets of numbers:
   (a) 6 and 15   (b) 12 and 6
   (c) 3, 5 and 6   (d) 4, 10 and 12
   (e) 3, 5 and 9

7. Write down all the factors of 20 and 24. Hence find the common factors and write down the HCF of 20 and 24.

8. Write down the multiples of 6 and 10 less than 61. Find the common multiples and hence find the LCM of 6 and 10.

9. Find the HCF of the following sets of numbers:
   (a) 10 and 30   (b) 6 and 15
   (c) 4, 10 and 12

10. Consider the numbers 2, 3, 4, 5, 6, 12, 18 and 24.
    (a) Which of them are factors of 12?
    (b) Which of them are multiples of 6?

11. Write down all the factors of
    (a) 12   (b) 36   (c) 60   (d) 100

12. Write down all the factors of 84 and check by pairing.

Miscellaneous Exercise 2

Section A

1. Write down two prime numbers whose sum is 16 and whose difference is 6.

2. List the following numbers:
   (a) Multiples of 8 less than 48.
   (b) The next three prime numbers greater than 23.

3. Find the smallest number that is divisible by 3, 4, 6 and 9.

4. Find the value of $13^2 - 12^2$.

5. Work out the values of $\sqrt{16 \times 49}$.

6. Find the sum of all the prime numbers between 10 and 20.

7. Consider the number sequence 1, 3, 6, 10, 15, 21, ... What is the next number in the sequence that has a square root which is a whole number?

8. What is the lowest common multiple of 3, 4 and 6?

Section B

1. Given the numbers 49, 55, 60, 63, 81 and 122 write down (a) the multiples of 5, (b) the multiples of 3, (c) the even numbers.

2. Express 360 as a product of prime factors.

3. Find the LCM of the following sets of numbers:
   (a) 10 and 15 (b) 12 and 15

4. Write down (a) all the positive whole numbers which are factors of 42, (b) all the multiples of 5 between 11 and 29.

5. Three prime numbers are 11, 13 and 37.
   (a) Find their sum. Is the sum a prime number?
   (b) Calculate their product. Is the product a prime number?

6. (a) Find the sum of 2^5 and 5^2.
   (b) Calculate the product of 3^2 and 2^3.

7. Consider the sequence 2, 3, 5, 8, 13, 21, 34, 55.
   (a) List all the even numbers in the sequence.
   (b) List all the prime numbers in the sequence.
   (c) Find the sum of the last four numbers.
   (d) Write down the next number in the sequence.

8. List the following numbers:
   (a) The prime factors of 100.
   (b) Multiples of 6 less than 55.
   (c) The next three prime numbers after 25.

Multi-Choice Questions 2

1. The sum of all the prime numbers between 10 and 20 is
   A 90   B 75   C 70   D 60

2. The value of 2^5 is
   A 32   B 16   C 10   D 3

3. The LCM of 3, 4, 6 and 9 is
   A 36   B 24   C 18   D 12

4. The difference between 3^4 and 4^3 is
   A 0   B 1   C 7   D 17

5. The number of prime numbers between 10 and 30 is
   A 2   B 5   C 6   D 8

6. The number 400 when expressed as the product of its prime factors is

    A $2^4 \times 5^2$ B $2^3 \times 5^3$

    C $3^2 \times 2^6$ D $2^3 \times 5 \times 10$

7. 243 expressed as a power of 3 is

    A 81   B 3^5 C 729   D 3^{81}

8. The LCM of 3, 4 and 6 is

    A 12   B 13   C 24   D 72

9. The HCF of 16 and 20 is

    A 2   B 4   C 80   D 320

10. What is the value of $4^2 + 7^2$?

    A 11   B 14   C 28   D 65

Mental Test 2

Try to answer the following questions without writing anything down except the answer.

1. Is 128 an even number?

2. Is 347 divisible by 3?

3. What is the square of 8?

4. Find the value of 2^4.

5. Is 529 an even number?

6. What is the square root of 36?

7. Find the value of $\sqrt{25 \times 64}$.

8. Find the difference between 2^3 and 3^2.

9. Is the number 425 divisible by 5?

10. Write down the factors of 24.

11. Write down all the multiples of 7 up to 41.

12. State the prime factors of 18.

13. What is the LCM of 2 and 5?

14. What is the HCF of 15 and 25?

15. Find the sum of the prime numbers in the following set: 3, 6, 8, 9, 11, 15, 19 and 21.

Fractions

Introduction

The circle shown in Fig. 3.1 has been divided into eight equal parts. Each of these parts is called one-eighth of the circle and is written $\frac{1}{8}$. The number below the line shows how many equal parts there are and it is called the **denominator**. The number above the line shows how many of these equal parts are taken and it is called the **numerator**. If five of the eight equal parts are taken then we have taken $\frac{5}{8}$ of the circle.

Fig. 3.1

$\frac{1}{8}$ of circle

Exercise 3.1

Fig. 3.2 shows some drawings of circles. Write down the fractions represented by the shaded portions.

Fig. 3.2

1.   2.   3.

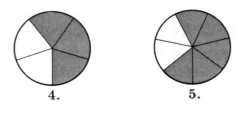

4.   5.

In a similar way to that shown in Fig. 3.2, sketch five circles and shade them to represent:

6. $\frac{7}{8}$ 7. $\frac{2}{5}$ 8. $\frac{5}{12}$ 9. $\frac{3}{10}$

10. $\frac{4}{7}$

Equivalent Fractions

In Fig. 3.3 we have shaded $\frac{1}{2}$ of a circle in diagram (a) and $\frac{4}{8}$ of a circle in diagram (b). From the diagrams we see that the same area has been shaded in both diagrams.

Hence

$$\frac{1}{2} = \frac{4}{8}$$

Fig. 3.3

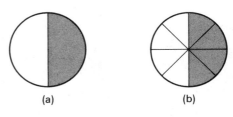

(a)   (b)

Two fractions are **equivalent** if they have the same value. We see that $\frac{1}{2}$ and $\frac{4}{8}$ are equivalent fractions and that to change $\frac{1}{2}$ into $\frac{4}{8}$ we have multiplied top and bottom by 4.

By drawing more sketches of fractions (Fig. 3.4) we find that if we multiply or divide the top and bottom of a fraction by the same number (provided it is not zero) we do not alter its value.

Fig. 3.4

$$\tfrac{3}{4} = \tfrac{6}{8} \qquad\qquad \tfrac{3}{6} = \tfrac{1}{2}$$

Example 1

(a) $\frac{3}{7}$ is equivalent to $\dfrac{3 \times 5}{7 \times 5} = \dfrac{15}{35}$

(b) $\frac{12}{20}$ is equivalent to $\dfrac{12 \div 4}{20 \div 4} = \dfrac{3}{5}$

Exercise 3.2

Draw suitable sketches to show that

1. $\frac{2}{3} = \frac{6}{9}$ 2. $\frac{4}{5} = \frac{8}{10}$ 3. $\frac{3}{4} = \frac{6}{8}$

4. $\frac{8}{14} = \frac{4}{7}$ 5. $\frac{6}{10} = \frac{3}{5}$

Copy the following fractions and fill in the missing numbers:

6. $\frac{3}{4} = \frac{}{20}$ 7. $\frac{2}{3} = \frac{}{12}$ 8. $\frac{3}{7} = \frac{}{28}$

9. $\frac{3}{8} = \frac{}{32}$ 10. $\frac{1}{6} = \frac{}{24}$ 11. $\frac{9}{12} = \frac{}{36}$

12. $\frac{4}{8} = \frac{}{4}$ 13. $\frac{9}{15} = \frac{}{5}$ 14. $\frac{35}{42} = \frac{}{6}$

15. $\frac{21}{49} = \frac{}{7}$

Reducing a Fraction to its Lowest Terms

When the numerator and denominator of a fraction possess no common factors, the fraction is said to be in its **lowest terms**.

The fraction $\frac{7}{16}$ is in its lowest terms because there is no whole number that will divide exactly into 7 and 16.

The fraction $\frac{6}{8}$ is not in its lowest terms because 2 will divide exactly into 6 and 8 to give the equivalent fraction $\frac{3}{4}$.

Example 2

Reduce $\frac{8}{12}$ to its lowest terms.

$\frac{8}{12}$ is equivalent to $\dfrac{8 \div 4}{12 \div 4} = \dfrac{2}{3}$

Exercise 3.3

Reduce each of the following fractions to its lowest terms:

1. $\frac{8}{16}$ 2. $\frac{15}{25}$ 3. $\frac{9}{15}$ 4. $\frac{8}{64}$

5. $\frac{12}{18}$ 6. $\frac{42}{48}$ 7. $\frac{22}{33}$ 8. $\frac{18}{27}$

9. $\frac{21}{24}$ 10. $\frac{35}{56}$

Types of Fractions

If the top number of a fraction is less than the bottom number then the fraction is called a **proper fraction**. Thus $\frac{2}{3}$, $\frac{5}{8}$ and $\frac{9}{20}$ are all proper fractions.

If the top number of a fraction is **greater** than the bottom number then the fraction is called an **improper fraction** or a **top-heavy fraction**. Thus $\frac{8}{5}$, $\frac{5}{2}$ and $\frac{19}{7}$ are all top-heavy fractions.

Every top-heavy fraction can be expressed as a whole number and a proper fraction. These are sometimes called **mixed numbers**. Thus $2\frac{3}{5}$, $5\frac{7}{8}$ and $9\frac{1}{2}$ are all mixed numbers.

Example 3

Change $\frac{7}{3}$ into a mixed number.

$$\frac{7}{3} \text{ may be written as } \frac{6+1}{3} = \frac{6}{3} + \frac{1}{3}$$
$$= 2 + \frac{1}{3}$$
$$= 2\frac{1}{3}$$

From this example we see that the rule for changing top-heavy fractions into mixed numbers is:

Divide the numerator (top number) by the denominator (bottom number). The answer becomes the whole-number part of the mixed number. The remainder becomes the numerator (top number) of the fractional part whose denominator is the denominator of the top-heavy fraction.

Example 4

Change $\frac{35}{4}$ into a mixed number.

$$35 \div 4 = 8 \text{ remainder } 3$$

Therefore $\frac{35}{4} = 8\frac{3}{4}$.

All mixed numbers can be changed into top-heavy fractions. The rule is:

To change a mixed number into a top-heavy fraction multiply the whole number by the denominator (bottom part) of the fractional part. Then add the numerator (top part) of the fractional part to this product. This gives the numerator of the top-heavy fraction whose denominator is the denominator of the original fractional part of the mixed number.

Example 5

Change $5\frac{2}{7}$ into a top-heavy fraction.

$$5\frac{2}{7} = \frac{(5 \times 7) + 2}{7}$$
$$= \frac{35 + 2}{7}$$
$$= \frac{37}{7}$$

Exercise 3.4

Change the following top-heavy fractions into mixed numbers:

1. $\frac{7}{2}$ 2. $\frac{19}{8}$ 3. $\frac{22}{9}$

4. $\frac{12}{11}$ 5. $\frac{21}{8}$

Change the following mixed numbers into top-heavy fractions:

6. $2\frac{3}{8}$ 7. $5\frac{1}{10}$ 8. $8\frac{2}{3}$

9. $6\frac{5}{7}$ 10. $4\frac{2}{9}$

Comparing the Values of Fractions

When putting fractions in order of size, first express each of the fractions with the same denominator. This common denominator should be the LCM of the denominators of the fractions to be compared.

Example 6

Arrange the fractions $\frac{3}{4}$, $\frac{5}{8}$, $\frac{7}{10}$ and $\frac{11}{20}$ in order of size, beginning with the smallest.

The LCM of the denominators 4, 8, 10 and 20 is 40.

We now change each of the given fractions so that it has a denominator of 40.

$\frac{3}{4}$ is equivalent to $\dfrac{3 \times 10}{4 \times 10} = \dfrac{30}{40}$

$\frac{5}{8}$ is equivalent to $\dfrac{5 \times 5}{8 \times 5} = \dfrac{25}{40}$

$\frac{7}{10}$ is equivalent to $\dfrac{7 \times 4}{10 \times 4} = \dfrac{28}{40}$

$\frac{11}{20}$ is equivalent to $\dfrac{11 \times 2}{20 \times 2} = \dfrac{22}{40}$

After the fractions have been expressed with the same denominator, all we have to do is compare their numerators.

Therefore the order is

$$\tfrac{22}{40}, \tfrac{25}{40}, \tfrac{28}{40} \text{ and } \tfrac{30}{40}, \text{ or } \tfrac{11}{20}, \tfrac{5}{8}, \tfrac{7}{10} \text{ and } \tfrac{3}{4}$$

Exercise 3.5

Arrange the following sets of fractions in order of size, beginning with the smallest:

1. $\tfrac{1}{2}, \tfrac{5}{6}, \tfrac{2}{3}$ and $\tfrac{7}{12}$ 2. $\tfrac{9}{10}, \tfrac{3}{4}, \tfrac{7}{8}$ and $\tfrac{3}{5}$

3. $\tfrac{3}{4}, \tfrac{5}{8}, \tfrac{9}{16}$ and $\tfrac{17}{32}$ 4. $\tfrac{1}{3}, \tfrac{1}{4}, \tfrac{1}{2}$ and $\tfrac{1}{5}$

5. $\tfrac{3}{8}, \tfrac{5}{9}, \tfrac{2}{6}$ and $\tfrac{5}{18}$

Addition of Fractions

The steps when adding fractions are as follows:

(1) Find the LCM of the denominators of the fractions to be added.

(2) Express each of the fractions with this common denominator.

(3) Add the numerators of the new fractions to give the numerator of the answer. The denominator of the answer is the LCM found in (1).

Example 7

Add $\tfrac{2}{7}$ and $\tfrac{3}{4}$.

The LCM of the denominators 7 and 4 is 28.

Expressing each fraction with a denominator of 28 gives

$$\tfrac{2}{7} \text{ is equivalent to } \frac{2 \times 4}{7 \times 4} = \frac{8}{28}$$

$$\tfrac{3}{4} \text{ is equivalent to } \frac{3 \times 7}{4 \times 7} = \frac{21}{28}$$

$$\tfrac{2}{7} + \tfrac{3}{4} = \tfrac{8}{28} + \tfrac{21}{28}$$
$$= \frac{8 + 21}{28}$$
$$= \frac{29}{28}$$
$$= 1\tfrac{1}{28}$$

The work may be set out as follows:

$$\tfrac{2}{7} + \tfrac{3}{4} = \frac{(2 \times 4) + (3 \times 7)}{28}$$
$$= \frac{8 + 21}{28}$$
$$= \tfrac{29}{28}$$
$$= 1\tfrac{1}{28}$$

If mixed numbers are present then add the whole numbers first.

Example 8

Add $5\tfrac{1}{2}$, $2\tfrac{2}{3}$ and $3\tfrac{2}{5}$.

The LCM of the denominators 2, 3 and 5 is 30.

$$5\tfrac{1}{2} + 2\tfrac{2}{3} + 3\tfrac{2}{5}$$
$$= 5 + 2 + 3 + \tfrac{1}{2} + \tfrac{2}{3} + \tfrac{2}{5}$$
$$= 10 + \frac{(1 \times 15) + (2 \times 10) + (2 \times 6)}{30}$$
$$= 10 + \frac{15 + 20 + 12}{30}$$
$$= 10 + \tfrac{47}{30}$$
$$= 10 + 1\tfrac{17}{30}$$
$$= 10 + 1 + \tfrac{17}{30}$$
$$= 11 + \tfrac{17}{30}$$
$$= 11\tfrac{17}{30}$$

Exercise 3.6

Work out:

1. $\frac{1}{2} + \frac{1}{3}$ 2. $\frac{2}{5} + \frac{9}{10}$

3. $\frac{3}{4} + \frac{3}{8}$ 4. $\frac{3}{10} + \frac{1}{4}$

5. $\frac{1}{2} + \frac{3}{4} + \frac{7}{8}$ 6. $\frac{1}{8} + \frac{2}{3} + \frac{5}{12}$

7. $1\frac{3}{8} + 3\frac{9}{16}$ 8. $7\frac{2}{3} + 6\frac{3}{5}$

9. $3\frac{3}{8} + 5\frac{2}{7} + 4\frac{3}{4}$ 10. $4\frac{1}{2} + 3\frac{5}{6} + 2\frac{1}{3}$

11. $7\frac{3}{8} + 2\frac{3}{4} + \frac{7}{8} + \frac{5}{16}$ 12. $7\frac{2}{3} + \frac{2}{5} + \frac{3}{10} + 2\frac{1}{2}$

Subtraction of Fractions

The method is similar to that used in addition. We first find the LCM of the denominators of the fractions and then express each fraction with this common denominator. We then subtract the numerators of the new fractions.

Example 9

Subtract $\frac{2}{5}$ from $\frac{5}{8}$.

The LCM of the denominators 5 and 8 is 40.

$$\frac{5}{8} - \frac{2}{5} = \frac{(5 \times 5) - (2 \times 8)}{40}$$

$$= \frac{25 - 16}{40}$$

$$= \frac{9}{40}$$

Example 10

Take $2\frac{1}{4}$ from $3\frac{7}{10}$.

The LCM of the denominators 4 and 10 is 20.

$$3\frac{7}{10} - 2\frac{1}{4} = 3 - 2 + \frac{7}{10} - \frac{1}{4}$$

$$= 1 + \frac{(7 \times 2) - (1 \times 5)}{20}$$

$$= 1 + \frac{14 - 5}{20}$$

$$= 1 + \frac{9}{20}$$

$$= 1\frac{9}{20}$$

Sometimes we have to 'borrow' from the whole number as shown in Example 11.

Example 11

Subtract $2\frac{3}{4}$ from $4\frac{2}{3}$.

The LCM of 4 and 3 is 12.

$$4\frac{2}{3} - 2\frac{3}{4} = 4 - 2 + \frac{2}{3} - \frac{3}{4}$$

$$= 2 + \frac{2}{3} - \frac{3}{4}$$

Because $\frac{3}{4}$ is larger than $\frac{2}{3}$ we cannot subtract and so we write

$$4\frac{2}{3} - 2\frac{3}{4} = 1 + 1 + \frac{2}{3} - \frac{3}{4}$$

$$= 1 + \frac{(1 \times 12) + (2 \times 4) - (3 \times 3)}{12}$$

$$= 1 + \frac{12 + 8 - 9}{12}$$

$$= 1 + \frac{11}{12}$$

$$= 1\frac{11}{12}$$

'Borrowing' can be avoided by changing mixed numbers into top-heavy fractions before attempting to subtract.

$$4\frac{2}{3} - 2\frac{3}{4} = \frac{14}{3} - \frac{11}{4}$$

$$= \frac{(14 \times 4) - (11 \times 3)}{12}$$

$$= \frac{56 - 33}{12}$$

$$= \frac{23}{12}$$

$$= 1\frac{11}{12}$$

Work out:

1. $\frac{1}{2} - \frac{1}{3}$ 2. $\frac{7}{8} - \frac{5}{6}$ 3. $\frac{1}{3} - \frac{1}{5}$

4. $\frac{2}{3} - \frac{1}{2}$ 5. $3\frac{3}{8} - 1\frac{1}{4}$ 6. $3 - \frac{5}{7}$

7. $5\frac{7}{16} - 3\frac{1}{5}$ 8. $5\frac{3}{8} - 2\frac{9}{10}$ 9. $4\frac{3}{8} - 3\frac{2}{5}$

10. $2\frac{2}{3} - 1\frac{3}{4}$

Combined Addition and Subtraction

When a problem contains both addition and subtraction it is often best to make each of the mixed numbers into a top-heavy fraction.

Example 12

Work out $5\frac{3}{8} - 1\frac{1}{4} + 2\frac{1}{2} - \frac{7}{16}$.

$$5\frac{3}{8} - 1\frac{1}{4} + 2\frac{1}{2} - \frac{7}{16}$$
$$= \frac{43}{8} - \frac{5}{4} + \frac{5}{2} - \frac{7}{16}$$
$$= \frac{(43 \times 2) - (5 \times 4) + (5 \times 8) - 7}{16}$$
$$= \frac{86 - 20 + 40 - 7}{16}$$
$$= \frac{(86 + 40) - (20 + 7)}{16}$$
$$= \frac{126 - 27}{16}$$
$$= \frac{99}{16}$$
$$= 6\frac{3}{16}$$

Work out:

1. $2\frac{1}{2} + 3\frac{1}{4} - 4\frac{3}{8}$

2. $5\frac{1}{10} - 3\frac{1}{2} - 1\frac{1}{4}$

3. $4\frac{3}{8} - 2\frac{1}{2} + 5$

4. $6\frac{1}{2} - 3\frac{1}{6} + 2\frac{1}{12} - 4\frac{3}{4}$

5. $1\frac{3}{16} - 2\frac{2}{5} + 3\frac{3}{4} + 5\frac{5}{8}$

6. $12\frac{3}{4} - 6\frac{7}{8} - 2\frac{13}{16} + 5\frac{1}{2}$

7. $3\frac{9}{20} + 1\frac{3}{8} - 2\frac{7}{10} - \frac{3}{4}$

8. $1\frac{1}{12} - 4\frac{3}{4} - 3\frac{3}{3} + 8\frac{3}{8}$

Multiplication of Fractions

To multiply two or more fractions together, first multiply all the top numbers together and then multiply all the bottom numbers together. Mixed numbers must always be made into top-heavy fractions before attempting to multiply.

Example 13

Multiply $\frac{5}{8}$ by $\frac{3}{7}$.

$$\frac{5}{8} \times \frac{3}{7} = \frac{5 \times 3}{8 \times 7}$$
$$= \frac{15}{56}$$

Example 14

Multiply $1\frac{3}{8}$ by $1\frac{1}{4}$.

$$1\frac{3}{8} \times 1\frac{1}{4} = \frac{11}{8} \times \frac{5}{4}$$
$$= \frac{11 \times 5}{8 \times 4}$$
$$= \frac{55}{32}$$
$$= 1\frac{23}{32}$$

Work out:

1. $\frac{2}{3} \times \frac{4}{5}$ 2. $\frac{3}{4} \times \frac{5}{7}$ 3. $\frac{2}{9} \times 1\frac{2}{3}$

4. $\frac{5}{9} \times 2\frac{3}{4}$ 5. $1\frac{2}{5} \times 3\frac{1}{2}$ 6. $2\frac{1}{2} \times 2\frac{1}{3}$

7. $1\frac{2}{9} \times 1\frac{2}{5}$ 8. $1\frac{7}{8} \times 1\frac{4}{7}$

Cancelling

Example 15

Work out $\frac{2}{3} \times 1\frac{7}{8}$.

$$\frac{2}{3} \times 1\frac{7}{8} = \frac{2}{3} \times \frac{15}{8}$$

$$= \frac{2 \times 15}{3 \times 8}$$

$$= \frac{30}{24}$$

$$= \frac{5}{4}$$

$$= 1\frac{1}{4}$$

The step of reducing $\frac{30}{24}$ to its lowest terms has been achieved by dividing 6 into both the numerator and the denominator. The work can often be made easier by cancelling the common factors before multiplying:

$$\frac{\overset{1}{\cancel{2}}}{\underset{1}{\cancel{3}}} \times \frac{\overset{5}{\cancel{15}}}{\underset{4}{\cancel{8}}} = \frac{1 \times 5}{1 \times 4}$$

$$= \frac{5}{4}$$

$$= 1\frac{1}{4}$$

We have divided 2 into 2 (a top number) and into 8 (a bottom number). Also we have divided 3 into 15 (a top number) and into 3 (a bottom number). You will see that we have divided the top and bottom numbers by the same amount. Note carefully that we can only cancel between a top number and a bottom number.

Example 16

Work out $3\frac{1}{8} \times \frac{7}{10} \times \frac{2}{21}$.

$$3\frac{1}{8} \times \frac{7}{10} \times \frac{2}{21} = \frac{\overset{5}{\cancel{25}}}{\underset{4}{\cancel{8}}} \times \frac{\overset{1}{\cancel{7}}}{\underset{2}{\cancel{10}}} \times \frac{\overset{1}{\cancel{2}}}{\underset{3}{\cancel{21}}}$$

$$= \frac{5 \times 1 \times 1}{4 \times 2 \times 3}$$

$$= \frac{5}{24}$$

Sometimes in calculations involving fractions the word 'of' appears. It should always be taken as meaning multiply.

Example 17

Find $\frac{4}{5}$ of 20.

$$\frac{4}{5} \text{ of } 20 = \frac{4}{\underset{1}{\cancel{5}}} \times \frac{\overset{4}{\cancel{20}}}{1}$$

$$= \frac{4 \times 4}{1 \times 1}$$

$$= \frac{16}{1}$$

$$= 16$$

Exercise 3.10

Work out:

1. $\frac{3}{4} \times 1\frac{7}{9}$ 2. $5\frac{1}{5} \times \frac{10}{13}$

3. $1\frac{5}{8} \times \frac{7}{26}$ 4. $1\frac{1}{2} \times \frac{2}{5} \times 2\frac{1}{2}$

5. $\frac{5}{8} \times \frac{7}{10} \times \frac{2}{21}$ 6. $2 \times 1\frac{1}{2} \times 1\frac{1}{3}$

7. $3\frac{3}{4} \times 1\frac{3}{5} \times 1\frac{1}{8}$ 8. $\frac{15}{32} \times \frac{8}{11} \times 24\frac{1}{5}$

9. $\frac{3}{4}$ of 16   10. $\frac{5}{7}$ of 140

11. $\frac{2}{3}$ of $4\frac{1}{2}$ 12. $\frac{4}{5}$ of $2\frac{1}{2}$

Division of Fractions

To divide by a fraction, invert it and multiply.

Example 18

(a) Divide $\frac{3}{5}$ by $\frac{2}{7}$.

$$\frac{3}{5} \div \frac{2}{7} = \frac{3}{5} \times \frac{7}{2}$$

$$= \frac{3 \times 7}{5 \times 2}$$

$$= \frac{21}{10}$$

$$= 2\frac{1}{10}$$

(b) Divide $1\frac{4}{5}$ by $2\frac{1}{3}$.

$$1\frac{4}{5} \div 2\frac{1}{3} = \frac{9}{5} \div \frac{7}{3}$$

$$= \frac{9}{5} \times \frac{3}{7}$$

$$= \frac{9 \times 3}{5 \times 7}$$

$$= \frac{27}{35}$$

Exercise 3.11

Work out:

1. $\frac{4}{5} \div 1\frac{1}{3}$
2. $2 \div \frac{1}{4}$
3. $\frac{5}{8} \div \frac{15}{32}$
4. $3\frac{3}{4} \div 2\frac{1}{2}$
5. $2\frac{1}{2} \div 3\frac{3}{4}$
6. $5 \div 5\frac{1}{5}$
7. $3\frac{1}{15} \div 2\frac{5}{9}$
8. $2\frac{3}{10} \div \frac{3}{5}$

Operations with Fractions

The sequence of operations when dealing with fractions is the same as that used with whole numbers:

(1) Work out the contents of brackets.

(2) Multiply and divide.

(3) Add and subtract.

Example 19

Simplify $(2\frac{1}{2} - 1\frac{1}{3}) \div 1\frac{5}{9}$.

$$(2\frac{1}{2} - 1\frac{1}{3}) \div 1\frac{5}{9} = (\frac{5}{2} - \frac{4}{3}) \div \frac{14}{9}$$

$$= \frac{5 \times 3 - 4 \times 2}{6} \div \frac{14}{9}$$

$$= \frac{7}{6} \div \frac{14}{9}$$

$$= \frac{\cancel{7}^{1}}{\cancel{6}_{2}} \times \frac{\cancel{9}^{3}}{\cancel{14}_{2}}$$

$$= \frac{1 \times 3}{2 \times 2}$$

$$= \frac{3}{4}$$

Exercise 3.12

Work out:

1. $\frac{1}{4} \div (1\frac{1}{8} \times \frac{2}{5})$
2. $1\frac{2}{3} \div (\frac{3}{5} \div \frac{9}{10})$
3. $(1\frac{7}{8} \times 2\frac{2}{5}) - 3\frac{2}{3}$
4. $(2\frac{2}{3} + 1\frac{1}{5}) \div 5\frac{4}{5}$
5. $3\frac{2}{3} \div (\frac{2}{3} + \frac{4}{5})$
6. $\frac{2}{5} \times (\frac{2}{3} - \frac{1}{4}) + \frac{1}{2}$
7. $2\frac{8}{9} \div (1\frac{2}{3} + \frac{1}{2})$
8. $(2\frac{1}{2} - 1\frac{3}{8}) \times 1\frac{1}{3}$

Practical Applications of Fractions

The examples which follow are all of a practical nature. They depend upon fractions for their solution.

Example 20

(a) A girl spends $\frac{3}{4}$ of her pocket money and has 90p left. How much did she have to start with?

The whole amount of her pocket money is represented by 1, so the amount left is represented by

$$1 - \frac{3}{4} = \frac{1}{4}$$

$\frac{1}{4}$ represents 90p

1 represents $4 \times 90p = 360p$

The girl had £3.60 to start with.

(b) A group of school children went to a hamburger bar. $\frac{2}{5}$ of them bought hamburgers, $\frac{1}{4}$ bought chips and the remainder bought drinks.

(i) What fraction bought food?

(ii) What fraction bought drinks?

(i) Fraction who bought food $= \frac{2}{5} + \frac{1}{4}$

$$= \frac{8 + 5}{20}$$

$$= \frac{13}{20}$$

(ii) The whole group of school children is a whole unit, i.e. 1. Therefore

Fraction who bought drinks

$$= 1 - \frac{13}{20}$$

$$= \frac{20 - 13}{20}$$

$$= \frac{7}{20}$$

Exercise 3.13

1. Calculate $\frac{3}{16}$ of £800.

2. Jane takes $5\frac{3}{4}$ minutes to iron a blouse. How many blouses can she iron in 23 minutes?

3. At a Youth Club $\frac{2}{5}$ of those present were playing darts and $\frac{1}{4}$ were playing other games.

   (a) What fraction were playing games?

   (b) What fraction were not playing games?

4. A watering can holds $12\frac{1}{2}$ litres. It is filled 11 times from a tank containing 400 litres. How much water is left in the tank?

5. A school has 600 pupils. $\frac{1}{5}$ are in the upper school, $\frac{1}{4}$ in the middle school and the remainder in the lower school. How many pupils are in the lower school?

6. A boy spends $\frac{5}{8}$ of his pocket money and has 60p left. How much money did he have to start with?

7. During 'bob a job week' a boy scout decided to earn money by cleaning shoes. It takes him $2\frac{1}{2}$ minutes to clean one pair. At one house he was given 12 pairs to clean. How long did it take him to complete the task?

8. The profits of a business are £29 000. It is shared between two partners A and B. If A receives $\frac{9}{20}$ of the profits, how much money does B receive?

Miscellaneous Exercise 3

Section A

1. Express $\frac{3}{8} \times \frac{5}{6} \times \frac{4}{15}$ as a single fraction.

2. Find the value of $\frac{2}{3} - \frac{4}{9}$.

3. Find the next three terms of the sequence $1\frac{3}{4}, 2, 2\frac{1}{4}, \ldots$

4. Work out $6 - 1\frac{7}{8}$.

5. Make $3\frac{1}{5} + 1\frac{5}{8}$ into a single fraction.

6. What is the square root of $\frac{16}{49}$?

7. Which is the smallest of the fractions $\frac{2}{3}, \frac{11}{15}, \frac{7}{10}$ and $\frac{5}{6}$?

8. What is $\frac{2}{3}$ of 60?

9. Write down the next two terms of the sequence $2, 1\frac{2}{3}, 1\frac{1}{3}, \ldots$

10. Write down the next term of the sequence $\frac{1}{2}, \frac{3}{4}, \frac{9}{8}, \ldots$

Section B

1. Consider the fractions $\frac{3}{5}, \frac{3}{4}, \frac{7}{10}$ and $\frac{13}{20}$.

   (a) Which fraction is the largest?

   (b) What is the difference between the largest and smallest fractions?

   (c) Calculate the product of $\frac{3}{5}$ and $\frac{3}{4}$.

   (d) Divide $\frac{3}{5}$ by $\frac{13}{20}$.

2. Give the answers to the following in their lowest terms:

   (a) $\frac{3}{5} \times \frac{10}{21}$ (b) $\frac{5}{8} \div 2\frac{3}{4}$

   (c) $\frac{7}{8} - \frac{2}{3}$ (d) $1\frac{1}{3} + 2\frac{3}{4}$

3. Find $\frac{3}{4}$ of $\frac{2}{3}$.

4. Find the next three terms in the sequence $2\frac{1}{4}, 2\frac{5}{8}, 3, \ldots$

5. Find $\frac{5}{8}$ of 32.

6. Write down the next fraction in the sequence $\frac{3}{4}, \frac{9}{16}, \frac{3}{8}, \ldots$

7. Express $\frac{1}{2} - \frac{3}{4} + \frac{5}{8} - \frac{7}{16} + \frac{19}{32}$ as a single fraction.

8. Express as a single fraction $(2\frac{1}{2} + 1\frac{1}{4}) \div 2\frac{1}{2}$.

9. A man left $\frac{3}{8}$ of his money to his wife and half the remainder to his son. If he left £8000, how much did his son receive?

10. Find $\frac{3}{4}$ of $7\frac{1}{3}$.

Multi-Choice Questions 3

1. The value of $(\frac{3}{4} \times 1\frac{1}{3}) - \frac{8}{15}$ is

   A $\frac{17}{60}$ B $\frac{7}{15}$ C $\frac{11}{20}$ D $\frac{3}{5}$

2. In its simplest terms $\frac{3}{5} \times \frac{5}{12} =$

   A $\frac{8}{60}$ B $\frac{1}{4}$ C $\frac{5}{17}$ D $\frac{15}{12}$

3. Which of the fractions $\frac{5}{6}, \frac{3}{4}, \frac{2}{3}$ and $\frac{4}{5}$ has the greatest value?

   A $\frac{5}{6}$ B $\frac{3}{4}$ C $\frac{2}{3}$ D $\frac{4}{5}$

4. $(\frac{1}{4})^2$ is equal to

   A $\frac{1}{16}$ B $\frac{1}{8}$ C $\frac{2}{8}$ D $\frac{1}{2}$

5. The value of $3 \times (\frac{1}{2} \times \frac{1}{4})$ is

   A 1   B $1\frac{3}{4}$ C $\frac{3}{8}$ D $3\frac{3}{4}$

6. Find the value of $\frac{3}{4} \div \frac{8}{9}$ expressed in its lowest terms.

   A $\frac{24}{36}$ B $\frac{2}{3}$ C $\frac{27}{32}$ D $\frac{3}{2}$

7. The mixed number $3\frac{5}{6}$ is equal to

   A $\frac{5}{18}$ B $\frac{15}{18}$ C $\frac{15}{6}$ D $\frac{23}{6}$

8. One of the following fractions is equal to $\frac{4}{9}$. Which one?

   A $\frac{4}{36}$ B $\frac{20}{90}$ C $\frac{12}{27}$ D $\frac{36}{4}$

9. The top-heavy fraction $\frac{104}{14}$ when written as a mixed number in its lowest terms is

   A $7\frac{3}{7}$ B $\frac{364}{49}$ C $7\frac{6}{14}$ D $\frac{52}{7}$

10. $(1\frac{2}{3})^2$ is equal to

    A $2\frac{7}{9}$ B $2\frac{4}{9}$ C $1\frac{2}{3}$ D $1\frac{4}{9}$

11. $(3\frac{1}{3} - 2\frac{1}{2}) \div \frac{5}{12}$ in its simplest form is

    A $\frac{5}{12}$ B $\frac{5}{6}$ C 1   D 2

12. $(\frac{1}{3} - \frac{1}{6}) \div (\frac{1}{3} + \frac{1}{6})$ is equal to

    A $\frac{1}{6}$ B $\frac{1}{3}$ C $\frac{1}{2}$ D 2

Mental Test 3

Try to write down the answers to the following without writing anything else.

1. Work out $\frac{3}{8} + \frac{1}{2}$.

2. Find the sum of $\frac{3}{10}$ and $\frac{7}{20}$.

3. Take $\frac{3}{5}$ from $\frac{7}{10}$.

4. Add $\frac{1}{2}$, $\frac{1}{4}$ and $\frac{1}{8}$.

5. Find the product of $\frac{1}{2}$ and $\frac{1}{5}$.

6. Work out $\frac{2}{3} \times \frac{3}{7}$.

7. Work out $\frac{2}{5} \times \frac{5}{8}$.

8. What is $\frac{1}{4}$ of 20?

9. Work out $\frac{1}{5} \div 2$.

10. Work out $\frac{1}{4} \div \frac{1}{2}$.

11. Write down $\frac{12}{15}$ in its lowest terms.

12. Write $5\frac{2}{7}$ as a top-heavy fraction.

The Decimal System

Place Value

When we write the number 666 we mean
$600 + 60 + 6$. Reading from left to right,
each figure 6 is ten times the value of the
next 6.

We now have to decide how to deal with
fractional quantities, i.e. quantities whose
values are less than 1. If we regard 666.666
as meaning $600 + 60 + 6 + \frac{6}{10} + \frac{6}{100} + \frac{6}{1000}$
then the dot, called the **decimal point**,
separates the whole numbers from the
fractional parts.

Notice that in the fractional or decimal part
.666, each figure 6 has ten times the value of
the following 6. So $\frac{6}{10}$ has ten times the
value of $\frac{6}{100}$ and this has ten times the value
of $\frac{6}{1000}$.

Decimals, then, are fractions which have
denominators of 10, 100, 1000, etc., according to the position of the figure after the
decimal point.

If we want to write six hundred and five we
write 605, the zero keeping the place for the
missing tens. In the same way, to write
$\frac{3}{10} + \frac{5}{1000}$ we write .305, the zero keeping
the place for the missing hundredths. Also,
$\frac{4}{100} + \frac{7}{1000}$ would be written .047, the zero
keeping the place for the missing tenths.

When there are no whole numbers it is usual
to put a zero in front of the decimal point
so that, for instance, .35 would be written
0.35.

A table similar to the following may be
used to illustrate place value.

Hundreds	Tens	Units	.	Tenths	Hundredths	Thousandths	Number in decimal form
		9	.	6	5		9.65
	8	3	.	2	7	4	83.274
		0	.	3			0.3
		0	.	4	7		0.47
9	7	6	.	0	5	3	976.053

From the table we see that:

the first column after the decimal point
represents tenths;

the second column after the decimal point
represents hundredths;

the third column after the decimal point
represents thousandths and so on.

Example 1

(a) Write down the place value of the figure 7
in the number 54.374.

> Since the 7 occurs in the second column
> after the decimal point, its value is
> 7 hundredths, i.e. $\frac{7}{100}$.

(b) Find the difference between the actual
values of the two sevens in the number
679.73.

> The first seven has a value of 70.
>
> The second seven has a value of $\frac{7}{10}$.
>
> The difference between the actual
> values of the two sevens is
> $70 - \frac{7}{10} = 69\frac{3}{10} = 69.3$

Exercise 4.1

Write down the place value of

1. The figure 9 in
   (a) 34.79   (b) 5.369
   (c) 475.92   (d) 89.753

2. The figure 5 in
   (a) 579.2   (b) 0.059
   (c) 28.57   (d) 7.852

3. The figure 8 in
   (a) 8734.9   (b) 0.0083
   (c) 19.84   (d) 77.784

4. What is the numerical difference between the actual values of the two fours in the number 0.404?

5. What is the numerical difference between the actual values of the two nines in the number 594.93?

6. Read off as decimal numbers:
   (a) $\frac{7}{10}$ (b) $\frac{3}{10} + \frac{7}{100}$
   (c) $\frac{5}{10} + \frac{8}{100} + \frac{9}{1000}$ (d) $\frac{9}{1000}$
   (e) $\frac{3}{100}$ (f) $\frac{1}{100} + \frac{7}{1000}$

7. Read off the following with denominators (bottom numbers) 10, 100, 1000:
   (a) 0.2   (b) 4.6
   (c) 3.58   (d) 0.256
   (e) 0.004   (f) 0.036
   (g) 80.29   (h) 0.032

Number Line for Decimals

Fig. 4.1 shows a number line for decimal numbers. The numbers −1.3, 2.4 and 5.8 have been shown.

Fig. 4.1

Exercise 4.2

1. Make a copy of the scale shown in Fig. 4.1. Draw arrows and label the following points:
   (a) 7.2   (b) 3.7   (c) 4.2
   (d) 0.7   (e) −0.8   (f) −1.8

2. By looking at the number line (Fig. 4.1) decide which number is the larger:
   (a) 0.4 or 0.8   (b) 3.2 or 4.9
   (c) −2.0 or 2.0   (d) −0.8 or 0.3
   (e) −1.9 or 1.2

3. Rearrange the following numbers in order of size, largest first: −9, −29 and −3.

4. Write down in order of size, smallest first, the numbers −4, 0, −6 and 3.

5. Write the following numbers in order of size, smallest first: 5.205, −52.05, 5.502, 5.025 and −5.052.

The Decimal Point and Following Zeros

It is sometimes useful to add zeros at the end of numbers, particularly when adding and subtracting decimal numbers.

The number 3 means 3 units.

The number 3.0 means 3 units and no tenths.

The number 3.00 means 3 units, no tenths and no hundredths.

The number 8.6 means 8 units and six tenths.

The number 8.60 means eight units, six units and no hundredths.

Sometimes noughts are placed in front of integers. For instance, a bearing could be 036°, and James Bond is known as 007!

Adding and Subtracting Decimals

Adding and subtracting decimals is done in exactly the same way as for whole numbers. Care must be taken, however, to write the decimal points underneath one another. This makes sure that all the figures having the same place value fall in the same column. Putting zeros at the end of numbers so that all the numbers to be added have the same number of figures after the decimal point also helps.

Example 2

Add 11.362, 2.63 and 27.1.

$$
\begin{array}{r}
11.362 \\
2.630 \\
27.100\ + \\
\hline
41.092
\end{array}
$$

Therefore
$11.362 + 2.63 + 27.1 = 41.092$.

Most problems involving the addition of numbers will be done using a calculator. The method is as follows:

Input	Display
11.362	11.362
+	11.362
2.63	2.63
+	13.992
27.1	27.1
=	41.092

The addition may be checked by inputting the numbers in reverse order, i.e. $27.1 + 2.63 + 11.362$.

Example 3

Subtract 8.34 from 17.2.

$$
\begin{array}{r}
17.20 \\
8.34\ - \\
\hline
8.86
\end{array}
$$

Therefore $17.2 - 8.34 = 8.86$.

Using a calculator the programme is:

Input	Display
17.2	17.2
—	17.2
8.34	8.34
=	8.86

Subtraction may always be checked by adding the bottom two numbers on the calculation. If this sum equals the top number the subtraction is correct. Thus in Example 3:

$$8.34 + 8.86 = 17.20$$

Hence the answer is correct.

Many problems involve combined addition and subtraction. More often than not a calculator will be used, as shown in Example 4.

Example 4

Work out $9.24 - 35.7 - 58.32 + 6.53 + 75.6$

Input	Display
9.24	9.24
—	9.24
35.7	35.7
—	−26.46
58.32	58.32
+	−84.78
6.53	6.53
+	−78.25
75.6	75.6
=	−2.65

Hence $9.24 - 35.7 - 58.32 + 6.53 + 75.6 = -2.65$.

The work may be checked by inputting the numbers in reverse order.

Exercise 4.3

Find the value of:

1. $2.375 + 0.63$

2. $3.196 + 2.48 + 18.5$

3. $38.7 + 0.05 + 23$

4. $27.418 + 0.97 + 25 + 1.967$

5. $12.48 - 3.5$ 6. $0.867 - 0.039$

7. $5.48 - 0.0691$

8. $87.3 - 15.64 + 3.28 - 56.25$

9. $7.21 - 4.48 - 12.57 - 54.2 + 37.91 + 86.56$

10. $23.65 - 34.09 - 27.5 - 36.23 - 11.33 + 5.7$

Multiplying by Powers of 10

To multiply by 10 move all the figures one place to the left.

$$8.532 \times 10 = 85.32$$

$$0.083 \times 10 = 0.83$$

To multiply by 100 move all the figures two places to the left.

$$97.48 \times 100 = 9748$$

$$7.638 \times 100 = 763.8$$

To multiply by 1000 move all the figures three places to the left.

$$975.364 \times 1000 = 975\,364$$

$$0.007\,34 \times 1000 = 7.34$$

$$2.9 \times 1000 = 2900.0$$

$$= 2900$$

Dividing by Powers of 10

To divide by 10 move all the figures one place to the right.

$$57.48 \div 10 = 5.748$$

$$0.279 \div 10 = 0.0279$$

To divide by 100 move all the figures two places to the right.

$$598.65 \div 100 = 5.9865$$

$$4.738 \div 100 = 0.047\,38$$

To divide by 1000 move all the figures three places to the right.

$$9438 \div 1000 = 9.438$$

$$25.472 \div 1000 = 0.025\,472$$

Exercise 4.4

Multiply the following numbers by
(a) 10, (b) 100 and (c) 1000:

1. 0.45   2. 7.893   3. 0.058

4. 89.2346   5. 8.2643   6. 0.063

7. 0.000 28   8. 0.008

Divide each of the following numbers by
(a) 10, (b) 100 and (c) 1000:

9. 289   10. 28.17   11. 827.34

12. 0.04   13. 0.625   14. 0.0038

15. 6.52

Multiplying and Dividing Decimals

Multiplying and dividing longhand is very time consuming and it is best to use a calculator for these operations.

Example 5

Multiply 18.23 by 9.76.

Input	Display
18.23	18.23
$\times$	18.23
9.76	9.76
$=$	177.9248

Therefore $18.23 \times 9.76 = 177.9248$. To check the work, input the numbers in reverse order, i.e. 9.76×18.23.

Example 6

Divide 60.888 by 3.54.

Input	Display
60.888	60.888
$\div$	60.888
3.54	3.54
$=$	17.2

Therefore $60.888 \div 3.54 = 17.2$

Many problems occur that involve both multiplication and division together and these are best done using a calculator.

Example 7

Work out $\dfrac{12.88 \times 2.59}{4.6 \times 0.74}$

Input	Display
12.88	12.88
×	12.88
2.59	2.59
÷	33.3592
4.6	4.6
÷	7.252
0.74	0.74
=	9.8

Therefore $\dfrac{12.88 \times 2.59}{4.6 \times 0.74} = 9.8$

Exercise 4.5

Work out each of the following using a calculator:

1. 25.42×29.23
2. 0.361×2.63
3. 0.76×0.38
4. 3.025×2.45
5. 0.043×0.32
6. $44.8 \div 12.8$
7. $3.484 \div 0.26$
8. $3.914 \div 2.06$
9. $11.56 \div 2.72$
10. $0.0144 \div 0.32$
11. $\dfrac{53.04 \times 0.806}{3.4 \times 0.26}$
12. $\dfrac{1.6856 \times 14.9}{0.028}$
13. $\dfrac{3.122 \times 0.703}{11.15 \times 0.037}$
14. $\dfrac{65.178 \times 0.1674}{19.17 \times 0.62}$
15. $\dfrac{0.001\,35 \times 0.008\,37}{0.003 \times 0.027}$

Decimal Places

Example 8

Work out $15.187 \div 3.57$.

Using a calculator

$$15.187 \div 3.57 = 4.254\,061\,6$$

This is the limit of accuracy of the calculator

but it may not be the correct answer. It is the answer correct to 7 decimal places because this is the number of places following the decimal point.

For many purposes numbers can be approximated by stating them to so many decimal places (d.p.).

If the first figure to be discarded is 5 or more the previous figure is increased by 1.

Example 9

$$93.7257 = 93.726 \quad \text{correct to 3 d.p.}$$
$$= 93.73 \quad \text{correct to 2 d.p.}$$
$$= 93.7 \quad \text{correct to 1 d.p.}$$

Notice carefully how zeros must be kept to show the size of the number or to indicate that it is one of the decimal places.

Example 10

(a) $0.007\,362 = 0.007 \quad \text{correct to 3 d.p.}$
$$= 0.01 \quad \text{correct to 2 d.p.}$$

(b) $7.601 = 7.60 \quad \text{correct to 2 d.p.}$
$$= 7.6 \quad \text{correct to 1 d.p.}$$

Exercise 4.6

Write down the following numbers correct to the number of decimal places stated:

1. 19.372 (a) 2 d.p. (b) 1 d.p.
2. $0.007\,519$ (a) 5 d.p. (b) 3 d.p. (c) 2 d.p.
3. 4.9703 (a) 3 d.p. (b) 2 d.p.

Use a calculator to find the value of

4. $18.89 \div 14.2$ correct to 2 d.p.
5. $0.0396 \div 2.51$ correct to 3 d.p.
6. $7.21 \div 0.038$ correct to 2 d.p.
7. $\dfrac{184.3 \times 0.0063}{11.43 \times 0.7362}$ correct to 4 d.p.
8. $\dfrac{765.8 \times 0.000\,116}{178.2 \times 26.43}$ correct to 7 d.p.

Significant Figures

A second way of approximating a number is to use **significant figures.** In the number 2179 the 2 represents the most significant figure because it has the greatest value. Similarly 2 and 1 are the two most significant figures whilst 2, 1 and 7 are the three most significant figures. The rules regarding significant figures are as follows:

(1) If the first figure to be discarded is 5 or more, increase the previous figure by 1.

$$7.192\,53 = 7.1925 \quad \text{correct to 5 s.f.}$$
$$= 7.193 \quad \text{correct to 4 s.f.}$$
$$= 7.19 \quad \text{correct to 3 s.f.}$$
$$= 7.2 \quad \text{correct to 2 s.f.}$$

(2) Zeros must be kept to show the position of the decimal point or to indicate that zero is a significant figure.

$$35\,291 = 35\,290 \quad \text{correct to 4 s.f.}$$
$$= 35\,300 \quad \text{correct to 3 s.f.}$$
$$= 35\,000 \quad \text{correct to 2 s.f.}$$
$$0.0739 = 0.074 \quad \text{correct to 2 s.f.}$$
$$= 0.07 \quad \text{correct to 1 s.f.}$$
$$18.403 = 18.40 \quad \text{correct to 4 s.f.}$$
$$= 18.4 \quad \text{correct to 3 s.f.}$$

Unnecessary Zeros

A possible source of confusion is in deciding which noughts are necessary and which are not.

(1) Noughts are not needed after the last figure of a number unless the nought is a significant figure.

$$6.300 = 6.3; \quad 4.70 = 4.7; \quad 3.00 = 3$$

The number half way between 5.4 and 5.8 is 5.6.

The number half way between 5.9 and 6.1 is 6.0. In this case the nought is a significant figure and should be left.

(2) Noughts are needed to keep the places for any missing hundreds, tens, units, tenths, hundredths, etc. The noughts in these numbers are needed:

$$70; \quad 7.205; \quad 800; \quad 9.005$$

(3) One zero is usually put before the decimal point if there are no whole numbers, but this is not essential.

.837 is usually written 0.837

.058 is usually written 0.058

Number of Significant Figures in a Calculation

The answer to a calculation should not contain more significant figures than the least number of significant figures used amongst the given numbers.

Example 11

Find the value of $1.3 \times 7.231 \times 1.24$

The least number of significant figures amongst the given numbers is two (for the number 1.3). Hence the product should only be stated to two significant figures.

$$1.3 \times 7.231 \times 1.24 = 12$$
$$\text{correct to 2 s.f.}$$

Exercise 4.7

Write down the following numbers correct to the number of significant figures stated:

1. 24.935 82
   (a) to 6 s.f.   (b) to 4 s.f.
   (c) to 2 s.f.

2. 0.008 357 1
   (a) to 4 s.f.   (b) to 3 s.f.
   (c) to 2 s.f.   (d) to 1 s.f.

3. 17 359 285
    (a) to 5 s.f.   (b) to 2 s.f.
    (c) to 1 s.f.

4. 0.078 03
    (a) to 3 s.f.   (b) to 2 s.f.
    (c) to 1 s.f.

Write the following numbers without un-
necessary zeros:

5. 48.90   6. 4.000   7. 0.5000

8. 600.00   9. 108 070

Each number in the following questions is
correct to the number of significant figures
shown. Use a calculator to work out the
answers and state them to the correct number of significant figures:

10. 15.64×19.75 11. 14.6×8.7

12. $13.96 \div 0.42$

13. $43.5 \times 0.87 \times 1.23$

14. $\dfrac{15.76 \times 8.3}{9.725}$

Rough Estimates for Calculations

Before doing a calculation you should always
do a rough estimate to make sure that the
answer is sensible. You can often see if there
is a mistake by looking at the size of the
answer. In doing a rough estimate try to
select numbers which are easy to add, subtract and multiply. If division is needed try
to select numbers which will cancel.

Example 12

(a) Multiply 32.4 by 0.259.

For a rough estimate we will take
$$32 \times 0.25 = 32 \times \tfrac{1}{4}$$
$$= 32 \div 4$$
$$= 8$$

Accurate calculation:
$$32.4 \times 0.259 = 8.39$$
$$\text{(correct to 3 s.f.)}$$

(The rough estimate shows that the
answer is not 83.9 or 0.839.)

(b) Add 5.32, 0.925 and 17.81.

For a rough estimate we will take
$$5 + 1 + 18 = 24$$

Accurate calculation:
$$5.32 + 0.925 + 17.81 = 24.055$$

(c) Work out $\dfrac{47.5 \times 36.52}{11.3 \times 2.75}$

For a rough estimate we will take
$$\frac{\overset{5}{\cancel{50}} \times \overset{12}{\cancel{36}}}{\underset{1}{\cancel{10}} \times \underset{1}{\cancel{3}}} = \frac{5 \times 12}{1 \times 1}$$
$$= 60$$

Accurate calculation:
$$\frac{47.5 \times 36.52}{11.3 \times 2.75} = 55.8$$
$$\text{(correct to 3 s.f.)}$$

Exercise 4.8

Do a rough estimate for each of the following
and then, using a calculator, work out the
accurate answer:

1. $18.25 + 39.3 + 429.8$

2. $76.815 - 37.23 - 9.63 - 28.27$

3. 22×0.57

4. 41.35×0.26

5. $0.732 \times 0.098 \times 2.17$

6. $92.17 \div 31.45$

7. $0.092 \div 0.035$

8. $27.18 \times 29.19 \times 0.030$

9. $\dfrac{1.456 \times 0.0125}{0.0532}$

10. $\dfrac{29.92 \times 31.32}{10.89 \times 2.95}$

Fraction-to-Decimal Conversion

In Chapter 1 we found that the line separating numerator and denominator of a fraction takes the place of a division sign. Thus

$\frac{17}{80}$ is the same as $17 \div 80$

Example 13

(a) Convert $\frac{27}{32}$ to a decimal number.

$$\tfrac{27}{32} = 27 \div 32 = 0.843\,75$$

(by using a calculator)

(b) Convert $2\frac{9}{16}$ to a decimal number.

With mixed numbers we need only convert the fractional part to a decimal number. Thus

$$2\tfrac{9}{16} = 2 + \tfrac{9}{16} = 2 + (9 \div 16)$$
$$= 2 + 0.5625$$
$$= 2.5625$$

Recurring Decimals

Consider the calculation

$$\tfrac{3}{8} = 3 \div 8 = 0.375 \qquad 8\,\overline{)3.000}$$
$$\phantom{\tfrac{3}{8} = 3 \div 8 = 0.375 \qquad 8)}0.375$$

By adding three noughts after the decimal point we are able to complete the division and give an exact answer.

Now consider

$$\tfrac{2}{3} = 2 \div 3 = 0.666\,666\ldots \qquad 3\,\overline{)2.000\,000\ldots}$$
$$\phantom{\tfrac{2}{3} = 2 \div 3 = 0.666\,666\ldots \qquad 3)}0.666\,666\ldots$$

We see that we will continue to obtain sixes for evermore and we say that the 6 **recurs**.

Now consider

$$\tfrac{7}{11} = 7 \div 11 = 0.636\,363\ldots$$

$$11\,\overline{)7.000\,000\ldots}$$
$$0.636\,363\ldots$$

Here the 63 recurs. Sometimes it is one figure which recurs and sometimes it is a group of figures. If one figure or a group of figures continues to recur we are said to have a **recurring decimal**.

To save writing so many figures the dot notation is used. For example

$$\tfrac{1}{3} = 1 \div 3 = 0.\dot{3} \quad (\text{meaning } 0.333\,333\ldots)$$
$$\tfrac{1}{6} = 1 \div 6 = 0.16\dot{6} \quad (\text{meaning } 0.166\,666\ldots)$$
$$\tfrac{5}{11} = 5 \div 11 = 0.\dot{4}\dot{5} \quad (\text{meaning } 0.454\,545\ldots)$$

Note that when two figures have dots over them they both recur.

$$0.4\dot{0}2\dot{7} = 0.402\,702\,702\,7\ldots$$

For all practical purposes we never need recurring decimals; what we need is a decimal given to so many significant figures or decimal places. Thus

$$\tfrac{2}{3} = 0.67 \quad (\text{correct to } 2 \text{ s.f.})$$
$$\tfrac{7}{11} = 0.636 \quad (\text{correct to } 3 \text{ d.p.})$$

Converting Decimals into Fractions

It will be remembered that decimals are fractions with denominators (bottom numbers) of 10, 100, 1000, etc. Thus

$$0.53 = \tfrac{53}{100} \quad \text{and} \quad 0.625 = \tfrac{625}{1000} = \tfrac{5}{8}$$

Example 14

Find the difference between $2\frac{5}{16}$ and 2.3214.

$$2\frac{5}{16} = 2 + (5 \div 16)$$
$$= 2 + 0.3125$$
$$= 2.3125$$
$$\text{Difference} = 2.3214 - 2\frac{5}{16}$$
$$= 2.3214 - 2.3125$$
$$= 0.0089$$

Exercise 4.9

Using a calculator convert the following fractions to decimals:

1. $\frac{3}{4}$ 2. $\frac{1}{5}$ 3. $\frac{7}{8}$ 4. $\frac{13}{16}$

5. $1\frac{5}{8}$ 6. $2\frac{19}{32}$ 7. $3\frac{15}{64}$

Write down the following recurring decimals correct to 5 d.p.:

8. $0.\dot{5}$ 9. $0.\dot{8}$ 10. $0.1\dot{7}$

11. $0.4\dot{5}$ 12. $0.3\dot{5}$ 13. $0.\dot{2}\dot{1}$

14. $0.\dot{4}2\dot{8}$ 15. $0.\dot{5}67\dot{1}$

Convert each of the following into a recurring decimal:

16. $\frac{2}{9}$ 17. $\frac{2}{11}$ 18. $\frac{7}{9}$

19. $\frac{5}{11}$ 20. $\frac{4}{9}$

Write down the following decimals as fractions in their lowest terms:

21. 0.3   22. 0.65   23. 0.375

24. 0.4375   25. 2.62   26. 1.75

27. 9.185   28. 7.36

29. Find the difference between $\frac{3}{16}$ and 0.17.

30. What is the difference between $5\frac{3}{8}$ and 3.627?

31. Work out $\frac{13}{16} - 0.723$.

32. Find the difference between 0.281 35 and $\frac{9}{32}$.

33. Subtract 0.295 from $\frac{19}{64}$.

Estimation

Suppose that you want to put an extension on to your house. You would explain to the builder what you require and ask him to prepare an **estimate**. This estimate is made to give you some idea how much the extension is likely to cost. It may be a little more or a little less than the actual cost. (If the builder tells you the exact cost of the work he is giving you a quotation.)

To make his estimate the builder would have to work out roughly what he would need to charge for building work. His estimate might be as follows:

Building materials	£2500
Electrical work	£300
Plumbing work	£500
Labour	£3600
Final estimate	£6900

His costs would only be approximate (i.e. not correct but nearly correct). The costs given above might be correct to the nearest £100, and we say that the figures have been **rounded** to the nearest £100.

Rounding is another convenient way of approximating numbers. For instance

297 rounded to the nearest 100 is 300

648 rounded to the nearest 50 is 650

82 rounded to the nearest 10 is 80

Exercise 4.10

Round the following numbers to the nearest 100:

1. 698   2. 703   3. 870

4. 3620   5. 4870

Round the following numbers to the nearest 50:

6. 57   **7.** 237   **8.** 490

9. 880   **10.** 17 385

Round the following numbers to the nearest 10:

11. 94   **12.** 376   **13.** 693

14. 589   **15.** 604

Types of Numbers

We have already seen (in Chapter 1) that:

Counting or **natural** numbers are the numbers 1, 2, 3, 4, ...

Whole numbers are the numbers 0, 1, 2, 3, 4, ...

Integers are whole numbers but they include negative numbers. Thus the numbers ..., $-2, -1, 0, 1, 2, 3, ...$ are all integers.

A number with a sign in front of it is called a **directed** number. Thus -75 is a negative directed number or a negative integer.

$\frac{2}{3}$ is a **fractional** number and $-\frac{5}{6}$ is a negative fraction. 0.325 is a positive **decimal** number whilst -0.874 is a negative decimal number.

Fractional and decimal numbers, both positive and negative, as well as integers can be shown on a number line (Fig. 4.2).

Fig. 4.2

Rational numbers are numbers which can be expressed as a fraction. Thus 0.625 is a rational number because it can be expressed as the fraction $\frac{5}{8}$. Recurring decimals are also rational numbers because they can be expressed as fractions.

Irrational numbers cannot be expressed as fractions. For instance $\sqrt{2} = 1.414\,213\ldots$ and it is impossible to find an exact value for $\sqrt{2}$, and hence $\sqrt{2}$ is an irrational number.

Not all square roots are irrational. For instance

$$\sqrt{9} = 3 \quad \text{and} \quad \sqrt{2.25} = 1.5 = \tfrac{3}{2}$$

and these two numbers are rational.

Exercise 4.11

1. Which of the following numbers are positive integers:

    $5, -8, \frac{2}{3}, 1\frac{1}{4}, 2.75$ and 198?

2. Which of the following are negative integers:

    $8\frac{1}{2}, -9, 7, -\frac{1}{3}, 11$ and $-4\frac{3}{5}$?

3. Which of the following are rational numbers and which are irrational:

    $1.57, \frac{1}{7}, -5.625, \sqrt{16}, \sqrt{15}, 6.76$ and $-3\frac{1}{2}$?

4. Which of the following are whole numbers:

    $\frac{1}{4}, -2, 0, 6, 3.1$ and -0.4?

5. Arrange in order of size, smallest first, the numbers:

    $3, -4.7, \frac{5}{8}, -3\frac{3}{4}, 0, 7, -3.4$ and 2.5.

Finding the Square Root of a Decimal Number

Most calculators have a square-root key. Its use is shown in Example 15.

Example 15

Find the square root of 27.53 correct to 2 decimal places.

Input	Display
27.53	27.53
$\sqrt{}$	5.2469 ...

$\sqrt{27.53} = 5.25$ to 2 decimal places

Exercise 4.12

Find the square root of

1. 1.72　　　　2. 8.61　　　　3. 3.84

4. 12.61　　　5. 76.22　　　6. 90.41

7. 128　　　　8. 342

Miscellaneous Exercise 4

Section A

1. Work out $0.017 \div 0.027$ correct to 2 decimal places.

2. What is 0.081 778 rounded to the nearest thousandth?

3. By rounding off the numbers to the nearest number of tens, obtain the best estimate for $\dfrac{27.5 \times 60.52}{11.3 \times 20.51}$

4. Write $\frac{9}{11}$ as a recurring decimal.

5. Find the value of $\dfrac{0.8 \times 7 \times 1.3}{2.1 \times 4}$ correct to 1 decimal place.

6. In the number 52.058, find the numerical difference between the actual values of the two digits 5.

7. Write the number 738.0584 correct to

    (a) 2 decimal places

    (b) 1 decimal place

8. Change the following fractions to decimal numbers:

    (a) $\frac{15}{16}$　　　　(b) $9\frac{4}{25}$

9. Find the difference between $3\frac{7}{20}$ and 3.37.

10. Find the sum of 3.98, 0.745 and 52.073.

Section B

1. Find the exact value of $\dfrac{4.2 \times 0.0462}{0.077}$

2. Round off 0.049 49

    (a) correct to 3 significant figures

    (b) correct to 3 decimal places

3. Calculate

    (a) 7.3×2.1　　　　(b) $52.16 \div 3.2$

4. Find the exact value of $\dfrac{45}{0.9} + \dfrac{6.6}{0.55}$

5. Express 0.166 66 correct to 3 decimal places.

6. Express 0.026 66 correct to 2 significant figures.

7. Express $27 \div 81$ as

    (a) a fraction in its lowest terms

    (b) a decimal number correct to 3 decimal places

8. Work out

    (a) 87.4×1000　　　(b) $3.26 \div 100$

9. Work out $4 + 0.7 - 5.38$.

10. Rearrange the numbers 0.3, -0.7, $\frac{2}{3}$, -4.2 and 1.6 in order of size, starting with the greatest value.

11. Find the sum of 3.8, 273.09 and 0.265.

12. Express as decimal numbers

    (a) $\frac{7}{16}$　　　　(b) $\frac{1}{20}$

13. Express as fractions in their lowest terms

    (a) 0.95　　　　(b) 0.32

14. Find the square root of 86.32 correct to 4 significant figures.

Multi-Choice Questions 4

1. In the number 9.8392 the value of the digit 3 is

    A $\frac{3}{1000}$ B $\frac{3}{100}$ C $\frac{3}{10}$ D 300

2. 307.847 correct to 2 significant figures is

    A 30   B 31   C 308   D 310

3. $2\frac{1}{5}$ written in decimal form is

    A 2.01   B 2.02   C 2.1   D 2.2

4. The number 0.030 905 correct to 3 significant figures is

    A 0.030   B 0.0309
    C 0.031   D 0.3090

5. Correct to 2 decimal places, 3.0394 is

    A 3.0   B 3.03   C 3.039   D 3.04

6. 0.72 is equivalent to

    A $\frac{72}{1000}$ B $\frac{18}{25}$ C $\frac{7}{10}$ D $7\frac{1}{5}$

7. The largest of the numbers 0.22, 2.20, 0.022 and 2.02 is

    A 2.02   B 0.22   C 2.20   D 0.022

8. The number $0.1\dot{6}$ correct to 4 significant figures is

    A 0.1616   B 0.1617
    C 0.1666   D 0.1667

9. One of the numbers below is irrational. Which one?

    A $\frac{2}{9}$ B 9.2   C $\sqrt{9.2}$ D -9.2

10. The number 0.047 85 rounded off to the nearest thousandth is

    A 0.047   B 0.0478
    C 0.0479   D 0.048

Mental Test 4

Try to give the answers to the following without writing anything else.

1. Add 1.2 and 1.35.

2. Find the sum of 0.23 and 0.032.

3. Subtract 0.73 from 1.85.

4. Multiply 3.2 by 5.

5. Find the product of 0.3 and 0.2.

6. Write down the place value of the digit 9 in the number 0.394.

7. Read off the number 0.4 as a fraction in its lowest terms.

8. Read off $\frac{53}{100}$ as a decimal number.

9. Which number is the larger, -7.6 or 4.3?

10. Multiply 0.78 by 10.

11. Find the product of 0.357 and 1000.

12. Divide 53.21 by 100.

13. Write the number 2.467 correct to 2 decimal places.

14. Write the number 53 648 correct to 2 significant figures.

15. Write the number 17.504 correct to 3 significant figures.

16. Convert $\frac{4}{5}$ to a decimal number.

17. Write the recurring decimal $0.1\dot{7}$ correct to 3 significant figures.

18. Convert $4\frac{3}{4}$ to a decimal number.

19. Round off the number 3623 to the nearest hundred.

20. Is the number $-\frac{2}{3}$ rational?

Measurement

Measurement of Length

In the metric system the standard unit of length is the metre (abbreviation: m). For some purposes the metre is too large a unit and it is therefore split up into smaller units as follows:

$$1 \text{ metre (m)} = 10 \text{ decimetres (dm)}$$
$$= 100 \text{ centimetres (cm)}$$
$$= 1000 \text{ millimetres (mm)}$$

In dealing with large distances the metre is too small a unit and large distances are measured in kilometres such that

$$1 \text{ kilometre (km)} = 1000 \text{ metres (m)}$$

Because the metric system is essentially a decimal system it is easy to convert from one unit to another by multiplying or dividing by 10, 100 or 1000.

Example 1

(a) Convert 28.35 m into millimetres.

$$28.35 \text{ m} = 28.35 \times 1000 \text{ mm}$$
$$= 28\,350 \text{ mm}$$

(b) Convert 879 cm into metres.

$$879 \text{ cm} = 879 \div 100 \text{ m} = 8.79 \text{ m}$$

(c) Convert 734 mm into centimetres.

$$734 \text{ mm} = 734 \div 10 \text{ cm} = 73.4 \text{ cm}$$

(d) Convert 87 652 m into kilometres.

$$87\,652 \text{ m} = 87\,652 \div 1000 \text{ km}$$
$$= 87.652 \text{ km}$$

Exercise 5.1

This exercise should be done mentally.
Convert the following to millimetres:

1. 5 cm
2. 4.8 cm
3. 0.9 cm
4. 2 m
5. 8.3 m
6. 28.1 m
7. 7 dm
8. 19.4 dm

Convert the following to centimetres:

9. 8 m
10. 48 m
11. 0.8 m
12. 28 dm
13. 3.25 dm
14. 800 mm
15. 8964 mm
16. 5.4 mm

Convert the following to metres:

17. 8000 mm
18. 25 634 mm
19. 54.3 mm
20. 300 cm
21. 37 864 cm
22. 9.35 cm
23. 18 km
24. 6.2 km

Convert the following to kilometres:

25. 9000 m
26. 460 m
27. 58 m
28. 1800 cm
29. 750 cm
30. 470 000 mm

In the imperial system lengths are measured in inches, feet, yards and miles.

$$12 \text{ inches (in)} = 1 \text{ foot (ft)}$$
$$3 \text{ feet (ft)} = 1 \text{ yard (yd)}$$
$$1760 \text{ yards (yd)} = 1 \text{ mile}$$

Example 2

(a) Change 468 inches into feet.

$$468 \text{ in} = 468 \div 12 \text{ ft} = 39 \text{ ft}$$

(b) Change 1548 inches into yards.

$$1548 \text{ in} = 1548 \div 36 \text{ yd} = 43 \text{ yd}$$

(c) Change 22 352 yards into miles.

$$22\,352 \text{ yd} = 22\,352 \div 1760 \text{ miles}$$
$$= 12.7 \text{ miles}$$

(d) Change 32 yards into inches.

$$32 \text{ yd} = 32 \times 36 \text{ in}$$
$$= 1152 \text{ in}$$

Exercise 5.2

A calculator should be used to work out the answers.

Change to feet:

1. 48 in   2. 30 in   3. 39 in

4. 74.4 in   5. 8 yd   6. 9.2 yd

7. 813 yd   8. 39.4 yd

Change to yards:

9. 216 in   10. 158.4 in

11. 586.8 in   12. 2021.4 in

13. 36 ft   14. 8.7 ft

15. 67.5 ft   16. 1209 ft

Change to inches:

17. 8 ft   18. 23 ft   19. 15.8 ft

20. 31.9 ft   21. 7 yd   22. 5.1 yd

23. 6.8 yd   24. 54.2 yd

Change to miles:

25. 5280 yd   26. 11 264 yd

27. 26 400 ft   28. 49 104 ft

Measurement of Mass

In the metric system light objects are weighed using grams or milligrams whilst heavier objects are weighed using kilograms or tonnes.

$$1 \text{ gram (g)} = 1000 \text{ milligrams (mg)}$$
$$1 \text{ kilogram (kg)} = 1000 \text{ grams (g)}$$
$$1 \text{ tonne (t)} = 1000 \text{ kilograms (kg)}$$

Example 3

(a) Change 900 g into kilograms.

$$900 \text{ g} = 900 \div 1000 \text{ kg} = 0.9 \text{ kg}$$

(b) Change 8 kg to grams.

$$8 \text{ kg} = 8 \times 1000 \text{ g}$$
$$= 8000 \text{ g}$$

(c) Change 47 000 mg to grams.

$$47\,000 \text{ mg} = 47\,000 \div 1000 \text{ g}$$
$$= 47 \text{ g}$$

(d) Change 43 500 kg to tonnes.

$$43\,500 \text{ kg} = 43\,500 \div 1000 \text{ t}$$
$$= 43.5 \text{ t}$$

Exercise 5.3

This exercise should be done mentally.

Change to kilograms:

1. 5000 g   2. 6980 g

3. 8 t   4. 11.2 t

Change to grams:

5. 9 kg   6. 39 kg

7. 0.75 kg   8. 45 000 mg

9. 1830 mg   10. 85 mg

Change to milligrams:

11. 7 g   12. 4.9 g

13. 0.3 g   14. 0.009 g

Change to tonnes:

15. 50 000 kg   **16.** 8570 kg

17. 453 kg   **18.** 45 kg

In the imperial system light objects are weighed in ounces. Heavier objects are measured in pounds whilst very heavy objects are measured in hundredweights or tons.

$$16 \text{ ounces (oz)} = 1 \text{ pound (lb)}$$

$$112 \text{ pounds (lb)} = 1 \text{ hundredweight (cwt)}$$

$$20 \text{ hundred-}$$
$$\text{weight (cwt)} = 1 \text{ ton}$$

$$1 \text{ ton} = 2240 \text{ lb}$$

Example 4

(a) How many ounces are there in 2 pounds?

$$2 \text{ lb} = 2 \times 16 \text{ oz}$$
$$= 32 \text{ oz}$$

(b) Change 419.2 ounces to pounds.

$$419.2 \text{ oz} = 419.2 \div 16 \text{ lb}$$
$$= 26.2 \text{ lb}$$

(c) Change 817.6 pounds into hundred-weight.

$$817.6 \text{ lb} = 817.6 \div 112 \text{ cwt}$$
$$= 7.3 \text{ cwt}$$

(d) How many pounds are there in 8.6 tons?

$$8.6 \text{ tons} = 8.6 \times 2240 \text{ lb}$$
$$= 19 264 \text{ lb}$$

Exercise 5.4

A calculator should be used to work out the answers.

Change to ounces:

1. $\frac{1}{4}$ lb   **2.** 5 lb

3. $4\frac{1}{2}$ lb   **4.** 8.3 lb

Change to pounds:

5. 112 oz   **6.** 4 oz

7. 12 oz   **8.** 52 oz

9. 4 cwt   **10.** 5.2 cwt

11. 6 tons   **12.** 7.3 tons

Change to hundredweights:

13. 448 lb   **14.** 436.8 lb

15. 8 tons   **16.** 12.4 tons

Change to tons:

17. 13 440 lb   **18.** 1344 lb

19. 80 cwt   **20.** 184 cwt

Measurement of Capacity

Fluids of various kinds are usually stored in tins, bottles and tanks. The amount of fluid that a container will hold is known as its **capacity**.

In the metric system capacities are measured in litres, centilitres and millilitres such that

$$1 \text{ litre } (\ell) = 100 \text{ centilitres (c}\ell)$$
$$= 1000 \text{ millilitres (m}\ell)$$

Example 5

(a) How many centilitres are there in 0.7 litre?

$$0.7 \ell = 0.7 \times 100 \text{ c}\ell$$
$$= 70 \text{ c}\ell$$

(b) Change 45 000 millilitres into litres.

$$45 000 \text{ m}\ell = 45 000 \div 1000 \ell$$
$$= 45 \ell$$

Exercise 5.5

Change to centilitres:

1. 6 ℓ   2. 4.3 ℓ   3. 0.25 ℓ

4. 500 mℓ   5. 8000 mℓ

Change to millilitres:

6. 2 ℓ   7. 0.3 ℓ

8. 0.005 ℓ   9. 50 cℓ

10. 3 cℓ

Change to litres:

11. 550 cℓ   12. 40 cℓ

13. 8 cℓ   14. 3250 mℓ

15. 90 mℓ

In the imperial system capacities are measured in fluid ounces, pints and gallons such that

$$20 \text{ fluid ounces (fl oz)} = 1 \text{ pint (pt)}$$
$$8 \text{ pints (pt)} = 1 \text{ gallon (gal)}$$

Example 6

(a) Change 7 gallons into pints.

$$7 \text{ gal} = 7 \times 8 \text{ pt}$$
$$= 56 \text{ pt}$$

(b) Change 54 fluid ounces into pints.

$$54 \text{ fl oz} = 54 \div 20 \text{ pt}$$
$$= 2.7 \text{ pt}$$

Exercise 5.6

Change into pints:

1. 60 fl oz   2. 25 fl oz

3. 45 fl oz   4. 9 gal

5. 6.25 gal

Change into fluid ounces:

6. 4 pt   7. 1.3 pt   8. 0.45 pt

9. 2 gal   10. 0.3 gal

Change into gallons:

11. 24 pt   12. 5.2 pt   13. 760 pt

14. 320 fl oz   15. 4000 fl oz

Conversion of Imperial and Metric Units

Conversions from metric units to imperial units and vice versa are often needed. Sometimes an accurate conversion is required but often an approximate conversion is good enough.

Metric Equivalents
1 inch = 2.5 centimetres = $2\frac{1}{2}$ cm approx. (i.e. 2 in ≈ 5 cm)
1 foot = 30.48 centimetres = 30 cm approx.
1 yard = 0.91 metres = 1 m approx.
1 kilometre = 0.621 miles = $\frac{5}{8}$ mile approx. (i.e. 5 miles ≈ 8 km)
1 kilogram = 2.205 pounds = $2\frac{1}{4}$ lb approx. (i.e. 4 kg ≈ 9 lb)
1 fluid ounce = 28.41 millilitres = 30 mℓ approx.
1 litre = 1.760 pints = $1\frac{3}{4}$ pt approx. (i.e. 4ℓ ≈ 7 pt)
1 gallon = 4.546 litres = $4\frac{1}{2}$ ℓ approx. (i.e. 2 gal ≈ 9 ℓ)

Example 7

(a) Given that 1 kg = 2.205 lb, find in pounds the weight of a 50 kg sack of potatoes.

$$50 \text{ kg} = 50 \times 2.205 \text{ lb} = 110.25 \text{ lb}$$

(b) 1 kilometre $= \frac{5}{8}$ mile approx. Find the distance in kilometres of 135 miles.

$$135 \text{ miles} = 135 \div \frac{5}{8} \text{ kilometres}$$

$$= \frac{\overset{27}{\cancel{135}}}{1} \times \frac{8}{\underset{1}{\cancel{5}}} \text{ kilometres}$$

$$= 27 \times 8 \text{ kilometres}$$

$$= 216 \text{ kilometres}$$

Exercise 5.7

1. If $1 \text{ kg} = 2\frac{1}{4} \text{ lb}$, find how many pounds there are in 64 kg.

2. Taking 1 fluid ounce to be 30 millilitres, how many fluid ounces are equivalent to 180 millilitres?

3. If 1 kilometre $= 0.621$ miles, find to the nearest mile the distance equivalent to 32 km.

4. If 8 kilometres = 5 miles approximately, find the distance in kilometres of 25 miles.

5. A woman buys 2 kg of tomatoes. How many pounds of tomatoes does she buy? (Take $4 \text{ kg} = 9 \text{ lb}$.)

6. A shopper buys 5 lb of apples. How many kilograms of apples does she buy if $1 \text{ kg} = 2.2 \text{ lb}$?

7. A consignment of steel weighing 5 tonnes is delivered to a factory. Taking $1 \text{ kg} = 2.205 \text{ lb}$, find the weight of the steel in tons.

8. Taking 1 litre $= 1\frac{3}{4}$ pints, convert 49 pints into litres.

Addition and Subtraction of Metric Quantities

Metric quantities are added and subtracted in the same way as decimal numbers. However, it is important that all the quantities should be in the same units.

Example 8

Add 15.2 m, 39.2 cm and 150.2 mm. State the answer in metres.

$$15.2 \text{ m} + 39.2 \text{ cm} + 150.2 \text{ mm}$$

$$= 15.2 \text{ m} + 0.392 \text{ m} + 0.1502 \text{ m}$$

$$= 15.7422 \text{ m}$$

Example 9

Subtract 158 mm from 7.3 m. State the answer in metres.

$$7.3 \text{ m} - 158 \text{ mm} = 7.3 \text{ m} - 0.158 \text{ m}$$

$$= 7.142 \text{ m}$$

Exercise 5.8

Add the following lengths, giving the answer in metres:

1. 47 cm, 5.83 m and 15 mm

2. 93 km, 462 m and 5 cm

3. 0.185 m, 7.36 cm and 8.2 mm

4. Add together 792 g, 15 000 mg and 1.265 kg, giving the answer in kilograms.

5. A greengrocer starts the day with 127 kg of apples. He sells $3\frac{1}{2}$ kg, 450 g and 25 kg. What weight of apples has he left?

6. A length of ribbon 2.5 m long has the following lengths cut from it: 25 cm, 863 mm and 70 cm. What length of ribbon, in centimetres, remains?

7. Subtract 15.2 cm from 0.78 m, giving the answer in millimetres.

8. Subtract 25 mm from 7.5 cm, giving the answer in centimetres.

Multiplication and Division of Metric Quantities

Metric quantities are multiplied and divided in exactly the same way as decimal numbers.

Example 10

57 lengths of wood each 95 cm long are required by a builder. What total length of wood, in metres, is needed?

$$57 \times 95 \, \text{cm} = 57 \times 0.95 \, \text{m}$$
$$= 54.15 \, \text{m}$$

Hence total length of wood needed is 54.15 m.

Example 11

Frozen peas are packed in bags containing 450 grams. How many full bags can be obtained from 2000 kg of peas?

$$450 \, \text{grams} = 0.45 \, \text{kilograms}$$
$$\text{Number of bags filled} = 2000 \div 0.45$$
$$= 4444$$

Exercise 5.9

1. 95 lengths of steel bar, each 127 cm long, are required by a toy manufacturer. Find the total length of bar, in metres, needed.

2. 209 lengths of cloth each 135 cm long have to be cut from a bale containing 300 m. What length of cloth, in metres, remains?

3. How many lengths of string, each 53 cm long, can be cut from a ball containing 25 m? What length of string remains?

4. How many pieces of ribbon each 36 cm long can be cut from a reel containing 30 m?

5. Butter is packed in 250-gram packs. How many packs can be obtained from 8 tonnes of butter?

6. A certain spice is packed in jars containing 32 grams. How many jars can be filled from 15 kg of spice?

7. How many doses of 5 millilitres can be obtained from a full medicine bottle with a capacity of 24 centilitres?

8. A certain type of tablet has a weight of 8.2 milligrams. How much, in kilograms, do 5000 of these tablets weigh?

Money

The British system of currency uses the pound as the basic unit. The only sub-unit used is pence, such that

$$100 \, \text{pence} = 1 \, \text{pound}$$

The abbreviation p is used for pence and £ is used for pounds. A decimal point is used to separate the pounds from the pence, for instance £3.58 means 3 pounds and 58 pence. pence.

There are two ways of expressing amounts less than £1. For example 74 pence may be written as 74p or £0.74.

5 pence may be written as 5p or £0.05.

Adding and Subtracting Sums of Money

The addition and subtraction of sums of money is done in exactly the same way as the addition and subtraction of decimal numbers.

Example 12

(a) Add £8.94, £3.28 and 95p.

When amounts are given in pence it is best to convert these to pounds.

$$£8.94 + £3.28 + 95p = £8.94 + £3.28$$
$$+ £0.95$$
$$= £13.17$$

(b) Subtract 47p from £5.00.

$$£5.00 - 47p = £5.00 - £0.47$$
$$= £4.53$$

Exercise 5.10

1. Add £2.15, £7.28 and £6.54.

2. Add 27p, 82p, 97p and 31p.

3. Add £15.36, £10.42, 75p, 86p and £73.25.

4. Subtract £5.49 from £12.62.

5. Subtract 79p from £3.27.

6. A lady spends amounts of £43.64, £59.76 and £87.49 in a departmental store. How much did she spend altogether?

7. A man deposited £540 in his bank account but withdrew amounts of £138.26, £57.49 and £78.56 before depositing a further £436. How much money has he in his account?

8. Emma had £10. She spent amounts of £3.58, £4.29 and £1.35. How much money had she left?

Multiplication and Division with Money

The methods used when multiplying and dividing sums of money are very similar to those used with decimal numbers.

Example 13

(a) Find the cost of 35 articles if each costs 82p.

Since 82p = £0.82

cost of 35 articles = £0.82 × 35
= £28.70

(b) If 37 similar articles cost £34.78, how much does each article cost?

Cost of each article = £34.78 ÷ 37
= £0.94

Exercise 5.11

1. Find the cost of 12 articles if each costs 12p.

2. How much do 87 articles at 6p each cost?

3. Calculate the cost of 23 articles if each costs £9.46.

4. Bread rolls cost 7p each. 140 are needed for a party. How much will they cost?

5. 12 similar articles cost £6.24. How much does each cost?

6. 41 similar articles cost £102.09. How much does each cost?

7. I bought 3 loaves at 43p each, $\frac{1}{2}$ kg of butter at £1.80 per kg and $\frac{1}{4}$ kg of cheese at £9.60 per kg. How much was the bill?

8. An electricity bill was £74.36 for a quarter (which is 13 weeks). How much was the cost of electricity per week?

Miscellaneous Exercise 5

Section A

1. What is the total cost of renting a television set for six months at £18.32 per month?

2. Calculate the total cost of three pairs of jeans at £18.83 per pair and four pairs of socks at 84p per pair.

3. How much change would I get from £5 when I buy 4 packets of coffee at 92p per packet.

4. (a) Change 12.34 kg into grams.
   (b) Change 123.4 cm into metres.
   (c) Change 1234 mℓ to litres.

5. How many pieces of string, each 30 cm long, can be cut from a piece of string 4.2 m long.

6. If $4\,\text{kg} = 9\,\text{lb}$ approximately, work out in pounds the weight of a bag of flour weighing 3 kg.

7. A large piece of cheese weighs $2\frac{1}{4}$ kg. Smaller pieces weighing 525 g, 485 g and 370 g are cut from it. What weight of cheese, in grams, is left?

8. A bottle contains 20 cℓ of medicine. How many 5 mℓ doses can be obtained from it?

9. $1\,\text{km} = \frac{5}{8}$ of a mile approximately. How far, in miles, is a distance of 160 km?

Section B

1. Express each of the following in metres:
   (a) 2 km
   (b) 2 km 56 m
   (c) 672 cm
   (d) 28 400 mm

2. How many 25-millilitre doses can be obtained from a bottle containing $1\frac{1}{2}$ litres.

3. How many (a) milligrams in 3 grams, (b) centimetres in 9 metres, (c) litres in 3500 millilitres?

4. How many lengths of tape, each 75 cm long, can be cut from a reel 10 m long? What is the length of the piece left over?

5. Express 45 cm as a fraction of 1 m. Express the fraction in its lowest terms.

6. Which is the heaviest: 80 g, 0.8 kg, 800 mg or 0.88 g?

7. A bank delivers £50 worth of 2 pence pieces to a supermarket. Calculate the weight of these coins in kilograms, if each coin weighs 6.92 grams.

8. Calculate the exact cost of 375 toys at £3.21 each.

9. Find the cost of taking 34 children on an outing at £2.65 each.

10. A cup costs 48p and a saucer 31p. How much change should be obtained from a £10 note after purchasing 8 cups and 8 saucers?

Multi-Choice Questions 5

1. 42 600 m expressed in kilometres is
   A 4.26 B 42.6 C 426 D 4260

2. 650 mm expressed in metres is
   A 0.065 B 0.65 C 6.5 D 65

3. Four packets have weights marked 2 kg, 500 g, $3\frac{1}{2}$ kg and 250 g. Their total weight in kilograms is
   A 5.25 B 6 C 6.25 D 75.55

4. A metre rule is broken into four equal pieces. The length of each piece is
   A 10 cm B 20 cm
   C 25 cm D 50 cm

5. A piece of string is 4 m long. $\frac{1}{5}$ of the string's length in centimetres is
   A 0.4 cm B 0.8 cm
   C 20 cm D 80 cm

6. The difference between 2 kg and 1800 grams, in grams, is
   A 1.6 B 2 C 20 D 200

7. If $1\,\text{km} = 0.6$ miles then 60 miles is equivalent to
   A 36 km B 100
   C 360 km D 1000 km

8. Calculate the cost of 400 g of bacon if 4 kg costs £12.20.
   A 30p B 61p
   C £1.22 D £12.20

9. Calculate the number of packets of chemical, each containing 240 g, that can be filled from a sack containing 120 kg.
   A 5 B 50 C 500 D 5000

10. The value of £6.92 − 75p + £0.96 is

A £5.21   B £5.23

C £7.13   D £7.27

Mental Test 5

Try to answer the following questions without writing anything down except the answer.

1. Convert 4 m into centimetres.

2. Change 5 m into millimetres.

3. Change 20 mm into centimetres.

4. Change 7 cm into millimetres.

5. Change 6000 m into kilometres.

6. Convert 9.2 km into metres.

7. Change 36 inches into feet.

8. Change 72 in into yards.

9. How many feet are there in 1 mile?

10. Change 4000 g into kilograms.

11. How many grams are there in 8 kg?

12. Convert 7500 mg into grams.

13. Convert 9.2 g into milligrams.

14. How many ounces are there in 20 lb?

15. How many pounds are there in 3 cwt?

16. Change 80 fluid ounces into pints.

17. How many fluid ounces are there in 5 pints?

18. How many centilitres are there in 4 litres?

19. Change 6 gallons into pints.

20. If $1\ \text{in} = 2\frac{1}{2}\ \text{cm}$, how many centimetres are there in 4 inches?

21. If $1\ \text{fl oz} = 30\ \text{m}\ell$, calculate the number of fluid ounces in 90 millilitres.

22. Add 500 cm and 8 m, giving the answer in metres.

23. Add 7 cm and 50 mm, giving the answer in centimetres.

24. Subtract 50 mm from 2 metres, giving the answer in millimetres.

25. How many doses of 5 millilitres can be obtained from a medicine bottle holding 1 litre?

26. Add £3.35 and 40p.

27. Subtract 27p from £1.50.

28. Find the cost of ten articles if each one costs 57p.

29. If 20 articles cost £10, how much does each one cost?

Ratio and Proportion

Ratio

Concrete is made by mixing sand and cement in the ratio 3 : 1 (three to one) and then adding water. You could put three sacks of sand to one sack of cement or three bucketfuls of sand to one bucketful of cement or three cups of sand to one cup of cement. These ratios are all 3 : 1.

As long as sand and cement are kept in the same proportion when mixed with water they will make concrete. The actual amount of sand and cement used only affects the amount of concrete made.

Example 1

Blackcurrant jam is made by mixing 4 kg of blackcurrants to 2 ℓ of water and 6 kg of sugar. This will make 8 kg of jam.

(a) If 4 kg of jam are required, work out the amount of each ingredient needed.

(b) If 12 kg of blackcurrants are used, how much sugar is required?

(c) If 4 ℓ of water is used how much sugar and how many blackcurrants are needed?

  (a) Since half the amount of jam is required, we need only half of the ingredients. Therefore 2 kg of blackcurrants, 1 ℓ of water and 3 kg of sugar are needed.

  (b) Since three times as many blackcurrants are used we need three times as much sugar: 18 kg.

(c) Since twice as much water is used we need twice as much of the other ingredients. Hence we need 8 kg of blackcurrants and 12 kg of sugar.

Model trains are made to a scale of 1 : 72. Every measurement on the model is $\frac{1}{72}$nd of the real measurement. The model measurements and the real measurements are in the ratio of 1 : 72.

Exercise 6.1

1. A recipe for a cake to serve four people is as follows:

    400 g butter   320 g castor sugar
    1 egg separated   800 g self-raising flour
    200 g currants   60 g mixed peel
    1 pinch of salt

    (a) Write down the ingredients needed to make a cake to serve eight people.
    (b) If 1200 g of butter is used, how much sugar is needed?
    (c) If 3200 g of flour is used, how much mixed peel is needed?
    (d) If 800 g of currants are used, how much flour is required?

2. A mortar mixture is made of cement, sand and water in the ratio 1 : 2 : 4. If 80 kg of cement is used, how much sand and water is required?

3. A metal alloy is made by mixing copper and zinc in the ratio of $5:1$. If 8 kg of zinc is used, how much copper is required?

4. Solder is made by mixing tin and lead in the ratio of $3:1$. If 5 kg of lead is used, work out
   (a) the weight of tin required
   (b) the total weight of solder made

5. A suitable batter for pancakes uses 120 grams of flour, 1 egg and 250 millilitres of milk. This is sufficient for 12 pancakes.
   (a) Work out the amount of each ingredient if 24 pancakes are to be made.
   (b) If 360 grams of flour is used, how many eggs should be used?
   (c) If 1 litre of milk is used, how many grams of flour should be used?

Simplifying Ratios

Problems are often made easier by putting the ratios into their simplest terms.

The ratios $1:3, 3:7$ and $9:4$ are in their simplest terms because there is no number which will divide exactly into both sides.

The ratio $8:6$ is not in its simplest terms because 2 will divide into both sides to give $4:3$, which is the same as $8:6$.

Example 2

Put the ratio $72:84$ in its simplest terms.

> 12 divides into 72 and 84, hence
>
> $72:84$ is the same as
> $72 \div 12:84 \div 12$. That is
> $72:84$ is the same as $6:7$,
> which is the ratio $72:84$
> in its simplest terms.

Exercise 6.2

Express each of the following ratios in its simplest terms:

1. $5:20$
2. $4:12$
3. $42:49$
4. $35:42$
5. $64:56$
6. $60:48$
7. $7500:5000$
8. $50:60:70$
9. $25:30:45$
10. $20:25:30:35$

Exercise 6.2 should bring out the similarities between ratios and fractions. Compare, for instance

$$42:49 = 6:7 \quad \text{with} \quad \frac{42}{49} = \frac{6}{7}$$

and $\quad 5:20 = 1:4 \quad \text{with} \quad \frac{5}{20} = \frac{1}{4}$

Exercise 6.3

Express the following ratios as fractions in their lowest terms:

1. $9:7$
2. $5:10$
3. $14:7$
4. $12:15$
5. $16:24$
6. $80:100$
7. $3:9$
8. $15:18$
9. $21:24$
10. $45:81$

To simplify a ratio such as $1\frac{1}{4}:\frac{1}{3}$ change the mixed number into a top-heavy fraction and then express each fraction with the same denominator (bottom number). The common denominator should be the LCM of the original denominators.

$1\frac{1}{4}:\frac{1}{3}$ is the same as $\frac{5}{4}:\frac{1}{3}$

The LCM of 4 and 3 is 12. Expressing each fraction with a denominator of 12 gives

$\frac{5}{4}:\frac{1}{3}$ is the same as $\frac{15}{12}:\frac{4}{12}$

We now multiply each side of the ratio by 12:

$\frac{15}{12}:\frac{4}{12}$ is the same as $15:4$

This is the ratio $1\frac{1}{4}:\frac{1}{3}$ expressed in its simplest terms.

Example 3

Express the ratio $1\frac{1}{3} : 2\frac{1}{2}$ in its simplest terms.

$$1\frac{1}{3} : 2\frac{1}{2} = \frac{4}{3} : \frac{5}{2} = \frac{8}{6} : \frac{15}{6} = 8:15$$

Exercise 6.4

Simplify the following ratios:

1. $\frac{3}{4} : \frac{1}{3}$ 2. $\frac{2}{5} : \frac{3}{10}$ 3. $\frac{1}{2} : \frac{3}{8}$

4. $\frac{1}{6} : \frac{5}{12}$ 5. $2\frac{1}{4} : \frac{1}{4}$ 6. $3\frac{1}{2} : 1\frac{1}{2}$

7. $\frac{5}{6} : 1\frac{2}{3}$ 8. $2\frac{3}{4} : 1\frac{1}{2}$ 9. $1\frac{1}{3} : 3\frac{1}{4}$

10. $2\frac{2}{5} : 5\frac{3}{20}$

Simplifying Ratios with Units

We can simplify the ratio $30\,\text{cm} : 80\,\text{cm}$ by first removing the units (because they are the same) and then dividing both sides by 10.

$$30\,\text{cm} : 80\,\text{cm} \text{ is the same as } 3:8$$

The ratio $5\,\text{cm} : 2\,\text{m}$ can be simplified by making the units the same and then removing them.

$$5\,\text{cm} : 2\,\text{m} = 5\,\text{cm} : 200\,\text{cm}$$
$$= 5:200 = 1:40$$

(by dividing both sides by 5)

Exercise 6.5

Simplify the following ratios:

1. $400\,\text{cm} : 100\,\text{cm}$ 2. $500\,\text{cm} : 1\,\text{m}$

3. $60\,\text{g} : 2\,\text{kg}$ 4. £1.50 : 50p

5. $5\,\text{kg} : 250\,\text{g}$ 6. $400\,\text{mm} : 4\,\text{m}$

7. $2\frac{1}{2}\,\text{km} : 500\,\text{m}$ 8. $3\,\text{cm} : 9\,\text{mm}$

9. £8.00 : 25p   10. $3\,\text{kg} : 90\,\text{g}$

Proportional Parts

The line AB (Fig. 6.1), whose length is 15 cm, has been divided into two parts in the ratio $2:3$. The line has been divided into its **proportional parts** and, as can be seen from the diagram, the line has been divided into a total of 5 parts. The length AC contains 2 of these parts and the length BC contains 3 of them. Each part is $15\,\text{cm} \div 5 = 3\,\text{cm}$ long. Hence

$$\text{AC} = 2 \times 3\,\text{cm} = 6\,\text{cm} \text{ and}$$
$$\text{BC} = 3 \times 3\,\text{cm} = 9\,\text{cm}.$$

Fig. 6.1

The problem of dividing the line into two parts in the ratio $2:3$ could be tackled in this way:

Total number of parts $= 2 + 3 = 5$

Length of each part $= 15\,\text{cm} \div 5 = 3\,\text{cm}$

Length of AC $= 2 \times 3\,\text{cm} = 6\,\text{cm}$

Length of BC $= 3 \times 3\,\text{cm} = 9\,\text{cm}$

Example 4

Divide £240 in the ratio $5:4:3$.

Total number of parts $= 5 + 4 + 3 = 12$

Amount of each part $= £240 \div 12 = £20$

Amount of first part $= £20 \times 5 = £100$

Amount of second part $= £20 \times 4 = £80$

Amount of third part $= £20 \times 3 = £60$

Direct Proportion

Example 5

Two lengths are in the ratio $8:5$. If the first length is 120 metres, what is the second length?

> The first length is represented by 8 parts. Each part $= 120$ metres $\div 8 = 15$ metres. The second length is represented by 5 parts. Therefore the second length is 5×15 metres $= 75$ metres.

Alternatively the problem may be tackled by using fractions:

> The second length $= \frac{5}{8}$ of the first length
>
> $\qquad = \frac{5}{8} \times 120$ metres
>
> $\qquad = 75$ metres

Exercise 6.6

1. Divide £800 in the ratio $5:3$.

2. Divide 160 kg in the ratio $7:3$.

3. Divide 120 m in the ratio $2:3:5$.

4. A line 1.68 metres long is to be divided into three parts in the ratio $2:7:11$. Find, in millimetres, the length of each part.

5. A sum of money is divided into two parts in the ratio $5:7$. If the smaller amount is £200, find the larger amount.

6. An alloy consists of copper, zinc and tin in the ratio $2:3:5$. Find the amount of each metal in 75 kg of the alloy.

7. A sum of money is shared into three parts in the ratio $2:4:5$. If the largest share is £40, what is the total amount of money shared?

8. Four friends contribute sums of money to a charitable organization in the ratio $2:4:5:7$. If the largest amount contributed is £1.40, calculate the total amount contributed by the four people.

Direct Proportion

Two quantities are said to **vary directly**, or be in **direct proportion**, if they increase or decrease at the same rate.

Thus the quantity of petrol used and the distance travelled by a motor car are in direct proportion.

Again, if we buy potatoes at 20 pence for 2 kilograms then we expect to pay 40 pence for 4 kilograms and 10 pence for 1 kilogram. If we double the amount bought we double the cost; if we halve the amount bought we halve the cost. The amount of potatoes bought and their cost are in direct proportion.

Example 6

If 7 pens cost 56p, how much will 5 pens cost?

> *Method 1* (the unitary method)
>
> 7 pens cost 56p
>
> 1 pen costs 56p $\div 7 = 8$p
>
> 5 pens cost 5×8p $= 40$p
>
> *Method 2* (the fractional method)
>
> 7 pens cost 56p
>
> 5 pens cost $\frac{5}{7} \times \frac{56}{1} = 40$p

You should decide for yourself which method, unitary or fractional, to use when solving a problem.

Exercise 6.7

1. If 74 exercise books cost £11.84, how much do 53 cost?

2. If 40 articles cost £35, how much do 55 articles cost?

3. Eggs cost 70 pence for 10. How much will 25 eggs cost?

4. A car travels 205 km on 20 litres of petrol. How much petrol will be needed for a journey of 369 km?

5. If 9 metres of stair carpet cost £21, how much will 96 metres cost?

6. A train travels 200 km in 4 hours. If it travels at the same rate, how long will it take to complete a journey of 350 km?

7. A machine makes 15 articles in $\frac{1}{2}$ hour. How many articles will it make in 3 hours?

8. 4 books weigh 800 grams. What is the weight, in kilograms, of 20 similar books?

Inverse Proportion

If an increase (or decrease) in one quantity produces a decrease (or increase) in a second quantity in the same ratio, the two quantities are in inverse proportion.

Example 7

Five men building a wall take 20 days to complete it. How long would it take four men to complete it?

5 men take 20 days

1 man takes 5×20 days $= 100$ days

(1 man takes longer so multiply by 5)

4 men take $100 \div 4 = 25$ days

(4 men take less time so divide by 4)

Alternatively: 4 men will take longer than 5 men so the time ratio is $5:4$. Hence

time taken by 5 men $= \frac{5}{4} \times \frac{20}{1}$

$= 25$ days

Exercise 6.8

1. 20 men working in a factory produce 3000 articles in 12 working days. How long would it take 15 men to produce the articles?

2. A farmer employs 12 men to harvest his potato crop. They take 9 days to do the job. If he had employed only 8 men, how long would it have taken them?

3. A bag contains sweets. When divided amongst 8 children each child receives 9 sweets. If the sweets are divided amongst 12 children, how many sweets would each child receive?

4. 4 people can clean an office in 6 hours. How many people would be needed to clean the office in 4 hours?

5. A farmer employs 15 women to pick his fruit. It takes them 3 days. How long would it take 10 women to pick the fruit?

6. Seven ladies take 30 minutes to make 21 toys. How long would it take 20 women to make the toys?

7. 8 women can do a piece of work in 60 hours. If the work is to be completed in 20 hours, how many women are needed?

Exercise 6.9

1. A car does 8 km per litre of petrol. How far will it go on $5\frac{1}{2}$ litres?

2. 7 men take 1 hour to make 21 model cars. How long would it take 21 men?

3. 10 men dig up a row of potatoes in 40 minutes. How long would it take 5 men?

4. A car travels 80 miles in 2 hours. If it keeps the same speed, how far will it travel in **(a)** 1 hour,   **(b)** 5 hours?

5. If 100 g of sweets costs 44p, find the cost of 300 g.

6. 8 people take 5 hours to pick a row of raspberries. How long would it take 4 people to do the work?

7. It took 4 men 2 hours to dig a hole in a road. How long would 12 men take?

8. 15 kg of potatoes costs £3.00. Calculate the cost of **(a)** 5 kg,   **(b)** 40 kg.

Measures of Rate

If a car travels 8 km on 1 litre of fuel we say that its fuel consumption is 8 km per litre of petrol. This is the rate at which the car consumes fuel.

The flow of water from a pipe or a tap is usually measured in gallons per minute or litres per minute. This is the rate of flow of the water.

When a car has a speed of 40 miles per hour this gives the rate at which it travels, that is 40 miles in one hour.

Example 8

(a) A car has a fuel consumption of 7 km per litre. How much fuel will be needed for a journey of 105 km?

$$\text{Fuel required} = \frac{\text{Length of journey}}{\text{Fuel consumption}}$$

$$= \frac{105}{7} \text{ litres}$$

$$= 15 \text{ litres}$$

(b) The flow of water from a pipe is 20 litres per minute. How long will it take to fill a container with a capacity of 70 litres?

$$\text{Time taken} = \frac{\text{Capacity of container}}{\text{Rate of flow}}$$

$$= \frac{70}{20} \text{ minutes}$$

$$= 3.5 \text{ minutes}$$

We see from this example that if we know the rate of flow we can work out the time taken to fill a container of known capacity. We can also work out the amount of water delivered in a given time. If butter, for instance, is sold at £1.60 per kg then this is the rate at which butter is sold. We can, knowing the cost per kilogram, easily work out the cost of any amount of butter.

Exercise 6.10

1. Potatoes are sold at £3.30 per 55 kg bag.
   (a) Calculate the cost per kilogram.
   (b) How much will 12 kg cost?

2. The flow of water from a tap is measured as 60 litres in 5 minutes.
   (a) Work out the rate of flow in litres per minute.
   (b) Calculate the time needed to fill a container with a capacity of 45 litres.
   (c) How much water will flow in 7 minutes if this rate of flow is maintained?

3. A car travels a distance of 60 miles on 2 gallons of petrol.
   (a) Calculate the fuel consumption of the car in miles per gallon.
   (b) How far would the car be expected to travel on 5 gallons of fuel?
   (c) How many gallons of fuel would be needed for a journey of 120 miles?

4. A car travels at an average speed of 50 km/h.

   (a) How far will the car travel in 3 hours if this speed is maintained?

   (b) How long will it take the car to travel 175 km?

5. The density of aluminium is 2700 kg per cubic metre. How much will 5 cubic metres of aluminium weigh?

6. The density of nylon is 1.16 grams per cubic centimetre. How many cubic centimetres of nylon weigh 8.12 grams?

7. A schoolboy found that he walked 100 metres in 50 seconds.

   (a) Calculate his speed in metres per second.

   (b) If he continues to walk at this speed, how long will it take him to walk a further 350 metres?

Rate of Exchange

Every country has its own monetary system. If there is to be trade and travel between any two countries there must be a rate at which the money of one country can be converted into money of the other country. This rate is called the **rate of exchange.**

Foreign Exchange Rates at June 1985

Country	Rate of exchange
Belgium	77.75 francs = £1
France	11.73 francs = £1
Germany	3.85 marks = £1
Greece	170 drachmas = £1
Italy	2450 lire = £1
Spain	220 pesetas = £1
United States	$1.27 = £1

Example 9

(a) If 220 Spanish pesetas = £1, find to the nearest penny the value in British money of 5000 pesetas.

Method 1 (the unitary method)

$$220 \text{ pesetas} = £1$$

$$1 \text{ peseta} = £\frac{1}{220}$$

$$5000 \text{ pesetas} = £\frac{1}{220} \times \frac{5000}{1}$$

$$= £\frac{5000}{220}$$

$$= £22.73$$

Method 2 (the fractional method)

$$5000 \text{ pesetas} = £\frac{5000}{220}$$

$$= £22.73$$

(b) A tourist changes traveller's cheques for £50 into French francs at 11.73 francs to £1. How many francs does he get?

$$£50 = 50 \times 11.73 \text{ francs}$$

$$= 587 \text{ francs}$$

Exercise 6.11

Use a calculator and where necessary state the answers correct to 2 decimal places.

Using the exchange rates given above find:

1. The number of German marks equivalent to £15.

2. The number of Spanish pesetas equivalent to £25.

3. The number of United States dollars equivalent to £8.

4. The number of pounds equivalent to 225 dollars.

5. The number of Belgian francs equivalent to £98.50.

6. The number of pounds equivalent to 8960 lire.

7. A transistor set costs £52.60 in London. An American visitor wishes to buy a set but wants to pay in dollars. How much will he pay in dollars?

8. A tourist changes traveller's cheques for £100 into Greek currency at 70 drachmas to £1. He spends 6240 drachmas and changes the remainder back into British money at £1 = 68 drachmas. How much will the tourist get for his drachmas?

Map Drawing and Model Scales

When drawings or models of large objects are to be made they are usually made to a scale, for example 1 cm = 10 m. This means that an actual distance of 10 m on the object will be represented by 1 cm on the drawing or model.

Example 10

(a) A house is drawn to a scale of 1 cm = 2 m. On the drawing the kitchen has a length of 4 cm. What is the actual length of the kitchen?

$$\text{Actual length of kitchen} = 4 \times 2 \text{ m}$$
$$= 8 \text{ m}$$

(b) A model of a train is made $\frac{1}{72}$nd full size. If the actual length of the locomotive is 108 ft, how long will the model be?

$$\text{Length of model} = \frac{1}{72} \times \frac{108}{1} \text{ ft}$$
$$= 1.5 \text{ ft}$$

On maps and drawings scales are often expressed as a ratio, for instance 1:1000. This means that a distance of 1000 m on the ground would be represented by 1 m on the map or drawing.

Example 11

(a) A road map is drawn to a scale of 2 km to 1 cm. Express this scale as a ratio.

$$2 \text{ km} = 2 \times 1000 \text{ m}$$
$$= 2 \times 1000 \times 100 \text{ cm}$$
$$= 200\,000 \text{ cm}$$

Since the map scale is 2 km to 1 cm, 1 cm on the map represents 200 000 cm on the ground. Therefore the map scale expressed as a ratio is 1:200 000.

(b) A map is drawn to a scale of 1:1 000 000. How many miles is represented by 1 inch?

1 inch on the map represents 1 000 000 inches on the ground.

$$1\,000\,000 \text{ inches} = \frac{1\,000\,000}{12} \text{ feet}$$
$$= \frac{1\,000\,000}{12 \times 5280} \text{ miles}$$
$$= 15.8 \text{ miles}$$

So the map scale is 15.8 miles to 1 inch (approximately 16 miles to the inch).

Exercise 6.12

1. A drawing of a house is made to a scale of 1 cm = 2 m. The length of the lounge measures 4 cm on the drawing. What is the actual length of the lounge?

2. The scale of a map is 1:20 000. What distance, in metres, does 8 cm on the map represent?

3. The model of a car is 50 times smaller than the real car.

    (a) What length on the real car does 4 cm on the model represent?
    (b) What length on the model would represent 3 m on the real car? Give your answer in centimetres.

4. The scale of a map is 1:100 000. The actual distance between two towns shown on the map is 25 km. What is the distance between the two towns, in centimetres, on the map?

5. A model aeroplane is built 72 times smaller than the real aeroplane. What is the real measurement represented by (a) 3 cm, (b) 0.5 cm?

6. On a map 1 cm represents 5 km.

    (a) What distance on the ground is represented by 0.6 cm on the map?

    (b) What length on the map represents 7.5 km on the ground?

7. The distance between two islands is measured as 8.1 cm on a map drawn to a scale of 1 : 5 000 000. What is the actual distance between the two islands?

8. A model car is made to a scale of 1 : 25. If the actual length of the car is 12 ft, how long is the model?

Miscellaneous Exercise 6

Section A

1. If 5 pencils cost 45p, work out the cost of 12 pencils.

2. A car does 7 km per litre of petrol. How far will it go on $4\frac{1}{2}$ litres?

3. If a man walks 3 km in 40 minutes, how many kilometres will he have walked in 60 minutes if he keeps going at the same rate?

4. At a wedding reception for 120 people it was estimated that one bottle of wine would be sufficient for 8 people.

    (a) How many bottles would supply the 120 guests?

    (b) At a cost of £2.90 per bottle, what would be the total bill for wine?

5. Simplify the ratio 175 : 200.

6. Simplify the ratio 2 km : 500 m.

7. Express the ratio 8 : 64 as a fraction in its lowest terms.

8. Is 100 g of cornflour at 70p a better buy than 160 g at £1.20?

Section B

1. In a school the ratio of the number of pupils to the number of teachers is 18 : 1. If the number of pupils is 540, how many teachers are there?

2. A model of an aeroplane is made to a scale of 1 : 78. If the wing span of the model is 35 cm find, in metres, the wing span of the actual aeroplane.

3. £979 is to be divided into three parts in the ratio 6 : 3 : 2. Calculate the value of the smallest part.

4. A car goes 12 km per litre of petrol. How many whole litres of petrol will be needed to be sure of completing a journey of 100 km?

5. A councillor gets a car allowance of 17.4 pence per mile. How much will he be paid for a journey of 20 miles?

6. A vending machine needs 20 litres of orangeade to fill 50 cups. How many litres are needed to fill 60 cups?

7. Change £150 into dollars when the exchange rate is $1.25 to £1.

8. If two men can paint a fence in 6 hours, how long will it take three men to paint it?

9. 2 dratt equal 1 assam and 1 dratt equals 4 yeda. How many yeda equal 1 assam?

10. Taking £1 as 12.06 francs, express 5.50 francs as pence correct to the nearest penny.

11. A car which travels 10 km on a litre of petrol requires $22\frac{1}{2}$ litres for a journey. Find the number of litres which would be required for the same journey by a car which has a fuel consumption of 4 km per litre.

12. The annual rent of a field amounts to £93.50. The rent is shared by two farmers in the ratio 15 : 7. Find the difference between the amounts of their shares.

13. Twelve bottles of claret cost £48.96. How many bottles can be purchased for £61.20?

14. When a petrol tank is $\frac{7}{8}$ full it contains $31\frac{1}{2}$ litres of petrol. How much petrol does the tank hold when full?

15. A car runs 4 km on a litre of petrol which costs 32 pence. Find the cost of petrol for a journey of 392 km.

Multi-Choice Questions 6

1. The ratio of A's share to B's share in the profits of a business is $5:4$. If the total profits are £4500, then A's share is

   A £900   B £2000
   C £2250   D £2500

2. If £120 is shared in the ratio $2:3$ then the smallest share is

   A £40   B £48   C £72   D £80

3. Change £150 into dollars when the exchange rate is $\$2.00 = £1$.

   A $75   B $125
   C $300   D $600

4. When £48 is divided in the ratio $5:7$, the smallest share is

   A £12   B £14   C £20   D 24

5. Given that $£1 = \$1.98$, then $33 is equivalent to

   A £6.53   B £16.67
   C £33.98   D £65.34

6. If $£1 = \$2.80$ and $£1 = 9.92$ francs, how many francs can be exchanged for $700?

   A 1692   B 1695
   C 2480   D 6944

7. If £90 is divided in the ratio $4:5:6$, the amount of the largest share is

   A £22.50   B £33.75
   C £36   D £45

8. If £1 is shared in the ratio $7:13$, the largest share is

   A 65p   B 35p   C 30p   D 13p

9. A mortar mixture is made of cement, sand and water mixed by weight in the ratio $1:3:5$. What weight of sand is contained in 63 kg of the mixture?

   A 7 kg   B 21 kg
   C 35 kg   D 42 kg

10. Which one of the following is not in the ratio $45:72$?

   A 5:9   B 10:16
   C 25:40   D 65:104

Mental Test 6

Try to answer the following questions without writing anything down except the answer.

1. An alloy is made by mixing copper and zinc in the ratio $5:2$. If 4 kg of zinc is used, how much copper is used?

2. Put the ratio $6:15$ in its simplest terms.

3. Express the ratio $15:20$ as a fraction in its lowest terms.

4. Simplify the ratio £2:50p.

5. Divide £200 in the ratio $2:3$.

6. If 8 rubbers cost 72p, how much do 2 rubbers cost?

7. 2 men digging a ditch take 4 days to complete it. How long would 8 men take?

8. A car has a fuel consumption of 10 miles per gallon. How many gallons are needed for a journey of 50 miles?

9. If the exchange rate is 10 francs = £1, how many francs are equivalent to £50.

10. A map is drawn to a scale of 1 cm : 5 km. A road is 20 km long on the ground. What length, in centimetres, will represent it on the map?

Percentages

Introduction

When comparing fractions it is often convenient to express them with a denominator of 100. Thus

$$\tfrac{1}{2} = \tfrac{50}{100} \quad \text{and} \quad \tfrac{2}{5} = \tfrac{40}{100}$$

Fractions with a denominator of 100 are called **percentages**. Thus

$$\tfrac{1}{4} = \tfrac{25}{100} = 25 \text{ per cent}$$
$$\tfrac{3}{10} = \tfrac{30}{100} = 30 \text{ per cent}$$

The symbol % is usually used instead of the words per cent. Thus

$$\tfrac{3}{20} = \tfrac{15}{100} = 15\%$$

Changing Fractions and Decimals into Percentages

To convert a fraction or a decimal into a percentage multiply it by 100.

Example 1

(a) Convert $\tfrac{17}{20}$ into a percentage.

$$\frac{17}{20} = \frac{17}{\cancel{20}_1} \times \frac{\cancel{100}^5}{1}\,\%$$

$$= \frac{17 \times 5}{1 \times 1}\,\%$$

$$= 85\%$$

(b) Convert 0.3 into a percentage.

$$0.3 = \frac{3}{10}$$

$$= \frac{3}{\cancel{10}_1} \times \frac{\cancel{100}^{10}}{1}\,\%$$

$$= \frac{3 \times 10}{1 \times 1}\,\%$$

$$= 30\%$$

Alternatively, multiplying the decimal number by 100

$$0.3 = 0.3 \times 100\% = 30\%$$

(c) Convert 0.56 into a percentage.

$$0.56 = 0.56 \times 100\% = 56\%$$

Exercise 7.1

Convert the following fractions into percentages:

1. $\tfrac{7}{10}$ 2. $\tfrac{4}{5}$ 3. $\tfrac{11}{20}$ 4. $\tfrac{9}{25}$

5. $\tfrac{31}{50}$ 6. $\tfrac{1}{4}$ 7. $\tfrac{9}{10}$ 8. $\tfrac{1}{5}$

Change the following decimal numbers into percentages:

9. 0.2 10. 0.34 11. 0.73

12. 0.68 13. 0.813 14. 0.927

15. 0.333

Not all percentages are whole numbers. For instance

$$\frac{3}{8} = \frac{3}{\cancel{8}_2} \times \frac{\cancel{100}^{25}}{1}\%$$

$$= \frac{3 \times 25}{2}\%$$

$$= \frac{75}{2}\%$$

$$= 37\tfrac{1}{2}\%$$

Also $\qquad \dfrac{1}{3} = \dfrac{1}{3} \times \dfrac{100}{1}\%$

$$= \frac{100}{3}\%$$

$$= 33\tfrac{1}{3}\%$$

Banks, business houses and building societies frequently state their interest rates as mixed numbers.

For instance $12\tfrac{1}{2}\%$ and $7\tfrac{1}{4}\%$.

Changing Percentages into Fractions and Decimals

To convert a percentage into a fraction or a decimal divide by 100.

Example 2

(a) Convert 45% into a fraction in its lowest terms.

$$45\% = \frac{45}{100}$$

$$= \frac{9}{20}$$

(b) Convert $7\tfrac{1}{2}\%$ into a fraction in its lowest terms.

$$7\tfrac{1}{2}\% = \frac{15}{2} \div 100$$

$$= \frac{\cancel{15}^3}{2} \times \frac{1}{\cancel{100}_{20}}$$

$$= \frac{3 \times 1}{2 \times 20}$$

$$= \frac{3}{40}$$

(c) Convert 3.9% into a decimal number.

$$3.9\% = 3.9 \div 100 = 0.039$$

Note that all we have done is to move the decimal point 2 places to the left.

Exercise 7.2

Convert the following percentages into fractions in their lowest terms:

1. 32%   2. 24%   3. 30%
4. 45%   5. 6%   6. $37\tfrac{1}{2}\%$
7. $66\tfrac{2}{3}\%$ 8. $7\tfrac{1}{4}\%$ 9. $8\tfrac{1}{3}\%$
10. $62\tfrac{1}{2}\%$

Convert the following percentages into decimals:

11. 78%   12. 31%   13. 48.2%
14. 2.5%   15. 1.25%   16. 3.95%
17. 20.1%   18. 1.96%

The following table gives corresponding fractions, decimals and percentages. Copy the table and write in the figures which should be placed in each of the spaces filled with a question mark. Put fractions into their lowest terms.

	Fraction	Decimal	Percentage
	$\frac{1}{4}$	0.25	25
19.	$\frac{11}{20}$	?	?
20.	$\frac{17}{50}$	?	?
21.	$\frac{7}{8}$	?	?
22.	?	0.76	?
23.	?	0.08	?
24.	?	0.375	?
25.	?	?	15
26.	?	?	27
27.	?	?	45
28.	?	?	$5\frac{1}{4}$

Percentage of a Quantity

To find the percentage of a quantity we must first express the percentage as a fraction or a decimal.

Example 3

(a) What is 10% of 40?

    Method 1 (the fractional method)

$$10\% \text{ of } 40 = \frac{10}{100} \text{ of } 40$$

$$= \frac{10}{100} \times \frac{40}{1}$$

$$= 4$$

    Method 2 (the decimal method)

    Since $10\% = 10 \div 100 = 0.1$

$$10\% \text{ of } 40 = 0.1 \times 40 = 4$$

If a calculator is to be used the decimal method is to be preferred.

(b) What is $7\frac{1}{2}\%$ of £50?

    Since $7\frac{1}{2}\% = 7.5 \div 100 = 0.075$

$$7\frac{1}{2}\% \text{ of } £50 = £50 \times 0.075 = £3.75$$

(c) 22% of a certain length is 55 cm. What is the complete length?

    Method 1 (the unitary method)

    22% of the length $= 55$ cm

      1% of the length $= 55 \div 22$ cm

$$= 2.5 \text{ cm}$$

The complete length will be 100%, hence

100% of the length $= 100 \times 2.5$ cm

$$= 250 \text{ cm}$$

Therefore the complete length is 250 cm.

    Method 2 (the fractional method)

    22% of the complete length $= 55$ cm

$$\text{Complete length} = \frac{55}{1} \times \frac{100}{22} \text{ cm}$$

$$= 250 \text{ cm}$$

(d) What percentage is 37 of 264? Give the answer correct to 5 significant figures.

$$\text{Percentage} = \frac{37}{264} \times \frac{100}{1}$$

$$= \frac{37 \times 100}{264}$$

$$= 14.015\%$$
$$\text{(correct to 5 s.f.)}$$

Exercise 7.3

1. What is:
    (a) 20% of 50?   (b) 30% of 80?
    (c) 5% of 120?   (d) 12% of 20?
    (e) 20.3% of 105?   (f) 3.7% of 68?

2. What percentage is:
    (a) 25 of 200?   (b) 30 of 150?
    (c) 24 of 150?   (d) 29 of 178?
    (e) 15 of 33?

Where necessary give the percentage correct to 5 significant figures.

3. In a test a girl scores 36 marks out of 60.

    (a) What is her percentage mark?

    (b) The percentage needed to pass the test is 45%. How many marks are needed to pass?

4. If 20% of a length is 23 cm, what is the complete length?

5. Given that 13.3 cm is 15% of a certain length, what is the complete length?

6. What is
    (a) 9% of £80
    (b) 12% of £110
    (c) 75% of £250

7. 27% of a consignment of fruit is bad. If the consignment weighs 800 kg, how much fruit is good?

8. In a certain county the average number of children eating lunches at school was 29 336, which represented 74% of the total number of children attending school. Calculate the total number of children attending school in the county.

Percentage Change

An increase of 5% in a number means that the number has been increased by $\frac{5}{100}$ of itself. Thus if the number is represented by 100, the increase is 5 and the new number is 105. The ratio of the new number to the old number is 105 : 100.

Example 4

An increase of 10% in salaries makes the wage bill for a factory £55 000.

(a) What was the wage bill before the increase?

(b) What was the amount of the increase?

(a) If 100% represents the wage bill before the increase then 110% represents the wage bill after the increase.

$$\text{Wage bill before the increase} = \frac{100}{110} \times £55\,000$$

$$= £50\,000$$

(b) The amount of the increase $= 10\%$ of £50 000

$$= 0.1 \times £50\,000$$

$$= £5000$$

Example 5

When a sum of money is decreased by 20% it becomes £320. What was the original sum?

If 100% represents the original sum then the sum after the decrease of 20% is represented by 80%. Therefore

$$\text{Original sum} = \frac{100}{80} \times £320$$

$$= £400$$

Exercise 7.4

1. The duty on an article is 20% of its value. If the price of the article after the duty has been paid is £960, find the price exclusive of tax.

2. When a sum of money is decreased by 10% it becomes £18. What was the original sum?

3. A man sells a car for £850 thus losing 15% of what he paid for it. How much did the car cost him?

4. During an epidemic 40% of the people in a certain place in Africa died and 1200 were left. How many people died?

5. The value of a machine depreciates by 15% of its value at the beginning of the year. If its value at the end of the year was £1360, what was its value at the beginning of the year?

6. 25% of a consignment of fruit was bad. If 1500 kg of the consignment was good, how much did the consignment weigh?

7. A man pays 20% of his salary in income tax. If his salary net of tax is £6400 per annum, what was his gross salary?

Miscellaneous Exercise 7

Section A

1. Change to fractions in their lowest terms
   (a) 48%   (b) 6%

2. Change to decimal numbers
   (a) 64%   (b) 7%

3. Below is a table of corresponding fractions, decimals and percentages. Write down the figures which should be placed in the spaces marked a, b, c, d, e and f.

Fraction	Decimal	Percentage
$\frac{1}{2}$	0.5	50
a	0.85	b
$\frac{7}{20}$	c	d
e	f	12

4. Find the value of
   (a) 9% of £20   (b) 38% of £70

5. A coat has been reduced by 15% in a sale. The old price was £80.
   (a) How much was the reduction in price?
   (b) How much will the customer pay for the coat?

6. In a class of 30 children, 40% are boys.
   (a) How many boys are there in the class?
   (b) What percentage of the class are girls?

Section B

1. Express $\frac{132}{150}$ as a percentage.

2. Find 5% of £260.

3. After prices have risen by 8% the new price of an article is £70.47. Calculate the original price.

4. Next year a man will receive a 12% wage increase and his weekly wage will then be £161.28. What is his present weekly wage?

5. The number of people working for a company at the end of 1984 was 1210. This was an increase of 10% on the number working for the company at the beginning of 1984. How many people worked for the company at the beginning of 1984?

6. The entry fee for an examination was £5.00 in 1984 and it rose to £6.40 in 1985. Express the increase in the fee as a percentage of the fee in 1984.

7. Calculate $7\frac{1}{2}$% of £160.

8. After 6% of a woman's wages have been deducted she receives a net amount of £98.70. Calculate the amount deducted.

9. A girl scores 66 marks out of 120 in her maths examination. What is her percentage mark? If the percentage required for a pass is 45%, how many marks must be obtained for a pass?

10. 8% of a sum of money is equal to £9.60. Find:
    (a) 1% of the sum of money
    (b) the sum of money
    (c) 92% of the sum of money

Multi-Choice Questions 7

1. $\frac{9}{10}$ expressed as a percentage is

    A 0.9%   B 9%   C 90%   D 99%

2. A boy scored 70% in a test. If the maximum mark was 40, then the boy's mark was

    A 4   B 28   C 30   D 35

3. For his holidays a man put aside 10% of his weekly wage of £150 for 40 weeks in the year. How much did he save for his holiday?

    A £150   B £300   C £450   D £600

4. A girl bought a record for 75p and sold it for 60p. Her loss as a percentage of the cost price is

    A 15%   B 20%   C 60%   D 75%

5. $\frac{1}{4}$% of £480 is

    A £1.20   B £12
    C £19.20   D £120

6. Written as a fraction $12\frac{1}{2}$% is

    A $\frac{1}{25}$ B $\frac{2}{25}$ C $\frac{1}{8}$ D $\frac{1}{3}$

7. 28 expressed as a percentage of 112 is

    A $\frac{1}{4}$%   B 4%   C 25%   D 28%

8. The price of a machine was £2000 but it is increased in price by 10%. What is the new price of the machine?

    A £2010   B £2020
    C £2100   D £2200

9. If 40% of a sum of money is £400 then the whole sum of money is

    A £40   B £1000
    C £4000   D £20 000

Mental Test 7

Try to answer the following questions without writing anything down except the answer.

1. Convert $\frac{3}{10}$ into a percentage.

2. Convert 0.4 into a percentage.

3. Convert 70% into a fraction in its lowest terms.

4. Convert 24% into a decimal number.

5. What is 10% of 50?

6. What percentage is 8 of 50?

7. Calculate 25% of £80.

8. 50% of a length is 20 cm. What is the whole length?

9. What is 7% of 200?

10. Change $66\frac{2}{3}$% into a fraction.

Wages and Salaries

Introduction

Everyone who works for an employer receives a wage or salary in return for his or her labour. However, the payments may be made in several different ways. Wages are usually paid weekly and salaries monthly.

Payment by the Hour

Many people are paid a certain amount of money for each hour that they work. They usually work a fixed number of hours known as the **basic week**. It is this basic week which determines the hourly (or basic) rate of pay. The basic week and the basic rate of pay are often fixed by negotiation between the employer and the trade union which represents the workers.

Example 1

A man works a basic week of 35 hours and his weekly wage is £140. Find the hourly rate of payment.

$$\text{Hourly rate of pay} = \frac{\text{Weekly wage}}{\text{Basic week}}$$

$$= \frac{£140}{35}$$

$$= £4.00$$

Example 2

A man works a basic week of 38 hours and his hourly rate of pay is £3. Calculate his weekly wage.

$$\text{Weekly wage} = 38 \text{ hours} \times £3 \text{ per hour}$$

$$= £114$$

Exercise 8.1

1. A woman works a basic week of 35 hours and her basic rate is £3 per hour. Calculate her weekly wage.

2. A man works a basic week of 40 hours for which he is paid £200. Calculate his hourly rate.

3. Calculate the weekly wage for a 35-hour week if the basic rate is

   (a) £2   (b) £2.50   (c) £4

4. The basic week is 42 hours and the weekly wage is £147. What is the basic rate?

5. If the basic rate is £2.92 and the basic week is 35 hours, what is the weekly wage?

Overtime

Hourly paid workers are usually paid extra for working more hours than the basic week requires. These extra hours of work are called **overtime**. Overtime is usually paid at time-and-a-quarter ($1\frac{1}{4}$ times the basic rate), time-and-a-half ($1\frac{1}{2}$ times the basic rate) or double time (twice the basic rate).

Example 3

A girl is paid a basic rate of £2 per hour. Calculate the hourly rate of overtime when this is paid at (a) time-and-a-quarter, (b) time-and-a-half,   (c) double time.

(a) Overtime rate at time-and-a-quarter

$$= 1\frac{1}{4} \times £2$$

$$= 1.25 \times £2$$

$$= £2.50 \text{ per hour}$$

(b) Overtime rate at time-and-a-half

$$= 1\frac{1}{2} \times £2$$

$$= 1.5 \times £2$$

$$= £3.00 \text{ per hour}$$

(c) Overtime rate at double time

$$= 2 \times £2$$

$$= £4 \text{ per hour}$$

Example 4

Peter Taylor works a 40-hour week for which he is paid £140. He works 5 hours overtime at time-and-a-half. What is his total wage for the week?

$$\text{Basic rate of pay} = \frac{\text{Weekly wage}}{\text{Basic week}}$$

$$= \frac{£140}{40}$$

$$= £3.50 \text{ per hour}$$

Overtime rate at time-and-a-half

$$= 1\frac{1}{2} \times £3.50$$

$$= 1.5 \times £3.50$$

$$= £5.25 \text{ per hour}$$

Payment for overtime

$$= 5 \times £5.25$$

$$= £26.25$$

Total wage for the week

$$= £140 + £26.25$$

$$= £166.25$$

Exercise 8.2

1. A man is paid a basic rate of £5 per hour. What is his overtime rate at (a) time-and-a-quarter, (b) time-and-a-half,   (c) double time?

2. Find the hourly overtime rate at time-and-a-half when the basic rate is

    (a) £3   (b) £4   (c) £7

3. A shop assistant works a basic week of 46 hours for which she is paid £121.44. During a certain week she works 4 hours overtime for which she is paid time-and-a-half. How much does she earn in overtime?

4. Michael Evans is paid £100.80 for a basic week of 40 hours. During a certain week he works 12 hours overtime which is paid at time-and-a-quarter. Calculate his total wage for the week.

5. A woman's basic hourly rate is £3.76. She works a basic week of 35 hours from Monday to Friday. On Sunday she works 7 hours for which she is paid double time.
    (a) What is her overtime rate?
    (b) How much money does she earn in overtime?
    (c) What is her total wage for the week?

Piecework

Some workers are paid a fixed amount for each article or piece of work that they make. This is called piecework. Often, if they can make more than a certain number of articles they are paid a bonus.

Example 5

A man is paid 5p for every handle he fixes to an electric iron up to 200 per day. For each handle over 200 that he fixes he is paid a bonus of 2p. If he fixes 250 handles in a day, calculate how much he earns.

$$\text{Money earned on first } 200 = 200 \times 5p$$
$$= £10.00$$
$$\text{Money earned on next } 50 = 50 \times (5 + 2)p$$
$$= 50 \times 7p$$
$$= £3.50$$
$$\text{Total amount earned} = £10.00 + £3.50$$
$$= £13.50$$

Exercise 8.3

1. A woman is paid 4p for each bag of sweets she packs up to a limit of 350. After that she earns a bonus of 1p per bag. Work out how much she earns if she packs 450 bags in a day.

2. Nancy Jones is paid 5p for each article she makes up to a limit of 200 per day. For each article over 200 that she completes she is paid a bonus of 2p. Calculate how much she earns in a day in which she completes 320 articles.

3. A man is paid 6p for each spot weld that he makes up to 300 per day. After that he is paid a bonus of 1p. If he makes 450 spot welds in a day, calculate how much he earns in the day.

4. For each article up to a limit of 150, a pieceworker is paid 8p. After that she is paid 10p per article. If she completes 220 articles in a day, how much will she earn?

5. Workers in a shirt factory are paid 50p for each shirt they complete up to a maximum of 25. After that they are paid 60p for each completed shirt. If a worker completes 35 shirts in a day, how much will she earn?

Commission

Shop assistants, salesmen and representatives are often paid commission on top of their basic wage. This is calculated as a small percentage of the value of the goods that they have sold.

Example 6

A salesman is paid a commission of 2% on the value of the goods which he has sold. Calculate his commission if he sells goods to the value of £5000.

$$\text{Commission} = 2\% \text{ of } £5000$$
$$= £5000 \times 0.02$$
$$= £100$$

Example 7

A shop assistant is paid a basic wage of £80. In addition she is paid 3% commission on the goods which she sells. During a certain week she sells goods to the value of £2000. What is her earnings during this week?

$$\text{Commission} = 3\% \text{ of } £2000$$
$$= £2000 \times 0.03$$
$$= £60$$
$$\text{Total earnings} = £80 + £60$$
$$= £140$$

Exercise 8.4

1. Find the commission at 2% on goods sold to the value of

    (a) £100   **(b)** £300

    (c) £2000   **(d)** £5000

2. A car salesman sells a car for £3500. He is paid commission of 2% on the sale. How much commission does he earn?

3. A representative sells farm machinery for £8000. He is paid commission of 3% on his sales. How much commission is he paid?

4. Furniture worth £900 is sold by a salesman who, in addition to his weekly wage of £70 per week, is paid commission of 3% on his sales. How much will he earn this week?

5. A sales assistant is paid a basic wage of £68 per week. In addition she is paid a commission of 4% on her sales. Work out her total wage for a week in which she sold £2500 of goods.

Salaries

People like teachers, civil servants, secretaries and company managers are paid a fixed amount each year. The money is usually paid monthly and they are not paid overtime or commission.

Example 8

A secretary is paid £7200 per annum. How much is she paid per month?

> Monthly salary = £7200 ÷ 12 = £600

Exercise 8.5

The following are the salaries of five people. How much is each paid monthly?

1. £5320   2. £6000   3. £8400

4. £6096   5. £7440

Deductions from Earnings

The wages and salaries which we have discussed previously are not the take-home pay or salary of the workers. A number of deductions are made first. The most important of these are

(1) Income tax

(2) National insurance

(3) Pension fund payments.

Before deductions the wage (or salary) is known as the **gross wage** (or **salary**).

After the deductions have been made the wage (or salary) is known as the **net wage** (or **salary**). It is also called the **take-home pay**.

National Insurance

Both employees and employers pay national insurance. Employees pay a certain percentage of their gross wage (9% in 1985). Married women and widows can pay at a reduced rate (3.85% in 1985). Men over 65, women over 60 and employees earning less than £34 per week do not pay national insurance.

The rates of national insurance are fixed by the Chancellor of the Exchequer and they vary from year to year. National insurance pays for such things as sick-pay and unemployment benefit.

Example 9

A man earns £12 000 per annum. His deductions for national insurance are 9% of his gross salary. How much does he pay in national insurance?

> Amount of
> national insurance = 9% of £12 000
>
> = 0.09 × £12 000
>
> = £1080

Income Tax

Taxes are levied by the Chancellor of the Exchequer to produce money to pay for such items as the armed services, the Civil Service and motorways. The largest producer of revenue is income tax. Every person who earns more than a certain amount pays income tax. Tax is not paid on the entire income. Certain allowances are made. In 1988-9 these were as follows:

(1) Single person's allowance   £2605
(2) Married couple's allowance   £4095
(3) Wife's earned income allowance   £2605

In certain cases other allowances such as an additional personal allowance for one parent families and blind persons are available. The residue of the income left over after all the allowances have been deducted is called the **taxable income**.

$$\text{Taxable income} = \text{Gross income} - \text{Allowances}$$

The amount of income tax payable per annum

$$= \text{Taxable income} \times \frac{\text{Rate of taxation}}{100}$$

In 1988-9 for taxable incomes up to £19 300 the rate of taxation was 25% but for any taxable income above £19 301 the rate of taxation was 40%.

Example 8

A woman's annual salary is £12 000. Her taxable income is calculated by deducting a personal allowance of £2605 and pension fund payments of £700 from her annual salary. She then pays income tax at 25% on her taxable income. How much does she pay in income tax per annum?

$$\text{Taxable income} = £12\,000 - £2605 - £700$$
$$= £8695$$

$$\text{Tax payable} = 25\% \text{ of } £8695$$
$$= £2173.75$$

Example 10

A man has a taxable income of £40 000. Work out the amount of income tax payable if the tax bands are 25% up to £19 300 and 40% above £19 301.

$$25\% \text{ of } £19\,300 = £4825$$

$$\text{Residue of taxable income} = £40\,000$$
$$- £19\,300$$
$$= £20\,700$$

$$40\% \text{ of } £20\,700 = £8280$$

$$\text{Total tax payable p.a.} = £4825$$
$$+ £8280$$
$$= £13\,105$$

PAYE

Most people pay tax by a method known as **pay-as-you-earn** or PAYE for short. The tax is deducted from their wages or salaries before they receive it. The taxpayer and his employer receive a notice of coding which gives the employee's allowances (based upon tax forms which he or she has previously completed) and sets a code number. The employer then knows from tax tables the amount of tax to deduct from the wages of an employee.

Pension Fund

Many firms operate their own private pension scheme to provide a pension in addition to that provided by the state. Usually, both employer and employee contribute, the employee's share being of the order of 5 or 6% of gross annual salary.

The amount of pension received depends upon the length of service and earnings at the time of retirement.

Example 12

An employee earns a salary of £8000 per annum. His pension fund payments amount to 6% of his annual salary. How much are his pension fund payments?

$$\text{Pension fund payments} = 6\% \text{ of } £8000$$
$$= 0.6 \times £8000$$
$$= £480$$

Example 13

A woman earns £9000 per annum. She pays £540 in national insurance contributions, £1950 in income tax and £456 in pension fund payments. Calculate her monthly take-home pay.

The deductions are   £540
                          £1950
                          £456
                          ‾‾‾‾‾
                          £2946

$$\text{Annual salary after deductions} = £9000$$
$$- £2946$$
$$= £6054$$
$$\text{Monthly take-home pay} = £6054 \div 12$$
$$= £504.50$$

Exercise 8.6

1. A person earns £8500 per annum. He pays 9% of his salary in national health contributions. How much does he pay?

2. Calculate the amount paid into a pension fund at a rate of 5% of the annual wage if this is £7000.

3. A woman earns £95 per week. Her deductions are: national insurance £3.66, income tax £19.36 and pension fund payments £5.70. Work out her take home pay.

4. A man's taxable income is £2400. If tax is paid at 30%, calculate the amount of income tax paid.

5. When income tax is levied at 30% a man pays £240 per annum in income tax. What is his taxable income?

6. A man earns £7500 per year. £1400 is deducted for income tax, £675 for national insurance and £450 for his pension fund.

    (a) Find his yearly take-home pay after deductions.

    (b) What is his monthly take-home pay to the nearest penny?

7. A single man has a personal allowance of £2205 and no other allowances. If he earns £8000 per annum, calculate

    (a) his taxable income

    (b) the amount of tax payable when this is levied at 30%

8. A married man earns £12 000 per year. His allowances to be set against his income are a married man's allowance of £3455 and pension fund payments of £720. Calculate his taxable income and hence find the amount of tax payable per year when this is levied at 30%.

9. The first £16 200 of a man's taxable income is taxed at 30%. The remainder of his taxable income, if any, is taxed at 50%. Calculate

    (a) the amount of tax payable on a taxable income of £21 000

    (b) the taxable income on which the tax payable is £8860

10. A woman is paid £3 per hour for a 40-hour week. She pays national insurance contributions at 3.85% of her weekly wage. How much does she pay in national insurance contributions?

Miscellaneous Exercise 8

Section A

1. Margaret worked for 40 hours. She is paid £2.90 per hour. What was her wage?

2. A man's taxable income is £3000. He pays income tax at 25%. How much tax did he pay?

3. A woman's total income is £8000 per annum. Her tax free allowances amount to £3250. How much is her taxable income?

4. A man is paid £200 for a 40-hour week. Calculate the man's basic hourly rate of pay.

5. A girl is paid £2.50 per hour. Overtime is paid at time-and-a-half. What is her overtime rate?

6. A salesman is paid commission of 3% on his weekly sales. During one week he sold goods to the value of £8000. How much commission did he earn?

7. A woman is paid 12p for every article she makes. During one week she made 750 articles. How much money did she make?

8. A trainee earned £78 for a week's work. He paid £9.36 in income tax. What percentage of his weekly wage was paid in income tax?

Section B

1. A man's income is £7000 per annum. He is allowed £3000 in tax-free allowances.

   (a) Calculate his taxable income.

   (b) If tax is levied at 30%, find the amount of tax payable in a year.

2. A man is paid £240 for a 40-hour week. Overtime is paid at time-and-a-half.

(a) Calculate the man's basic hourly rate of pay.

(b) Calculate his overtime rate of pay per hour.

(c) If he works 7 hours overtime during a certain week, work out his total wages for the week.

3. A salesman is paid a basic wage of £1100 per month. On top of this he is paid commission of $2\frac{1}{2}$% on goods that he sells. During one month he sells goods to the value of £20 000. Calculate his total wages for the month.

4. A woman is paid 8p for every article that she makes up to a limit of 150 per day. For each article made over this limit she is paid a bonus of 2p per article. In one day she made 220 articles. How much did she earn that day?

5. A woman earns £8000 per year. Deductions from her yearly pay are £720 for national insurance, £1850 for income tax and £450 for pension fund payments.

   (a) What are her total deductions for the year?

   (b) What is her net annual pay?

   (c) What is her monthly take-home pay?

6. A man has a taxable income of £23 000. For up to £16 200 of taxable income, tax is levied at 30%. After this the next £3000 is taxed at 40% and the remainder, if any, is taxed at 45%. Calculate the amount of tax payable per annum.

7. A woman is paid £3.50 per hour for a 40-hour week.

   (a) Calculate her gross weekly wage.

   (b) If she pays national insurance at 3.85% of her gross salary find how much she pays in national insurance contributions per week.

(c) If, in addition, she pays £18.30 per week in income tax, find her net wage for the week.

8. A man works a basic week of 40 hours at £3 per hour. He then earns overtime at time-and-a-half. Calculate his gross wage for a 44-hour week.

Multi-Choice Questions 8

1. An agent is paid $12\frac{1}{2}\%$ commission on each sale. If goods to the value of £64 are sold, his commission is

   A £1.92   B £2.24

   C £7.68   D £8.00

A trainee works a 40-hour week and is paid £1.50 per hour. Overtime is paid at time-and-a-half. Use this information to answer questions 2, 3, 4 and 5.

2. How much money does he earn in a week in which he works 52 hours?

   A £58   B £60   C £78   D £87

3. In another week his earnings are £73.50. How many hours has he worked?

   A 46   B 47   C 48   D 49

4. In a week in which his wages were £78 he paid £3.90 in income tax. What percentage of his wage was paid in income tax?

   A 2%   B 5%   C 20%   D 50%

5. If the trainee gains an 8% rise in wages, how much would he then earn for a 40-hour week?

   A £60   B £64.80

   C £72   D £72.50

6. When income tax is levied at 30%, a man pays £540 in income tax. His taxable income is

   A £162   B £1620

   C £1800   D £5400

7. A woman's basic pay for a 40-hour week is £80. Overtime is paid for at time-and-a-quarter. In a certain week she worked overtime and her total wage was £100. How many hours overtime did she work?

   A 4   B 8   C 10   D 25

8. A female employee in a factory works at a piecework rate of 21p for each item which she completes, up to 60 per day. After 60 per day she is paid a bonus of 6p extra for each item she completes. In one day she completes 82 items. How much did she earn that day?

   A £12.60   B £13.92

   C £17.22   D £18.54

Mental Test 8

Try to answer the following questions without writing anything down except the answer.

1. A man's basic wage is £120 for a 40-hour week. What is his basic hourly rate of pay?

2. A man earns £4 per hour. How much will he earn for a 40-hour week?

3. A girl earns £2 per hour and overtime is paid at time-and-a-half. What is her overtime rate?

4. A jobbing gardener is paid £3 per hour. How much will he earn for 5 hours work?

5. A woman is paid 20p for every article she completes. In one day she completed 50 articles. How much did she earn?

6. A man has a taxable income of £200. If tax is levied at 30%, how much tax is payable?

7. National insurance is paid at 9% of gross salary. A woman earns £5000 per year. How much does she pay in national insurance contributions?

8. Commission is paid at 5% to an agent who sold goods to the value of £500. How much commission did he earn?

9. A man earns £200 per week. His deductions are £20 for national insurance and £25 for income tax. What is his take-home pay for the week?

10. A woman employee is paid an annual salary of £6000 per year. What is her monthly salary?

Interest

Introduction

If you invest money with a bank or a building society then interest is paid to you for lending them the money. On the other hand, if you borrow money from the bank or building society, then you will be charged interest.

Example 1

I invest £1500 for 1 year at 12% interest. How much interest will I get at the end of the year?

$$\begin{aligned} \text{Amount of interest} &= 12\% \text{ of } £1500 \\ &= 0.12 \times £1500 \\ &= £180 \end{aligned}$$

Simple Interest

With simple interest the amount of interest earned is the same every year. If the money is invested for 2 years the amount of interest will be doubled; for 3 years the amount of interest will be multiplied by 3; for 6 months (i.e. $\frac{1}{2}$ year) the amount of interest will be halved.

Example 2

A man borrows £800 for 4 years. If the rate of simple interest is 15%, find the amount of interest that he pays.

$$\begin{aligned} \text{Amount of interest for 1 year} &= 15\% \text{ of } £800 \\ &= 0.15 \times £800 \\ &= £120 \\ \text{Amount of interest for 4 years} &= 4 \times £120 \\ &= £480 \end{aligned}$$

Exercise 9.1

Find the interest for 1 year on

1. £500 at 5% per annum.
2. £1100 at 10% per annum.
3. £5000 at 12% per annum.
4. £20 000 at 8% per annum.
5. £750 at 11% per annum.

Find the amount of simple interest on

6. £600 at 10% for 3 years.
7. £1200 at 8% for 5 years.
8. £600 at 12% for 6 months.
9. £850 at 7% for 2 years.
10. £900 at 15% for 7 years.

The Simple Interest Formula

From what has been said we can see that the simple interest can be calculated from

$$I = P \times R \times T$$

where I is the amount of simple interest, P is the principal (i.e. the amount of money invested or borrowed), R is the rate of interest as a decimal and T the time in years.

Since the rate R can be expressed as a fraction with a denominator of 100

$$I = \frac{P \times R \times T}{100}$$

Example 3

A person borrows £4500 for a period of 4 years at 20% simple interest. Calculate the amount of interest that the person must pay.

We are given that $P = 4500$, $R = 20$ and $T = 4$.

$$I = \frac{P \times R \times T}{100}$$

$$= \frac{4500 \times 20 \times 4}{100}$$

$$= £3600$$

The simple interest formula can be transposed to give

$$T = \frac{100 \times I}{P \times R}; \quad R = \frac{100 \times I}{P \times T}; \quad P = \frac{100 \times I}{R \times T}$$

The form to be used depends upon what we have to find.

Example 4

(a) £1500 is invested at 15% simple interest per annum. How long will it take for the investment to reach £2850? *FIND T*

Interest $= £2850 - £1500$

$$= £1350$$

We now have $I = 1350$, $P = 1500$ and $R = 15$. We have to find T.

$$T = \frac{100 \times I}{P \times R}$$

$$= \frac{100 \times 1350}{1500 \times 15}$$

$$= 6$$

Hence the investment will reach £2850 after 6 years.

(b) What principal is needed to earn interest of £96 if the money is invested for 5 years at 6% per annum?

We are given that $I = 96$, $R = 6$ and $T = 5$ and we have to find P.

$$P = \frac{100 \times I}{R \times T}$$

$$= \frac{100 \times 96}{6 \times 5}$$

$$= 320$$

Hence a principal of £320 is required.

In practice, for periods of more than 1 year, simple interest is virtually unknown.

Exercise 9.2

Find the simple interest on

1. £1400 for 4 years at 5% per annum.

2. £2000 for 3 years at 10% per annum.

3. £500 for 5 years at 8% per annum.

4. £1500 for 6 months at 12% per annum.

5. £800 for 4 years at 24% per annum.

Find the length of time for

6. £1000 to be the interest on £5000 invested at 5% per annum.

7. £480 to be the interest on £2000 invested at 8% per annum.

8. £1200 to be the interest on £3000 invested at 10% per annum.

Find the rate per cent per annum simple interest for   *FIND R.*

9. £420 to be the interest on £1200 invested for 5 years.

10. £72 to be the interest on £200 invested for 3 years.

11. £1200 to be the interest on £3000 invested for 4 years.

Find the principal required for

12. The simple interest to be £600 on money invested for 3 years at 5% per annum.

13. The simple interest to be £40 on money invested for 2 years at 10% per annum.

14. The simple interest to be £432 on money invested for 4 years at 9% per annum.

15. £500 is invested for 7 years at 10% per annum simple interest. How much will the investment be worth after this period?

16. £1000 is invested for 3 years and after this time the investment is worth £1240. Calculate the rate per cent per annum simple interest.

17. £2000 was invested at 12% per annum simple interest. If the investment is now worth £2960, find how long ago the money was invested.

Compound Interest

Compound interest is different from simple interest in that the interest which is added to the principal also attracts interest. Thus after 2 years there will be more interest than after 1 year because there is more capital to attract interest.

Example 5

£5000 is invested for 2 years at 15% compound interest. Calculate the value of the investment after this period.

	£
Principal	5000
1st year's interest	750
Value of the investment after 1 year	5750
2nd year's interest	862.50
Value of the investment at the end of 2 years	6612.50

The value of an investment is often called the amount. Thus in Example 5 we would say that the amount after 2 years was £6612.50.

Compound interest tables are available which give the amount to which a principal of £1 grows.

Years	5%	6%	7%	8%	9%
1	1.050	1.060	1.070	1.080	1.090
2	1.103	1.124	1.145	1.166	1.188
3	1.158	1.191	1.225	1.260	1.295
4	1.216	1.262	1.311	1.360	1.412
5	1.276	1.338	1.403	1.469	1.539
6	1.340	1.419	1.501	1.587	1.677
7	1.407	1.504	1.606	1.714	1.828
8	1.477	1.594	1.718	1.851	1.993
9	1.551	1.689	1.838	1.999	2.172
10	1.629	1.791	1.967	2.159	2.367

Years	10%	11%	12%	13%	14%
1	1.100	1.110	1.120	1.130	1.140
2	1.210	1.232	1.254	1.277	1.300
3	1.331	1.368	1.405	1.443	1.482
4	1.464	1.518	1.574	1.603	1.689
5	1.611	1.685	1.762	1.842	1.925
6	1.772	1.870	1.974	2.082	2.195
7	1.949	2.076	2.211	2.353	2.502
8	2.144	2.304	2.476	2.658	2.853
9	2.358	2.558	2.773	3.004	3.252
10	2.594	2.839	3.106	3.395	3.707

Example 6

Using the compound interest table, find the compound interest earned by £800 invested for 7 years at 12% per annum.

From the table, £1 becomes £2.211. To find to what £800 grows we multiply £2.211 by 800.

$$\text{Amount} = £2.211 \times 800$$

$$= £1768.80$$

$$\text{Interest} = \text{Amount} - \text{Principal}$$

$$= £1768.80 - £800$$

$$= £968.80$$

Exercise 9.3

Calculate the compound interest earned on

1. £300 invested for 2 years at 5% per annum.

2. £1000 invested for 3 years at 10% per annum.

3. £500 invested for 2 years at 15% per annum.

Calculate the amount when

4. £800 is invested for 3 years at 9% per annum.

5. £2000 is invested for 2 years at 12% per annum.

6. £20 000 is invested for 2 years at 15% per annum.

7. A woman borrows £1500 for 2 years at 15% per annum compound interest. How much will the woman have to pay back?

8. An investor can invest money at 10% simple interest or 8% compound interest per annum. If she invests £8000 for 3 years, calculate which is the most profitable investment and find the difference between the two.

Use the compound interest table on page 82 to calculate the amount when

9. £1500 is invested for 7 years at 9% per annum.

10. £2500 is invested for 9 years at 11% per annum.

11. £500 is invested for 6 years at 7% per annum.

Use the compound interest table on page 82 to find the amount of compound interest earned on

12. £700 invested for 5 years at 8% per annum.

13. £3000 invested for 8 years at 10% per annum.

14. £4500 invested for 8 years at 14% per annum.

Depreciation

When the value of an article decreases with age this reduction in its value is called **depreciation**. Examples of articles which depreciate in value are machinery, motor vehicles and typewriters.

Problems on depreciation are similar to those on compound interest except that with compound interest the interest is **added** to the amount of the previous year, whilst for depreciation the interest is **subtracted** from the amount of the previous year.

Example 7

A small business buys a computer costing £2000. The rate of depreciation is 20% per annum. What is its value at the end of 3 years?

Cost of computer	=	£2000
1st year's depreciation	=	20% of £2000
	=	£400
Value at end of 1st year	=	£2000 − £400
	=	£1600
2nd year's depreciation	=	20% of £1600
	=	£320
Value at end of 2nd year	=	£1600 − £320
	=	£1280
3rd year's depreciation	=	20% of £1280
	=	£256
Value at end of 3rd year	=	£1280 − £256
	=	£1024

Hence the computer is reckoned to be worth £1024 at the end of 3 years.

Exercise 9.4

1. A firm buys a machine for £1600. If its value depreciates at 10% per annum, calculate its value at the end of 3 years.

2. The rate of depreciation of office machinery bought for £60 000 is 15% per annum. How much will the machinery be worth at the end of 5 years?

3. A lorry costs £25 000. It depreciates in value at 20% per annum. Find its value at the end of 7 years.

4. A machine costs £40 000 when new. Its value depreciates at 25% per annum. Work out its value at the end of 3 years.

Miscellaneous Exercise 9

Section A

1. What is the interest for 1 year on £800 at 10% per annum?

2. What is the interest for six months on £500 at 8% per annum?

3. £200 is invested for 2 years at 12% per annum simple interest.

   (a) What is the interest at the end of the first year?

   (b) How much is the interest at the end of the two years?

4. A woman invests £200 for 2 years at 10% compound interest.

   (a) How much interest does the money earn in the first year?

   (b) At the end of the second year, how much interest has been earned?

   (c) How much money will she get at the end of the second year?

5. What sum of money must be invested to produce simple interest of £100 at the end of 1 year. The rate is 20% per annum.

Section B

1. Calculate the simple interest on £1600 invested for 1 year at 5% per annum.

2. A man invests £5000 at 8% compound interest per annum. Calculate the value of his investment at the end of 2 years.

3. A woman has £3400 to invest. She can buy savings bonds which pay 9% per annum simple interest or she can put the money into a building society account paying 8% compound interest per annum.

   (a) Calculate the value of her investment at the end of 3 years if she buys savings bonds.

   (b) If she puts the money into a building society account, calculate the amount standing in her account at the end of 3 years.

   (c) Which is the better investment and what is the difference in the amount of interest earned?

4. A man earned £12 000 during a certain year. He invested 15% of his earnings in a savings account paying 10% per annum compound interest. Work out the value of this investment at the end of two years.

5. A sum of money was invested at 12% per annum simple interest. After 1 year the amount of interest was £600. How much money was invested?

6. A machine which costs £5500 depreciates at a rate of 15% per annum. How much will the machine be worth at the end of 2 years?

Multi-Choice Questions 9

1. The simple interest on £1000 invested for 4 years at 7% per annum is

   A £64   B £120

   C £280   D £28 000

2. The simple interest on £2400 invested at 8% per annum for 6 months is

    A £64   B £96   C £120   D £128

3. The simple interest on £1600 invested for 5 years is £480. What is the rate of interest per annum?

    A 3%   B 5%   C 6%   D 7%

4. The simple interest on £1600 invested at 5% per annum over a number of years amounted to £320. For how long was the money invested?

    A 4 years   B 5 years
    C 6 years   D 7 years

5. What sum of money must be invested to produce simple interest of £60 if the rate is 12% per annum and the time is 2 years?

    A £400   B £360   C £300   D £250

6. £1000 invested for 2 years at 10% per annum compound interest becomes

    A £1200   B £1210
    C £1331   D £1400

7. A car depreciates in value by 10% per annum. It costs £4000 when new. Its value at the end of 4 years is

    A £1000   B £2500
    C £2624   D £3000

8. A sum of money was invested at 5% per annum simple interest. After 4 years the amount of interest earned was £100. How much money was originally invested?

    A £800   B £600   C £500   D £400

Mental Test 9

Try to answer the following questions without writing anything down except the answer.

1. Find the interest on £200 invested at 5% for 1 year.

2. What is the simple interest on £100 invested for 2 years at 10% per annum?

3. How much does £1000 become when it is invested at 20% per annum for 1 year?

4. Use the compound interest table on page 82 to find the amount accruing when £100 is invested for 5 years at 8% compound interest per annum.

5. A machine depreciates at 10% per annum. If it cost £1000 when new, how much is it worth at the end of 1 year?

Money in Business and the Community

Profit and Loss

When a dealer buys or sells goods, the cost price is the price at which he buys the goods and the selling price is the price at which he sells the goods. The profit is the difference between the selling price and the cost price. That is

$$\text{Profit } = \text{ Selling price} - \text{Cost price}$$

The profit per cent is usually calculated on the cost price, and

$$\text{Profit \%} = \frac{\text{Selling price} - \text{Cost price}}{\text{Cost price}} \times 100$$

The profit per cent is called the **mark-up**.

Example 1

A shopkeeper buys an article for £5.00 and sells it for £6.00. What is his profit per cent?

Cost price $=$ £5   and   Selling price $=$ £6

$$\text{Profit \%} = \frac{6-5}{5} \times 100$$

$$= \frac{1}{5} \times \frac{100}{1}$$

$$= 20\%$$

It is often easier for a retailer to calculate the profit as a percentage of the selling price, because the bill shows the amount of the takings and it is then easy to calculate the profit as a percentage of the takings. This is called the **margin**.

$$\text{Margin } = \frac{\text{Profit}}{\text{Selling price}} \times 100$$

If a loss is made the cost price is greater than the selling price. That is

$$\text{Loss } = \text{ Cost price} - \text{Selling price}$$

As with profit per cent, the loss per cent is usually calculated on the cost price. Thus

$$\text{Loss \%} = \frac{\text{Cost price} - \text{Selling price}}{\text{Cost price}} \times 100$$

Example 2

A man buys a car for £3200 and sells it for £2400. Calculate his percentage loss.

Cost price $=$ £3200   and

Selling price $=$ £2400

$$\text{Loss \%} = \frac{3200 - 2400}{3200} \times 100$$

$$= 25\%$$

Example 3

A dealer buys 20 similar articles for £5. He sells them for 30p each. What is his profit per cent?

Cost per article $=$ £5 ÷ 20

$$= £0.25$$

$$= 25\text{p}$$

$$\text{Profit \%} = \frac{30 - 25}{25} \times 100$$

$$= 20\%$$

Exercise 10.1

1. A shopkeeper buys an article for 40p and sells it for 50p. What is the profit per cent?

2. An article is bought for £5 and sold for £4. What is the loss per cent?

3. A greengrocer buys a box of 150 grapefruit for £12. He sells them for 12p each. What is his profit per cent?

4. A second-hand car is bought for £3100 and sold for £775. What is the percentage loss?

5. A dealer buys 200 pencils for £50 and sells them for 30p each. Calculate his profit per cent.

6. A retailer buys 30 pens at 8p each. Three are damaged and unsaleable but he sells the others at 10p each. What profit per cent does he make?

7. A motor bike is bought for £168 and sold for £126.

    (a) What is the loss?

    (b) What is the loss per cent?

8. A dealer buys a chair for £30 and sells it for £40.

    (a) Calculate his profit.

    (b) What is the profit per cent?

The selling price (sometimes called the marked price) may be calculated when the cost price and the profit per cent are given. This calculation is of great importance to people such as shopkeepers and dealers who sell things.

Example 4

A dealer buys a table for £20 and wants to make a profit of 30%. What should his selling price be?

If we let the cost price be 100, then since the profit is 30% the selling price is 130. That is

$$\text{Selling price} = \frac{130}{100} \times \frac{20}{1}$$

$$= £26$$

Example 5

A lady buys a typewriter for £200 and when she sells it she shows a loss of 40%. How much did she sell the typewriter for?

If the cost price is 100, then since the loss is 40% the selling price is 60.

$$\text{Selling price} = \frac{60}{100} \times \frac{200}{1}$$

$$= £120$$

When the selling price and the profit per cent are known, the cost price may be calculated by the method shown in Example 6.

Example 6

By selling a wardrobe for £375 a trader makes a profit of 25%. For how much did he buy the wardrobe (i.e. what was the cost price of the wardrobe)?

If the cost price is 100 then, since the profit is 25% the selling price is 125. Therefore

$$\text{Cost price} = \frac{100}{125} \times \frac{375}{1}$$

$$= £300$$

Exercise 10.2

1. A dealer buys an article for £30. He wishes to make a profit of 20%. How much should he sell the article for?

2. A dealer buys an armchair for £200. If his profit is to be 40%, for how much should he sell the armchair?

3. A bicycle is sold for £125 and the trader makes a profit of 25%. What was the cost price of the bicycle?

4. A car was sold for £21 000 and the garage that sold the car made a profit of 40%. Calculate the cost price of the car.

5. A chair was marked for sale at £260 thereby making a profit of 30% for the shopkeeper. How much did the shopkeeper pay for the chair?

6. A shopkeeper buys an article for £14.40. His marked selling price is 20% more than the cost price. What is his marked selling price?

7. Calculate the selling price of potatoes per kilogram if they are bought for £4 per 50-kg bag and the profit per cent is $12\frac{1}{2}\%$.

Discount

Sometimes a dealer will deduct a percentage of the selling price if the customer is prepared to pay cash. This is called **discount**.

Example 7

A piece of furniture is offered for sale at £270. A customer is offered a discount of 10% for cash. How much does the customer actually pay?

Method 1

Discount $=$ 10% of £270

$=$ £27

Amount actually paid $=$ £270 $-$ £27

$=$ £243

Method 2

Since a discount of 10% is given, the customer pays only 90% (100% $-$ 10%) of the cost.

Amount actually paid $=$ 90% of £270

$=$ 0.9 $\times$ £270

$=$ £243

Exercise 10.3

1. An armchair is marked for sale at £200. The trader offers a discount of 10% for cash. How much will a cash-paying customer actually pay for the armchair?

2. During a sale a clothing shop offers a $12\frac{1}{2}\%$ discount for cash. If the price of a suit is £64, how much will it sell for during the sale?

3. A furniture store offers a 7% discount for cash. How much discount will be allowed on furniture whose marked price is £1100?

4. A shop offers a discount of 5p in the pound (i.e. 5%) for cash. How much discount will the shop allow on a washing machine whose selling price is £170?

5. A grocer offers a discount of $2\frac{1}{2}\%$ to his customers provided they pay their grocery bill within one week. If a bill of £36.25 is paid within one week, how much discount will the grocer allow?

Value Added Tax

Value added tax or VAT for short is a tax on goods and services which are purchased. Some services and goods bear no tax, for instance food, books and insurance. The rate varies from time to time but at the moment it is levied at 15%.

Example 8

A man buys a lawnmower which is priced at £80 plus VAT. If the rate of tax is 15%, how much will he actually pay for the mower?

VAT $=$ 15% of £80

$=$ 0.15 $\times$ £80

$=$ £12

Total cost of mower $=$ £80 $+$ £12

$=$ £92

Example 9

A housewife buys a table for £138, the price includes VAT at 15%. What is the price of the table exclusive of VAT?

Let 100 be the price exclusive of VAT. Then

Price inclusive of VAT at 15% is 115

Price exclusive of VAT $= \dfrac{100}{115} \times \dfrac{138}{1}$

$= £120$

Exercise 10.4

1. A woman buys a washing machine whose price exclusive of VAT is £240. If VAT is charged at 15%, how much will the woman pay for the machine?

2. A chair is priced at £54 inclusive of VAT which is charged at 15%. What is its price exclusive of VAT?

3. A woman buys a refrigerator for £160 but on top of this VAT at 15% is charged. How much does she pay altogether?

4. A telephone bill is £40 excluding VAT. How much is the bill when VAT at 15% is added on?

5. A radio costs £69 including VAT at 15%. What is the cost without VAT?

6. A carpet costs £230 including VAT. How much does it cost without VAT which is charged at 15%?

7. Calculate the amount of VAT which is charged on a garage bill of £160 which includes VAT at 15%.

Rates

Every property in a town or city is given a **rateable value** which is fixed by the local district valuer. This rateable value depends upon the size, condition and location of the property.

The rates of the town or city are levied at so much in the pound of rateable value, for instance, 90p in the pound. The money raised by the rates is used to pay for such things as libraries, police, education, etc.

Yearly rates $=$ Rateable value $\times$ Rate in £1

Example 10

The rateable value of a house is £240. If the rate is £1.25 in the pound, how much must the householder pay in rates?

Yearly rates $= 240 \times £1.25$

$= £300$

Example 11

A householder pays rates of £270 on property which has a rateable value of £300. What is the rate in the pound?

For a rateable value of £300 rates payable are £270. For a rateable value of £1 rates payable are

$$£\dfrac{270}{300} = £0.90$$

$$= 90p$$

Hence the rate is 90p in the pound.

Example 12

A local authority needs to raise £9 020 000 to cover its expenditure for the current year. If the rateable value of all the property in the area is £8 200 000, what rate in the pound should be levied?

Rate chargeable in £1 $= \dfrac{9\,020\,000}{8\,200\,000}$

$= 1.10$

Hence a rate of £1.10 in £1 should be levied.

Most local authorities state on their rate demand the product of a penny rate. This is the amount that would be raised if the rate levied was 1p in the pound, that is £0.01 in £1.

Example 13

The rateable value of a town is £9 350 000. What is the product of a penny rate?

Product of a penny rate = £0.01 × 9 350 000

= £93 500

Example 14

The cost of highways and bridges in a town is equivalent to a rate of 9.28p in the pound. The rateable value of all the property in the town is £15 400 000. Calculate the amount of money available for spending on highways and bridges.

Amount available = £0.0928 × 15 400 000

= £1 429 120

Exercise 10.5

1. The rateable value of a house is £180. Calculate the yearly rates payable by the householder when the rate is £1.20 in the pound.

2. A householder pays £180 in rates when the rate is levied at £1.50 in the pound. What is the rateable value of the house?

3. A house is assessed at a rateable value of £450. The owner pays £405 in rates for the year. What is the rate in the pound?

4. What rate should a local authority charge if they need to raise £1 000 000 from a total rateable value of £4 000 000?

5. Calculate the total income from the rates of a town of rateable value £4 300 000 when the rates are £1.08 in the pound.

6. A town having a rateable value of £7 720 000 needs to raise £704 000 from the rates. What local rate should be levied?

7. The rateable value of the property in a city is £8 500 000. What is the product of a penny rate?

8. The rateable value of all the property in a city is £8 796 000.

   (a) Calculate the product of a penny rate.

   (b) The total expenses for running the city are £4 837 800. Calculate the rate that should be levied.

9. The total rateable value of the property in a locality is £850 000. Calculate the total cost of the public library if a rate of 4.6p in the pound must be levied for this purpose.

10. The expenditure of a town is £9 000 000 and its rates are 87p in the pound. The cost of its libraries is £300 000. What rate in the pound is needed for the up-keep of the libraries?

Insurance

Our future is something which is far from certain. We could become too ill to work or we could be badly injured or even killed in an accident. Our house could be burgled or burnt down. We might be involved in a car accident and be liable for injuries and damage. How do we take care of such eventualities? The answer is to take out insurance policies. The insurance company collects premiums from thousands of people who wish to insure themselves. It invests this money to earn interest and this money is then available to pay to people who claim for their loss.

Example 15

A householder values his house and its contents at £35 000. His insurance company charges a premium of £3 per £1000 insured. How much is the annual premium?

$$\text{Annual premium} = \frac{3 \times 35\,000}{1000}$$

$$= £105$$

Car Insurance

By law a vehicle must be insured, and the owner of a vehicle can be prosecuted for not having third-party insurance. That is, a policy must be taken out in case someone is injured or damage is caused by an accident which is the fault of the policy holder.

Third-party insurance covers only the other person; it does not cover the policy holder who will have to pay himself for any damage to his car. A fully comprehensive policy is needed to cover damage to the policy holder's car plus damage to other people and their property.

The size of the premium depends upon

(1) The value of the policy holder's vehicle.

(2) The engine size.

(3) The area in which the policy holder lives.

(4) Special risks (sports cars and young drivers).

(5) The use to which the vehicle is put (taxi, private, goods, etc.).

If a driver makes no claims during a year he gets a bonus (called a no-claims bonus) which means that he or she will pay a smaller premium next year.

Example 16

The insurance on a car is £180 but the owner is allowed a $33\frac{1}{3}$% no-claims bonus. How much is the insurance premium for the year?

$$\text{No-claims bonus} = 33\tfrac{1}{3}\% \text{ of } £180$$

$$= \tfrac{1}{3} \text{ of } £180$$

$$= £60$$

$$\text{Insurance premium} = £180 - £60$$

$$= £120$$

Cost of Running a Car

When calculating the cost of running a car, the costs of tax, insurance, petrol, maintenance and depreciation should all be taken into account. Every motor vehicle requires a Road Fund Licence which has to be paid when the vehicle is first registered and renewed periodically. It is also a legal requirement that every motor vehicle must be insured to protect a third party who may be injured in an accident. The vehicle will need maintaining and its value will depreciate with age.

Example 17

A car is bought for £3600 and used for one year. It is then sold for £2600. During this year it did 20 000 km averaging 12 km per litre of petrol which cost 45p per litre. Insurance cost £87, tax £100, repairs and maintenance £350. What is the total cost for a year's motoring and what is the cost per kilometre?

Depreciation	$= £1000$
Insurance	$= £87$
Tax	$= £100$
Repairs and maintenance	$= £350$
Cost of petrol	$= £\dfrac{20\,000}{12} \times \dfrac{45}{100}$
	$= £750$
Total cost for the year	$= £2287$
Cost per kilometre	$= \dfrac{£2287 \times 100\text{p}}{20\,000}$
	$= 11.44\text{p}$

Life Assurance

With this type of assurance a sum of money, depending upon the size of the premium etc., is paid to the dependents (wife or husband and children) of the policy holder upon his or her death. The size of the premium depends upon

(1) The age of the person (the younger the policy holder is, the less he pays because there is less risk of him or her dying suddenly).

(2) The amount of money the policy holder wants the dependents to receive (the greater the amount the larger the premium).

Example 18

A man aged 35 years wishes to assure his life for £7500. The insurance company quotes an annual premium of £13.90 per £1000 assured. Calculate the amount of the monthly premium.

$$\text{Annual premium} = £13.90 \times \frac{7500}{1000}$$

$$= £13.90 \times 7.5$$

$$= £104.25$$

$$\text{Monthly premium} = £104.25 \div 12$$

$$= £8.69$$

Endowment Assurance

This is very similar to life assurance but the person can decide for how long he or she is going to pay the premiums. At the end of the chosen period a certain sum of money will be paid to the policy holder. If, however, the policy holder dies before the end of the chosen period, the assured sum of money will be paid to his or her dependents. Some endowment and life policies are 'with profits', which means that the sum assured may increase over the period of the policy depending upon the profits made by the insurance company.

Example 19

A man aged 30 years wishes to take out an endowment 'with profits' assurance policy. He is quoted a price of £48.50 per £1000 assured over a period of 20 years. Calculate the monthly premiums if he wishes to assure himself for £6000.

$$\text{Annual premiums} = £48.50 \times \frac{6000}{1000}$$

$$= £48.50 \times 6$$

$$= £291.00$$

$$\text{Monthly premiums} = £291.00 \div 12$$

$$= £24.25$$

Exercise 10.6

1. A householder wishes to insure his house for £25 000. His insurance company charges a premium of 15p per £100 insured. Calculate the householder's annual premium.

2. An insurance company offers the following rates to customers: buildings £1.25 per £1000 insured; contents 25p per £100 insured. Calculate the annual premium paid by a householder if his house is valued at £27 000 and the contents at £6500.

3. The insurance on a car is £120 but a 20% no-claims bonus is allowed. How much is the yearly premium?

4. If the insurance premium is £20 on £5000, how much will the premium be on £8000?

5. A man aged 25 years wishes to assure his life for £7000. He is quoted a rate of £9.00 per £1000 assured per annum. How much in premiums will he pay monthly?

6. A car owner is quoted an annual premium of £90 but she is allowed a no-claims bonus of $33\frac{1}{3}\%$. Calculate the amount of her annual premium.

7. A person aged 36 years is quoted a rate of £13.90 per £1000 assured per annum. He wishes to assure his life for £8000 but he wishes to pay monthly. The company states that for monthly payments the premiums will be increased by 3%. How much per month will the policy holder actually pay?

8. For a monthly premium of £50 a woman aged 40 is guaranteed £6624 at the end of 10 years. However, the company states that the maturity value of the policy is likely to be £11 241.

    (a) Work out how much the woman will pay in premiums over the 10-year period.

    (b) Calculate how much profit she is likely to make on the policy.

9. A man drives 15 000 km in a year. If he averages 10 km per litre of petrol costing 40p per litre how much does petrol cost him?

10. A second-hand car is bought for £900. It is run for a year and then sold for £640. The cost of insurance is £63.20, tax £100, repairs £76 and maintenance £16. It is driven 12 000 km in the year and it averages 11 km per litre of petrol costing 42p per litre. What is the cost of a year's motoring and how much is the cost per kilometre?

11. A new car is bought for £3000 and it is sold two years later for £1900. During this time it travelled 42 000 km at an average petrol consumption of 14 km per litre. If petrol costs 40p per litre and other expenses were £210, calculate the cost per kilometre of running the car.

Miscellaneous Exercise 10

Section A

1. The cost price of a table is £120 and it is sold for £200.

    (a) How much is the profit?

    (b) What is the profit per cent?

2. A motor cycle is bought for £300 and sold for £150.

    (a) How much loss is made?

    (b) What is the loss per cent?

3. A dealer offers a discount of 25% on all the clothes in his shop. A coat is priced at £80. How much will a customer pay when he buys the coat?

4. Value added tax is charged at 15%. An electric fire is priced at £50 exclusive of tax.

    (a) How much tax will be charged?

    (b) How much will a customer pay for the fire?

5. A house has a rateable value of £400. Rates are levied at £1.25 in the pound. How much will the householder pay in rates?

6. A man wants to insure his house for £30 000. The insurance company quotes him £3 per £1000 insured. How much will the man pay in premiums per annum?

7. The insurance on a car is £120 but there is a 30% no-claims bonus. How much is the premium for the year?

8. A woman pays a 10% deposit on a house which she bought for £30 000. How much mortgage does she need?

Section B

1. The rateable value of Mr Richards' house is £280. He pays rates of 93p in the pound. How much does Mr Richards pay in rates for the year?

2. A car owner is quoted a premium of £80 per annum but he is allowed a no-claims bonus of 40%. How much does he actually pay?

3. Given that the value added tax on goods supplied is 15% of their value, calculate the value added tax payable on goods valued at £90.

4. By selling a chair for £31.05 a shopkeeper makes a profit of 15% on his cost price. Calculate his profit.

5. By selling an article for £18 a dealer made a profit of 44% on the cost price. At what price must it be sold to make a profit of 40%?

6. A woman sells her car for £2430 and as a result loses 10% of the price she paid for it. What price did she pay for the car?

7. A city treasurer estimated that the city would need a rate of 82.5p in the pound in order to provide revenue of £68 887 500 and that the cost to the rates of Family and Community Service would be 8.6p in the pound. Calculate (a) the product of a penny rate, (b) the amount of money to be raised for Family and Community Service.

8. Find the rateable value of a house on which the amount paid in rates was £192 when the rate was £1.60 in the pound.

9. The price of a coat was £68.25 but a 5% discount for cash is allowed. How much does a cash-paying customer actually pay for the coat?

10. A man aged 45 years takes out a 'with profits' insurance policy for £5000. He is charged premiums of 45p per £100 assured. If he pays the premiums monthly, how much will he pay?

Multi-Choice Questions 10

1. The rateable value of a house is £150. The rates are 80p in the pound. The amount paid annually is

   A £12   B £120   C £140   D £150

2. In a certain area the rateable value of all the property is £8 000 000. The product of a penny rate is

   A £8000   B £16 000
   C £20 000   D £80 000

3. A car owner is quoted a premium of £99.60 to insure his car. If his no-claims bonus is 40%, how much does he pay?

   A £39.84   B £59.76
   C £99.60   D £139.44

4. When a dealer sells an article for £18 he makes a profit of £3. His percentage profit is

   A $14\frac{1}{2}\%$ B $16\frac{2}{3}\%$
   C 20%   D 25%

5. A shopkeeper buys an article for £1.20. His marked selling price is 20% more than the cost price. His marked selling price is

   A £1.25   B £1.40
   C £1.44   D £1.60

6. When an article is sold for £180 a loss of 10% on its cost price is recorded. The cost price of the article is

   A £162   B £190   C £198   D £200

7. A shopkeeper decides to give 10% discount on all purchases in his store. How much would a customer pay for a television set that originally cost £240?

   A £226   B £218.18
   C £216.16   D £216

8. By selling goods for £56 a dealer made a profit of 12%. How much did he gain?

   A £12   B £6.72

   C £6   D £4.67

9. A girl bought a record for 75p and sold it for 60p. Her loss as a percentage of the cost price is

   A 15% B 20% C 60% D 72%

10. During a sale, a shop reduced the price of everything by 10%. What was the sale price of an article originally priced at £4.30?

    A £3.40   B £3.87

    C £3.97   D £4.73

Mental Test 10

Try to answer the following questions without writing anything down except the answer.

1. The cost price of an article is £8 and the selling price is £10. What is the profit?

2. The cost of a car is £3000 and it is sold for £2500. How much loss is made?

3. If the profit is £10 and the cost price is £50, what is the percentage profit?

4. The cost price of an article is £100 and the selling price is £120. What is the profit per cent?

5. A dealer offers a discount of 20% on all the goods in his shop. An article is priced at £200. How much will a customer pay for it?

6. Value added tax is charged at 15%. How much tax is payable on goods priced at £50 exclusive of tax?

7. A house has a rateable value of £300. If the rates are levied at £1.50 in the pound, how much will the house-holder pay in rates?

8. All the property in a small town has a rateable value of £500 000. What is the product of a penny rate?

9. A man wishes to insure the contents of his house for £5000. The insurance company quotes him £2 per £1000 insured. How much will the man pay in premiums per year?

10. A woman insures her car for £200 but a no-claims bonus of 20% is given. How much will she pay in insurance premiums?

Household Finance

Rent

Rent is a charge for accommodation and it is usually paid weekly or monthly to the landlord who owns the property. The biggest owners of rented property are local authorities.

The landlord may include the rates in his charge for accommodation but if he does not it is the tenant's responsibility to see that these are paid.

Example 1

A landlord charges a rent of £17.50 per week and the rates are £101.40 per annum. What weekly inclusive amount should the landlord charge the tenant?

$$\text{Rates per week} = £101.40 \div 52$$
$$= £1.95$$

$$\text{Inclusive charge per week} = £17.50 + £1.95$$
$$= £19.45$$

Mortgages

If someone buying a house or flat cannot pay outright, he or she arranges a loan called a mortgage from a building society. The society generally requires a deposit of 5 or 10% of the purchase price of the property, although sometimes 100% mortgages are available.

The balance of the loan plus interest is paid back over a number of years — perhaps 20 or 25. The interest rates charged by the building society vary from time to time.

Example 2

(a) A man wishes to buy a house costing £32 000. He pays a 10% deposit to the building society. How much mortgage will he require?

Method 1

$$\text{Deposit} = 10\% \text{ of } £32 000$$
$$= £3200$$

$$\text{Mortgage required} = £32 000 - £3200$$
$$= £28 800$$

Method 2

Since a deposit of 10% is paid the mortgage required is 90% of the purchase price.

$$\text{Mortgage required} = 90\% \text{ of } £32 000$$
$$= £28 800$$

(b) A building society quotes the repayments on a mortgage as being £12.69 per month per £1000 borrowed. What will be the monthly repayments on a mortgage of £25 000?

$$\text{Monthly repayments} = \frac{25 000}{1000} \times £12.69$$
$$= £317.25$$

Sometimes a combined mortgage and endowment assurance can be arranged. An endowment policy (see Chapter 10) is taken out for the value of the loan. Interest is paid on the full amount of the loan for the whole period of the endowment, after which the money received from the insurance policy is used to pay off the loan.

Example 3

A man wishes to borrow £20 000 to buy a house. The loan is to be covered by an insurance policy for a 20-year period which costs £3.88 per £1000 assured per month. The building society charges interest on the loan at 13%. What will be the total monthly payments?

$$\text{Annual loan interest} = 13\% \text{ of } £20\,000$$
$$= £2600$$

$$\text{Monthly payments of interest} = £2600 \div 12$$
$$= £216.67$$

$$\begin{matrix}\text{Endowment payments} \\ \text{per month}\end{matrix} = \frac{20\,000}{1000} \times £3.88$$
$$= £77.60$$

$$\begin{matrix}\text{Total monthly} \\ \text{payments}\end{matrix} = £216.67 + £77.60$$
$$= £294.27$$

Exercise 11.1

1. A woman wishes to buy a house costing £28 000. She pays a 10% deposit. How much mortgage does she require?

2. A house is rented for £44.16 per week. Its rateable value is £174 and rates are levied at £1.08 in the pound. What will be the weekly charge for rent and rates?

3. A flat is rented for £39.50 per week and the rates are £194.40 per annum. Calculate the weekly charge for rent and rates.

4. A man borrows £24 000 from a building society in order to buy a house. The society charges £12.20 per month per £1000 borrowed. How much are the monthly repayments?

5. A person borrows £16 000 from a building society to buy a house. The loan is covered by an insurance policy for the ten-year term of the loan, the premiums being £7.95 per £1000 assured. The building society charges interest at 10% per annum. Work out the total monthly outgoings of the borrower.

6. Calculate the monthly cost of owning a house, given that the expenditure is as follows: mortgage repayments £220 per month; insurances £330 per year; rates £120 per half year.

7. A woman took out an £8000 mortgage when the building society rate was $8\frac{1}{2}\%$ per annum. If the rate is now 11% per annum, calculate the increase in her monthly payments to the society.

Hire-Purchase

When we purchase goods and pay for them by instalments we are said to have purchased them on **hire-purchase**. Usually the purchaser pays a deposit. The remainder of the purchase price (called the balance) plus interest is repaid in a number of instalments.

Example 4

A woman buys some furniture for £280. She pays a deposit of 25%, and interest at 20% is paid on the outstanding balance. She pays back the balance plus interest in twelve monthly instalments. Calculate the amount of each instalment.

$$\text{Deposit} = 25\% \text{ of } £280$$
$$= £70$$
$$\text{Balance} = £280 - £70$$
$$= £210$$
$$\text{Interest} = 20\% \text{ of } £210$$
$$= £42$$
$$\text{Total to be repaid} = £210 + £42$$
$$= £252$$
$$\text{Monthly instalments} = £252 \div 12$$
$$= £21$$

Exercise 11.2

1. If the hire-purchase payments on a record player are £5 per month, how much will be paid at the end of 2 years?

2. A woman buys a washing machine for £150. She pays a deposit of £30 and is charged 10% interest on the balance.
   (a) How much is the balance?
   (b) Calculate the interest payable.
   (c) How much is the interest plus the balance?
   (d) If the balance plus interest is payable in ten equal monthly instalments, what is the amount of each instalment?

3. A man buys a television set for £320 and pays a deposit of £50. He is to pay the balance plus interest in twelve monthly instalments. If interest is charged at 10% on the balance for the full period of the loan, calculate the amount of each instalment.

4. A woman buys a suite of furniture for £640. A deposit of 20% is paid and interest at 12% per annum is charged on the balance for the full period of the loan. The balance is to be paid in four quarterly payments.
   (a) Work out the amount of the deposit.
   (b) What is the amount of the balance?
   (c) Calculate the interest payable.
   (d) Work out the total amount to be repaid.
   (e) Calculate the amount of each instalment.

5. The cash price of a table is £200. To buy it on hire-purchase a deposit of £50 is payable plus ten equal payments of £21.
   (a) Work out the amount paid for the furniture if it is bought on hire-purchase.
   (b) Find the difference between the cash price and the hire-purchase price.

Bank Loans

Many people take out personal loans from a bank. The bank will calculate the interest for the whole period of the loan and the loan plus interest is usually repaid in equal monthly periods.

Example 5

A woman borrows £300 from a bank. The bank charges 18% interest for the whole period of the loan. If the repayments are in twelve equal monthly instalments, calculate the amount of each payment.

$$\text{Interest} = 18\% \text{ of } £300$$
$$= £54$$
$$\text{Total amount to be repaid} = £300 + £54$$
$$= £354$$
$$\text{Amount of each instalment} = £354 \div 12$$
$$= £29.50$$

1. A man borrows £500 from a bank and interest amounts to £100. He repays the loan plus interest in ten equal monthly instalments. What is the amount of each instalment?

2. A bank lends a woman £1200. They charge interest at 20%. She pays the loan plus interest back in ten equal instalments How much does she pay each time?

3. An advertisement in a newspaper states that if you borrow £1000 over a period of 10 years, the repayments will be £20.70 per month.

   (a) A man borrows £4500 over 10 years. How much will be his monthly repayments?

   (b) How much will the man repay altogether over the 10-year period?

4. Over 5 years a man who borrows £7500 repays the loan at £210.07 per month. How much does the finance house charge per £1000 borrowed per month?

5. A man borrows £2000 from a bank and repays £2400 over a period of one year. What rate of interest does the bank charge?

Gas Bills

Gas is charged according to the number of therms used (1 therm = 100 cubic feet approximately). To find the cost of gas used, multiply the number of therms by the charge per therm. In addition, a standing charge may also be made.

Example 6

A customer uses 90 therms of gas in a quarter. She is charged 27.216p per therm plus a standing charge of £9.00. How much is her gas bill?

Cost of gas used = 90 × 27.216p = £24.49

Gas bill = Cost of gas used + Standing charge

= £24.49 + £9.00

= £33.49

Reading Meter Dials

The amount of gas used is always measured in cubic feet (often in hundreds of cubic feet) and this is changed into therms. In the ordinary way your gas meter will be read at regular intervals. However, there may be occasions when you wish to check your gas consumption yourself.

Fig. 11.1 shows the dials on a gas meter. To read the dials copy down the readings in the order in which they appear. Where the hand stands between two figures, put down the lower figure. However, if the hand stands between 9 and 0 you should put down 9.

Fig. 11.1

| Dial 1 pointer has passed 7, reading is 7 | Dial 2 pointer has passed 5, reading is 5 | Dial 3 pointer has passed 1, reading is 1 | Dial 4 pointer has passed 9, reading is 9 |

The dials in the diagram show 7519. If the previous meter reading was 7491, the difference would be 7519 − 7491 = 28 hundred cubic feet = 28 therms approx. This is the amount of gas used since the meter was last read.

1. Copy and complete the following table:

Charge per therm	Number of therms used	Cost of gas used	
		In pence	In pounds
40p	100		
42p	80		
48p	135		
35.3p	240		
37.5p	400		

2. A customer uses 90 therms of gas in a certain quarter. If the cost per therm is 37.0p and the standing charge is £9.90, how much is the gas bill?

3. The present meter reading of a householder is 1119 and the previous reading was 796 hundreds of cubic feet.

   (a) How many hundreds of cubic feet have been used?

   (b) If 1 therm = 100 cubic feet, how many therms have been used?

   (c) If gas is charged at 39.0p per therm and the standing charge is £8.75, how much will be the gas bill?

4. Copy the set of dials shown in Fig. 11.2 and put in hands to show

   (a) 1700   (b) 9423   (c) 4561

Fig. 11.2

5. Gas meters are read every quarter. If the reading was 350 at the beginning of the quarter and 481 hundreds of cubic feet at the end of the quarter:

   (a) How many therms have been used? (Take 1 therm = 100 cubic feet.)

   (b) If the charge is 37.4 pence per therm, calculate the cost of gas used.

   (c) If the standing charge is £8.00 per quarter, work out the total amount of the gas bill.

6. Write down the readings on the dials shown in Fig. 11.3.

Fig. 11.3

(b)

(c)

7. Fig. 11.4 shows the dials of a gas meter which gives readings in hundreds of cubic feet.

   (a) Read the dials.

   (b) If the previous reading was 4392, find the number of therms used.

   (c) Gas is charged at 38.3p per therm plus a standing charge of £7.90. Calculate the gas bill which the householder will receive.

Fig. 11.4

Electricity Bills

Electricity is charged according to the number of units used (1 unit = 1 kilowatt hour). There is also a fixed charge which is added to the bill.

Example 7

A user of electricity uses 1765 units in a quarter. If the standing charge is £6.84 and the price per unit is 5.37 pence, find the amount of the quarterly electricity bill.

1765 units at 5.37p per unit = 1765 × 5.37p

$$= £94.78$$

Quarterly electricity bill = £94.78 + £6.84

$$= £101.62$$

Reading Electricity Meters

When reading an electricity meter remember that adjacent dials revolve in opposite directions. The five dials are read from left to right. The method of reading an electricity meter is shown in Fig. 11.5.

(1) Always note down the number the pointer has passed. In Fig. 11.5, on the dial labelled 10 000 the pointer is between 4 and 5 so write down 4.

(2) If the pointer is directly over a figure write down that figure and underline it. On the dial marked 1000 in the diagram, the reading is 5 so write 5.

(3) The reading of the dials in the diagram gives 45 928.

(4) Now look at the figures underlined. If one of these is followed by a 9 reduce the underlined figure by 1. Thus the corrected reading is 44 928.

(5) When you have worked out the meter reading, subtract the previous reading to find the number of units of electricity used.

Fig. 11.5

DIAL 1
Pointer has passed 4, reading is 4
DIAL 2
Pointer is directly over 5,
write down 5 and underline 5
DIAL3
Pointer has passed 9, reading is 9
DIAL 4
Pointer has passed 2, reading is 2
DIAL 5
Pointer is directly
over 8, write down 8
and underline 8

Example 8

Fig. 11.6 shows the dials of an electricity meter.

(a) Read the meter.

(b) If the previous meter reading was 53 432, find the number of units used.

(c) If the charge per unit is 5.93p and the standing charge is £9.80 per quarter, calculate the amount of the electricity bill.

Fig. 11.6

(a) The meter reads 56 378

(b) Number of units used $=\ 56\,378$

$$-\,53\,432$$

$$=\ 2946$$

(c) Cost of electricity $=\ 2946 \times £0.0593$

$$+\ £9.80$$

$$=\ £174.70$$

$$+\ £9.80$$

$$=\ £184.50$$

Exercise 11.5

1. Copy and complete the table below:

Charge per unit	Number of units used	Cost of electricity used	
		In pence	In pounds
4.85	100		
5.17	600		
6.30	450		
7.98	800		
9.80	750		

2. In a certain area electricity is charged for at a fixed rate of 9.80p per unit used. A householder uses 780 units in a quarter. How much will his electricity bill be?

3. A householder uses 2500 units of electricity in a quarter. Each unit used costs 5.72p and there is a standing charge of £9.75 per quarter. Calculate the amount the householder will pay for his electricity.

4. The meter reading on 15 Feb was
   009 870 and on 17 May it was 009 986.
   How many units of electricity have
   been used? If each unit costs 5.93p
   and there is a standing charge of £8.70,
   find how much the electricity bill will
   be.

5. An economy tariff charges for electricity as follows: day rate 5.70p per
   unit; night rate 2.04p per unit; standing
   charge £2.66. Work out the amount of
   the electricity bill if 1325 units are used
   during the day and 932 units during the
   night.

6. Copy the set of dials shown in Fig. 11.7
   and put in arrows to show readings of
   (a) 81 372 **(b)** 56 325 **(c)** 73 454

Fig. 11.7

7. Read the dials shown in Fig. 11.8.

Fig. 11.8

8. A householder receives an electricity
   bill for £77.60. If electricity is charged
   for at 5.70p per unit and the standing
   charge is £9.20, find the number of
   units used by the householder.

Telephone Bills

The amount of a telephone bill depends
upon the number of calls made and the
charge for rental of the telephone itself.
The bills are paid quarterly and VAT at 15%
is charged.

Example 9

A householder dialled 500 calls in a certain
quarter. Each call was charged for at 4.70
pence. The quarterly rental charge is £17.25.
If VAT is charged at 15%, find the total
amount of the householder's telephone bill.

$$500 \text{ calls at } 4.70\text{p} = £23.50$$

$$\text{Cost less VAT} = £23.50 + £17.25$$

$$= £40.75$$

$$\text{VAT at } 15\% = 15\% \text{ of } £40.75$$

$$= £6.11$$

$$\text{Total cost} = £40.75 + £6.11$$

$$= £46.86$$

Exercise 11.6

Work out the total cost of the following
telephone bills:

1. Standing charge of £14.27.
   100 units at 4.8p per unit.

2. Standing charge of £16.15.
   180 units at 4.8p per unit.

3. Standing charge of £17.80.
   150 units at 4.3p per unit.

4. A telephone bill is made up of the
   following charges: rental £16.15 per
   quarter, 603 dialled units at 4.8p each
   and VAT at 15%.

   (a) Find the cost of the dialled units.

   (b) Calculate the total bill not including
   VAT.

   (c) Work out the total amount of the
   bill including VAT.

5. The telephone meter was read on 17th Feb as 007 963 and on 15th May as 009 215.

    (a) How many units have been used during this period?

    (b) If each unit is charged for at 4.9p, calculate the cost of the units used.

    (c) The standing charge is £17.29. What is the total bill?

    (d) VAT is charged at 15% on the total bill. How much is VAT?

    (e) Work out the total telephone bill including VAT.

Miscellaneous Exercise 11

Section A

1. Mrs Evans insured her jewellery for £2400. The premium is 25p per £100 of cover. What is her total premium?

2. A small computer is priced at £300 plus VAT at 15%.

    (a) How much is the VAT?

    (b) What is the total cost of the computer including VAT?

    (c) Tindalls offer the computer with £30 off its total price. Lyons offer the computer with 20% off its total price. Find the difference between the two offers.

    (d) Thompsons offer the computer on hire-purchase as follows: Deposit £90 plus 12 equal monthly payments of £18.75. What is the hire-purchase price of the computer?

3. (a) Fig. 11.9 shows the positions of the dials of an electricity meter at the beginning of a quarter. Write down the meter reading.

Fig. 11.9

(b) If during this quarter 760 units of electricity are used, write down the new meter reading.

(c) Copy the dials shown in Fig. 11.10 and show the new meter reading on the dials.

Fig. 11.10

(d) If electricity is charged at 5.7 pence per unit plus a standing charge of £8.50, calculate the total cost of the 760 units in the quarter.

4. Copy and complete the following telephone bill:

    Quarterly rental charge   £9.75

    Dialled units 600 at 4.7p per unit

    Total cost (exclusive of VAT)

    VAT at 15%

    Total payable

5. Fig. 11.11 shows the positions of the dials of a gas meter at the end of a quarter. The reading gives the number of hundreds of cubic feet used.

    (a) What is the reading on the meter?

    (b) If 1 therm = 100 cubic feet, how many therms have been used.

    (c) If the previous reading was 1013 therms, how many therms have been used in the quarter?

    (d) Gas is charged at 34.9p per therm plus a standing charge of £9.25 per quarter. Calculate the total amount of the quarterly gas bill.

Fig. 11.11

Section B

1. A woman borrows £2500 from a bank which charges interest at 15% for the entire period of the loan.

   (a) How much does the woman pay in interest?

   (b) How much does she pay in total?

   (c) If the loan plus interest is to be repaid in 12 monthly instalments, calculate the amount of each instalment.

2. A man purchases a house using a combined life insurance and mortgage.

   (a) He borrows £18 000 from the building society which charges an interest rate of 13% per annum. Calculate the amount of interest he has to pay per annum.

   (b) The insurance premium is £2.95 per month per £1000 of cover. What is the total premium per month?

   (c) Calculate the total monthly payments the man must make?

3. A woman wants to buy a house for £30 000. She pays a deposit of 10% to the building society.

   (a) How much mortgage does she require?

   (b) The monthly repayments are £11.60 per £1000 borrowed. How much are her monthly repayments?

4. A man wants to borrow £960 from a bank. The bank agrees to the loan at 12% per annum simple interest.

   (a) How much does he pay in interest if the loan lasts for two years?

   (b) The loan plus interest is to be repaid in 24 equal instalments. Work out the amount of each instalment.

5. A suite of furniture is priced at £700. Hire-purchase terms are available which are: deposit 25% of cash price and 18 equal monthly instalments of £36.75.

   (a) Work out the amount of the deposit.

   (b) Calculate the total hire purchase price.

   (c) Find the difference between the hire purchase price and the cash price.

Multi-Choice Questions 11

1. A debt of £160.80 is to be paid in equal instalments of £20.10. How many instalments are required?

   A 4   **B** 8   **C** 10   **D** 16

2. A telephone subscriber is charged a rental of £7.50 and 5p for each dialled call. If her telephone bill is £25, how many calls did she make?

   A 150   **B** 350   **C** 500   **D** 650

3. A householder is charged for electricity at 8p per unit for all the units used. If his bill is £48, how many units were used?

   A 48   **B** 384   **C** 600   **D** 4800

4. Gas is charged at 57p per therm. If 560 therms are used in a quarter, how much is the quarterly bill?

   A £5.60   **B** £57
   C £319.20   **D** £560

5. Mortgages cost £15 per month per £1000 borrowed. The annual repayments on a mortgage of £20 000 are

   A £180   **B** £300
   C £2000   **D** £3600

6. A woman wishes to buy a house for £30 000. She pays a 10% deposit. How much mortgage does she require?

   A £2700   **B** £3000
   C £27 000   **D** £30 000

7. A man wishes to borrow £480 from a bank. The bank agrees to the loan at 12% interest per annum. The repayments are to be in 24 equal monthly instalments. How much will the man repay altogether?

   A £240   B £480
   C £537.60   D £595.20

8. A workbench is priced at £88. It can be bought for cash or on hire-purchase. The hire-purchase terms are £20 deposit plus 12 equal monthly payments of £6.40. The rate of interest is

   A 3.4%   B 5%   C 10%   D 15.5%

Mental Test 11

Try to answer the following questions without writing anything down except the answer.

1. A landlord charges an annual rent of £480. What is the monthly rent?

2. A building society charges a person £20 per month per £1000 borrowed. If the person borrows £10 000 what are the monthly repayments?

3. A woman buys furniture priced at £800. She pays a deposit of 10%. What is her outstanding balance?

4. A young man buys a motor cycle on hire-purchase. He pays a deposit of £50 and 20 monthly instalments of £30 each. How much does he pay in total for his motor bike?

5. A man borrows £500 from his bank which charges 20% interest for the whole period of the loan, which he pays off in 10 equal monthly instalments. How much is each instalment?

6. A housewife uses 100 therms of gas per quarter which is charged for at 30p per therm. How much is her quarterly gas bill?

7. A householder uses 2000 units of electricity which costs 5p per unit. How much is his electricity bill?

Time, Distance and Speed

Measurement of Time

The units of time are:

$$60 \text{ seconds (s)} = 1 \text{ minute (min)}$$
$$60 \text{ minutes (min)} = 1 \text{ hour (h)}$$
$$24 \text{ hours (h)} = 1 \text{ day (d)}$$

Example 1

(a) Change 720 s into minutes.

$$720 \text{ s} = 720 \div 60 \text{ min}$$
$$= 12 \text{ min}$$

(b) Change 8 h into minutes.

$$8 \text{ h} = 8 \times 60 \text{ min}$$
$$= 480 \text{ min}$$

(c) How many seconds are there in 1 hour?

$$1 \text{ h} = 60 \text{ min}$$
$$= 60 \times 60 \text{ s}$$
$$= 3600 \text{ s}$$

The Clock

There are two ways of showing the time:

(1) With a 12-hour clock (Fig. 12.1), in which there are two periods each of 12 hours duration during each day.

The period between midnight and noon is called a.m. whilst the period between noon and midnight is called p.m. Thus 8.45 a.m. is a time in the morning whilst 8.45 p.m. is a time in the evening.

Fig. 12.1

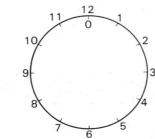

Example 2

Find the length of time between 2.15 a.m. and 8.30 p.m.

2.15 a.m. to 12 noon is	9 h 45 min
12 noon to 8.30 p.m. is	8 h 30 min +
Total length of time is	18 h 15 min

(2) With a 24-hour clock (Fig. 12.2) in which there is one period of 24 hours. This clock is used for railway and airline timetables. Times between midnight and noon are given times of 0000 hours to 1200 hours whilst times between noon and midnight are given times of 1200 hours to 2400 hours.

Fig. 12.2

3.30 a.m. is written 0330 hours whilst 3.30 p.m. is written 1530 hours and this avoids confusion between the two times.

Note that 24-hour-clock times are written with four figures. 1528 hours should be read as 'fifteen twenty-eight hours' and 1300 hours as 'thirteen hundred hours'.

Timetables

Timetables are usually written using the 24-hour clock. The timetable at the foot of the page is a railway timetable for the Paddington to Cheltenham service.

Example 3

Find the length of time between 0425 hours and 1812 hours.

1812 hours is	18 h 12 min
0425 hours is	4 h 25 min —
Total length of time is	13 h 47 min

Example 4

If I catch the 0935 from Paddington, at what time do I arrive at Stonehouse and how long will the journey take?

According to the timetable the 0935 from Paddington arrives at Stonehouse at 1121 hours.

1121 hours is	11 h 21 min
0935 hours is	9 h 35 min —
Time taken is	1 h 46 min

Exercise 12.1

1. How many minutes are there in
    (a) 4 hours   (b) $3\frac{1}{2}$ hours

2. How many seconds are there in
    (a) 9 min   (b) $2\frac{1}{4}$ min

3. Change 350 min into hours and minutes.

Find the length of time, in hours and minutes, between

4. 2.39 a.m. and 8.46 p.m.

5. 2.15 a.m. and 7.30 p.m.

6. 0.36 a.m. and 1.15 p.m.

7. 0049 hours and 1236 hours.

8. 0242 hours and 1448 hours.

9. 0836 hours and 2115 hours.

10. 0425 hours and 1712 hours.

Exercise 12.2

Using the Paddington to Cheltenham railway timetable answer the following questions:

1. I can catch either the 0638 or the 0730 train from Swindon. Which is the faster train to Cheltenham? What is the difference in the times taken for the journey?

2. What time does the 0725 from Paddington arrive at Gloucester? How long does the journey take?

3. Which is the faster train, the 0835 from Paddington to Swindon or the 0935 from Paddington to Swindon? How much shorter is the journey?

MONDAYS TO SATURDAYS

London, Paddington . . . dep	—	—	0725	0835	0935	1040	1100	1200	1405	1430	1535	1625	1650	1742
Swindon . . . . dep	0638	0730	0848	0942	1045	1134	1200	1305	1503	1542	1642	1738	1748	1835
Kemble. . . . . dep	0653	0745	0903	0957	1100	1148	1215	1320	1518	1557	1657	1755	1803	1849
Stroud . . . . dep	0709	0801	0919	1013	1116	1202	1231	1336	1534	1613	1713	1810	1819	1903
Stonehouse . . . . dep	0714	0806	0924	1018	1121	■	1236	1341	1539	1618	1718	1815	1824	1908
Gloucester . . . . arr	0727	0819	0937	1032	1135	1217	1249	1354	1552	1631	1731	1828	1838	1920
Cheltenham . . . . arr	0756	0833	1012	1046	1215	1231	1310	1415	1607	1718	1744	1845	1851	1935

4. I want to board the 1405 train from Paddington at Kemble. At what time is it scheduled to arrive? At what time does this train arrive at Cheltenham? How long is the journey?

5. Which is the fastest train from Stroud to Cheltenham? How long is the shortest journey?

Average Speed

The road from Campton to Burley is 50 miles long. It takes me two hours to drive from Campton to Burley. Obviously I do not drive at the same speed all the time. Sometimes I drive slowly and sometimes I have to stop at crossroads and traffic lights.

However, the time it takes me is exactly the same as if I were driving along a straight road at 25 miles per hour, i.e. 50 miles ÷ 2 hours. So we say that my average speed is 25 miles per hour.

$$\text{Average speed} = \frac{\text{Total distance travelled}}{\text{Total time taken}}$$

The unit of speed depends upon the unit of distance and the unit of time. If the distance is measured in kilometres and the time in hours then the average speed will be measured in kilometres per hour (km/h). If the distance is measured in miles and the time in hours then the average speed will be measured in miles per hour (mile/h). If the distance is measured in metres and the time in seconds then the average speed will be measured in metres per second (m/s).

Note carefully that

$$\text{Distance travelled} = \text{Average speed} \times \text{Time taken}$$

$$\text{Time taken} = \frac{\text{Distance travelled}}{\text{Average speed}}$$

Example 5

(a) A car travels a total distance of 400 km in 5 hours. What is the average speed?

$$\text{Average speed} = \frac{\text{Distance travelled}}{\text{Time taken}}$$

$$= \frac{400}{5} \text{ km/h}$$

$$= 80 \text{ km/h}$$

(b) A lorry travels at 35 miles per hour for 3 hours. How far does it travel?

$$\text{Distance travelled} = \text{Average speed} \times \text{Time taken}$$

$$= 35 \times 3 \text{ miles}$$

$$= 105 \text{ miles}$$

(c) A train travels 120 km at 40 km/h. How long does the journey take?

$$\text{Time taken} = \frac{\text{Distance travelled}}{\text{Average speed}}$$

$$= \frac{120}{40} \text{ hours}$$

$$= 3 \text{ hours}$$

Remember that average speed is defined as *total distance travelled* divided by *total time taken*.

Example 6

(a) A car travels 60 km at an average speed of 30 km/h and 30 km at an average speed of 40 km/h. Calculate its average speed for the entire journey.

Time taken to travel 60 km at 30 km/h

$$= \frac{60}{30} \text{ hours}$$

$$= 2 \text{ hours}$$

Time taken to travel 30 km at 40 km/h

$$= \frac{30}{40} \text{ hours}$$

$$= 0.75 \text{ hours}$$

Total time taken $= (2 + 0.75)$ hours

$\qquad = 2.75$ hours

Total distance travelled $= (60 + 30)$ km

$\qquad = 90$ km

Average speed $= \dfrac{90}{2.75}$ km/h

$\qquad = 32.7$ km/h

(b) A train travels between two towns 270 km apart in 9 hours. If on the return journey the average speed is reduced by 3 km/h, calculate the time taken for the return journey.

Average speed on outward journey

$\qquad = \dfrac{270}{9}$ km/h

$\qquad = 30$ km/h

Average speed on return journey

$\qquad = (30 - 3)$ km/h

$\qquad = 27$ km/h

Time taken for return journey

$\qquad = \dfrac{270}{27}$ hours

$\qquad = 10$ hours

Exercise 12.3

Find the time taken to travel

1. 10 miles at an average speed of 5 miles per hour.

2. 48 km at an average speed of 12 km/h.

3. 60 m at an average speed of 15 m/s.

Find the distance travelled if

4. Average speed is 20 miles/h and time is 3 h.

5. Average speed is 40 km/h and time is $3\frac{1}{2}$ h.

6. Average speed is 60 m/s and time is 8 s.

Find the average speed if

7. Distance is 30 miles and time is 2 hours.

8. Distance is 140 km and time is $3\frac{1}{2}$ hours.

9. Distance is 80 m and time is 5 seconds.

10. An aeroplane took $\frac{3}{4}$ hour to travel from one airport to another. If its average speed was 400 km/h, how far apart are the airports?

11. A boy walks 3 km at a speed of 6 km/h. He then cycles 6 km at a speed of 12 km/h. What is his average speed for the entire journey?

12. A car travels 136 miles at an average speed of 32 miles/h. On the return journey the speed is increased to 39 miles/h. Calculate the average speed for the complete journey.

13. For the first $1\frac{1}{2}$ hours of a 91 km journey the average speed was 30 km/h. If the average speed for the remainder of the journey was 23 km/h, calculate the average speed for the entire journey.

14. A motorist travelling at a steady speed of 90 km/h covers a section of motorway in 25 minutes. After a speed limit is imposed he finds that, when travelling at the maximum speed allowed, he takes 5 minutes longer than before to cover the same section. Calculate the speed limit.

15. In winter a train travels between two towns 264 km apart at an average speed of 72 km/h. In summer the journey takes 22 minutes less than in winter. Find the average speed in summer.

Miscellaneous Exercise 12

Section A

1. (a) A clock at a bus station shows the departure of an evening bus to be 8.15. How would this time be written in a timetable which uses the 24-hour-clock system?

   (b) The bus arrived at its destination at 2045 hours. For how long was the bus travelling?

2. An aeroplane left town A at 1535 hours taking 2 h 40 min for its flight to town B. At what time did the aeroplane reach town B?

3. A car leaves Baxton at 1325 and reaches Flighton at 1555 hours. The distance between the two towns is 125 km.

   (a) How long did the journey take?

   (b) What was the average speed of the car?

4. A boat leaves a harbour at 0235 hours and arrives at its destination at 0405 hours.

   (a) How long did the journey take?

   (b) If the distance travelled was 12 miles, calculate the average speed of the boat?

5. The distance between Cardiff and London is 150 miles. A car leaves Cardiff at 9.35 a.m. and arrives in London 5 hours later.

   (a) What time was it when the car arrived in London?

   (b) What was the average speed of the car?

Section B

1. The distance between two towns is 160 km.

   (a) How long will it take a motorist to make the journey if his average speed is 40 km/h?

   (b) If the motorist completes the return journey in $2\frac{1}{2}$ hours, what was his average speed?

2. A car leaves Bristol at 1055 hours and arrives in Hereford $1\frac{1}{4}$ hours later. The distance from Bristol to Hereford is 50 miles.

   (a) What time did it arrive in Hereford?

   (b) What was its average speed?

3. On a train journey of 117 km, the average speed for the first 27 km was 45 km/h.

   (a) Find the time taken for this part of the journey.

   (b) For the remainder of the journey the average speed was $37\frac{1}{2}$ km/h. Find the time taken for this part of the journey.

   (c) What was the total time for the whole journey?

   (d) Calculate the uniform speed at which the train would have to travel in order to cover the whole distance in the same time.

4. A motorway journey takes 3 hours at an average speed of 120 km/h.

   (a) How long, in kilometres, was the journey?

   (b) How long, in hours, will the journey take if the average speed is decreased to 80 km/h?

5. A greyhound runs 400 metres in 50 seconds.

   (a) Calculate its average speed in metres per second.

   (b) Work out its average speed in kilometres per hour.

Multi-Choice Questions 12

1. A walker covers a distance of 33 km at an average speed of 6 km/h. How long does the journey take?

    A 5 h 3 min   B 5 h 5 min

    C 5 h 20 min   D 5 h 30 min

2. A journey lasts 45 min and the average speed is 48 km/h. What is the length of the journey?

    A 64 km   B 36 km

    C 32 km   D 16 km

3. A motor car travels 40 km at 50 km/h. The time, in minutes, taken by the journey is

    A 20   B 48   C 75   D 180

4. An aeroplane flies non-stop for $2\frac{1}{4}$ h and travels a distance of 1620 km. Its average speed in kilometres per hour is

    A $364\frac{1}{2}$ B 720   C 800   D 3625

5. A boy takes 15 min to walk to school at a speed of 5 km/h. How long would it take if he cycled at 15 km/h?

    A 35 min   B 25 min

    C 10 min   D 5 min

6. A car travels 50 km at 50 km/h and 70 km at 70 km/h. Its average speed is

    A 58 km/h   B 60 km/h

    C 62 km/h   D 65 km/h

7. A car travels for 3 hours at a speed of 45 km/h and for 4 hours at a speed of 50 km/h. It has therefore travelled a distance of

    A $27\frac{1}{2}$ km   B 95 km

    C 335 km   D 353 km

Mental Test 12

Try to answer the following questions without writing anything down except the answer.

1. Change 240 seconds into minutes.

2. Change 4 minutes into seconds.

3. Change 180 minutes into hours.

4. Change 5 hours into minutes.

5. How many days in 120 hours?

6. Find the length of time between 9.30 a.m. and 11.45 a.m.

7. What is the length of time between 1520 hours and 1840 hours?

8. A train leaves Paddington at 1430 hours and arrives in Cheltenham at 1738 hours. How long does the journey take?

9. A motorist travels 160 km in 4 hours. What is his average speed?

10. A train travels 200 miles in 4 hours. Calculate its average speed.

11. A car travels for 3 hours at an average speed of 50 km/h. How far has it travelled?

12. A train travels 80 km at an average speed of 40 km/h. How long does the journey take?

13. An aeroplane flies at an average speed of 500 miles per hour. It flies for 3 hours. How far has it flown?

14. A car takes 5 hours to travel 240 km. What is its average speed?

15. A train travels 150 miles at an average speed of 50 miles per hour. How long does the journey take?

Use the bus timetable and route map shown below to answer the following questions:

16. How long does the bus take to travel from Stroud to Sapperton?

17. What time must I arrive at Chalford, Marle Hill to make sure of arriving in Stroud by the 1239 bus?

18. What is the number of the bus service from Stroud to Eastcombe?

19. Work out the time for the bus to travel between Bourne Bridge and Daglingworth.

20. Is there a bus from Cirencester to Stroud which leaves Cirencester at 10 minutes to 3 in the afternoon?

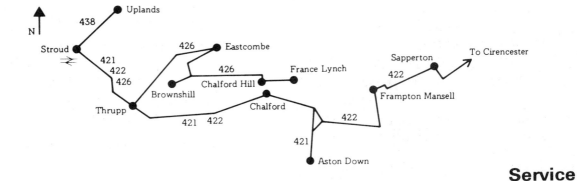

STROUD · CIRENCESTER via Chalford

Service 422

Service 422 via Bowbridge, Thrupp, Brimscombe, Chalford, Aston Down Turn, White Horse Inn, Frampton Mansell, Sapperton, Park Corner, Duntisbourne Rouse Turn, Daglingworth, Stratton.

Mondays to Fridays*

STROUD, Bus Station	0855	1055	1255	CIRENCESTER, Bus Station	0950	1150	1450
Thrupp, Brewery Lane	0903	1103	1303	Daglingworth	1000	1200	1500
Bourne Bridge	0907	1107	1307	Sapperton	1010	1210	1510
Chalford, Marle Hill.	0912	1112	1312	White Horse Inn	1017	1217	1517
White Horse Inn	0917	1117	1317	Chalford, Marle Hill.	1022	1222	1522
Sapperton	0924	1124	1324	Bourne Bridge	1027	1227	1527
Daglingworth	0934	1134	1334	Thrupp, Brewery Lane	1031	1231	1531
CIRENCESTER, Bus Station	0944	1144	1344	STROUD, Bus Station	1039	1239	1539

CODE
* — Not Bank Holiday Mondays, Good Friday or Boxing Day.

Basic Algebra

Introduction

In algebra, letters are used in place of numbers. Let us take two numbers and call them a and b. The sum of the two numbers is $a + b$. The difference of the two numbers is $a - b$. The product of the two numbers is $a \times b$ which is usually written ab (i.e. the multiplication sign is usually omitted in algebra).

The first number divided by the second number is $a \div b$ or $\dfrac{a}{b}$.

Example 1

(a) Translate into symbols six times a number x.

Six times a number x is $6 \times x = 6x$

(b) Translate into symbols 5 times a number y minus four.

Five times a number y minus four is

$$5 \times y - 4 = 5y - 4$$

Exercise 13.1

Translate each of the following into algebraic symbols:

1. Seven times a number a.

2. Five times a number b minus seven.

3. The sum of five times a number x and three times a number y.

4. The product of the three numbers x, y and z.

5. Eight times the product of the two numbers m and n.

6. Twice the number p divided by three times the number q.

7. Nine times a number r plus three times a number s.

8. The product of two numbers x and y divided by a third number p.

Substitution

Substitution is the process of assigning numbers to symbols in an algebraic expression.

Example 2

Find the value of $3x + 5y - 2z$ when $x = 6$, $y = 7$ and $z = 4$.

$$\begin{aligned}
3x + 5y - 2z &= 3 \times 6 + 5 \times 7 - 2 \times 4 \\
&= 18 + 35 - 8 \\
&= 45
\end{aligned}$$

(Remember that we must multiply before adding or subtracting.)

Exercise 13.2

If $x = 3$ and $y = 6$, find numerical values for the following algebraic expressions:

1. $x + 3y$
2. $3x - 4y$
3. $5x - 2y - 3x + 4y$
4. $x - 5$
5. $2x + 7$
6. $2x \div y$
7. $\dfrac{3y}{2x}$
8. $\dfrac{3y + 6}{4x}$

If $a = 3$, $b = 6$ and $c = -7$, find numerical values for the following expressions:

9. $b + 3$
10. $a - 3$
11. $12 - c$
12. $5a$
13. $3c$
14. bc
15. abc
16. $5ab$
17. $\dfrac{12}{a}$
18. $\dfrac{ab}{9}$
19. $a + 3b - 4c$
20. $a - 3b + 8c$
21. $\dfrac{c}{a}$
22. $\dfrac{b}{c}$

Powers

The quantity $a \times a \times a$ is usually written as a^3. The number 3 which shows the number of as to be multiplied together is called the **index** (plural: indices) and we say that a has been raised to the third power.

Example 3

(a) Find the value of y^6 when $y = 2$.

$$y^6 = 2^6$$
$$= 2 \times 2 \times 2 \times 2 \times 2 \times 2$$
$$= 64$$

(b) Find the value of a^3 when $a = -2$.

$$a^3 = (-2)^3$$
$$= (-2) \times (-2) \times (-2)$$
$$= -8$$

When dealing with expressions like $5ab^4$, note that it is only b which is raised to the power of 4. Thus

$$5ab^4 = 5 \times a \times b \times b \times b \times b$$

Example 4

Find the numerical value of $5b^2c^3$ when $b = 4$ and $c = 2$.

$$5b^2c^3 = 5 \times 4^2 \times 2^3$$
$$= 5 \times 4 \times 4 \times 2 \times 2 \times 2$$
$$= 640$$

Exercise 13.3

If $p = 2$, $q = -3$ and $r = 5$, find numerical values for the following algebraic expressions:

1. p^3
2. q^4
3. r^2
4. pq^4
5. $3p^4$
6. $4p^2q$
7. p^3r^2
8. pqr^3
9. $4pq^2r$
10. $p^2 + q^2$
11. $q^2 - r^2$
12. $p^4 - q^3$

Addition and Subtraction of Algebraic Terms

Like terms are numerical multiples of the same algebraic quantity. Thus $5x$, $8x$ and $-3x$ are three like terms.

The **numerical coefficient** is the number which appears before the symbols of a term. Thus in the term $8x$, the numerical coefficient is 8. If a term has no number in front of the symbols, for instance ab, the numerical coefficient is 1.

Like terms are added or subtracted by adding or subtracting their numerical coefficients.

Example 5

(a) $3x + 5x - 2x = (3 + 5 - 2)x = 6x$

(b) $3y - 5y = (3 - 5)y = -2y$

(c) $9b + b + 3b = (9 + 1 + 3)b = 13b$

Only like terms can be added or subtracted. $7a - 3b + c$ is an expression which contains three unlike terms. Hence it cannot be simplified further.

It is quite possible to have several sets of like terms in an expression, and each set may then be simplified by adding and/or subtracting.

Example 6

Simplify $8p + 2q - 5r + 6r - 4p + q - 3p + 5r$.

$$8p + 2q - 5r + 6r - 4p + q - 3p + 5r$$
$$= (8 - 4 - 3)p + (2 + 1)q$$
$$+ (-5 + 6 + 5)r$$
$$= 1p + 3q + 6r$$
$$= p + 3q + 6r$$

Note that it is usual in algebra to write, for instance, $1p$ as p, the 1 being understood.

Exercise 13.4

Simplify each of the following by adding or subtracting like terms:

1. $3x + 5x$
2. $7y - 3y$
3. $8p - 11p$
4. $q - 4q$
5. $3x - 5x$
6. $8p + 5p - 3p$
7. $-3x - 4x + x$
8. $5m + 8m - 15m$
9. $5x - 4y + 2c + 7x + 5c - 2y - 6y + 8c + x + 3y$
10. $a - 3b + 2c - 5a - 6c + 8b - 7b + 9a - c$

Multiplication of Algebraic Terms

The rules for the multiplication of algebraic terms are the same as those for directed numbers. Thus

$$(+x) \times (+y) = +xy = xy$$
$$(-x) \times (+y) = -xy$$
$$(+x) \times (-y) = -xy$$
$$(-x) \times (-y) = +xy = xy$$

Example 7

(a) $2x \times (-3y) = 2 \times (-3) \times x \times y$
$$= (-6) \times x \times y$$
$$= -6xy$$

(b) $5a \times 2b \times (-3c) = -(5 \times 2 \times 3)$
$$\times a \times b \times c$$
$$= -30abc$$

Note that in writing algebraic expressions such as those in Example 7, we have used the order: numbers first then letters in alphabetical order. It is not wrong to write terms in any other order but it looks better if we write the numbers first and then the letters in alphabetical order. Thus we would write $5xyz$ rather than $x5zy$.

Note also that it is bad practice to have two signs following one another, for instance $2x \times -3y$. A bracket should be used to avoid this. Thus $2x \times -3y$ would be written $2x \times (-3y)$.

When multiplying expressions containing the same symbols, indices are used.

Example 8

(a) $3m \times 5m = 3 \times m \times 5 \times m$
$$= 15m^2$$

(b) $3mn \times 2n^2 = 3 \times m \times n \times 2 \times n \times n$
$$= 6mn^3$$

Exercise 13.5

Simplify each of the following by making them into single terms:

1. $2y \times 3p$
2. $3m \times (-2p)$
3. $(-4a) \times 5b$
4. $3a \times 2b \times 5c$
5. $(-2x) \times (-3y) \times 2z$
6. $7m \times (-2n) \times p$
7. $3a \times 2b \times (-c)$
8. $(-3) \times (-5y) \times 2z \times (-4q)$
9. $2m \times (-3m)$
10. $2p \times 3q \times (-q) \times 4p$

Brackets

As in arithmetic, brackets are used for convenience in grouping terms together. When 'removing' a bracket each term within the bracket is multiplied by the term outside the bracket.

Example 9

(a) $5(2m + 3p) = (5 \times 2m) + (5 \times 3p)$
$$= 10m + 15p$$

(b) $-2(3a - 2b) = (-2 \times 3a)$
$$+ [-2 \times (-2b)]$$
$$= -6a + 4b$$

When a bracket has a minus sign outside it, it means that each term within the bracket has to be multiplied by -1 when the bracket is removed.

$$-(p + q) = (-1 \times p) + (-1 \times q)$$
$$= -p - q$$

Note that all the signs inside the bracket are changed when the bracket is removed.

Example 10

(a) $-(4m - 3n) = -4m + 3n$

(b) $-5(a - 2b) = -5a + 10b$

When simplifying expressions containing brackets remove the brackets first and then add and/or subtract like terms.

Example 11

Simplify $5(3a + 2b) - 3(2a - 6b)$.

$$5(3a + 2b) - 3(2a - 6b) = 15a + 10b$$
$$-6a + 18b$$
$$= 9a + 28b$$

Exercise 13.6

Remove the brackets and simplify:

1. $3(a + 2b)$
2. $5(2x - 3y)$
3. $5(4p - 3q)$
4. $-(p - q)$
5. $-2(3x - 4y)$
6. $-5(4m - 3n)$
7. $3(a - 2b) + 5(2a - 3b)$
8. $-(p + 2q) - (3p + 5q)$
9. $2(3m + 5n) - 3(2m - 5n)$
10. $5(2x - 3y) - 3(x - 2y)$

Division of Algebraic Terms

When dividing algebraic expressions, cancellation between numerator (top part) and denominator (bottom part) is often possible. Note that cancelling is equivalent to dividing top and bottom by the same quantity.

Example 12

(a) $$\frac{pq}{p} = \frac{\not{p} \times q}{\not{p}} = q$$

(b) $$\frac{3x^2y^3z}{xyz} = \frac{3 \times \not{x} \times x \times \not{y} \times y \times y \times \not{z}}{\not{x} \times \not{y} \times \not{z}}$$
$$= 3xy^2$$

Simplify the following:

1. $12x \div 6$

2. $(-3a) \div 3b$

3. $(-5x) \div (-5y)$

4. $4a \div 2b$

5. $4ab \div 2a$

6. $12x^2yz^2 \div 4xz^2$

7. $(-12a^2b) \div 6a$

8. $8a^2bc^2 \div 4ac^2$

9. $7a^2b^2 \div ab$

10. $2x^3y^2z^4 \div x^2yz^2$

Factorising

The expression $3x + 3y$ has the number 3 common to both terms.

$$3x + 3y = 3 \times (x + y) = 3(x + y)$$

3 and $(x + y)$ are said to be the **factors of** $3x + 3y$.

Example 13

$$2x + 6 = 2(x + 3)$$

Note that $2x$ and 6 have a common factor of 2. This common factor is placed outside the bracket. To find the terms inside the bracket, divide each of the terms making up the original expression by the common factor. Thus

$$\frac{2x}{2} = x \quad \text{and} \quad \frac{6}{2} = 3$$

Factorise the following:

1. $3x + 6$

2. $4y - 8$

3. $2p - 4$

4. $5m + 10$

5. $2x + 2y$

6. $3p - 3q$

7. $5x + 15y$

8. $4x - 6y$

9. $9x - 6y$

10. $3p + 9q$

Multiplying Algebraic Fractions

As with arithmetic fractions, to multiply algebraic fractions we first multiply the numerators and then multiply the denominators. We then cancel common factors.

Example 14

$$\frac{3ab}{5c^2} \times \frac{cd}{7a^2} = \frac{3ab \times cd}{5c^2 \times 7a^2}$$

$$= \frac{3 \times \cancel{a} \times b \times \cancel{c} \times d}{5 \times \cancel{c} \times c \times 7 \times \cancel{a} \times a}$$

$$= \frac{3 \times b \times d}{5 \times c \times 7 \times a}$$

$$= \frac{3bd}{35ac}$$

Dividing Algebraic Fractions

As with arithmetic fractions, when dividing algebraic fractions we invert the second fraction and then proceed as in multiplication.

Example 15

$$\frac{5pq^2}{8m^2n} \div \frac{3pq}{7mn} = \frac{5pq^2}{8m^2n} \times \frac{7mn}{3pq}$$

$$= \frac{5pq^2 \times 7mn}{8m^2n \times 3pq}$$

$$= \frac{5 \times \cancel{p} \times \cancel{q} \times q \times 7 \times \cancel{m} \times \cancel{n}}{8 \times \cancel{m} \times m \times \cancel{n} \times 3 \times \cancel{p} \times \cancel{q}}$$

$$= \frac{5 \times q \times 7}{8 \times m \times 3}$$

$$= \frac{35q}{24m}$$

Exercise 13.9

Simplify each of the following:

1. $\dfrac{6a}{b^2} \times \dfrac{b}{3a^2}$

2. $\dfrac{9x^2}{6y^2} \times \dfrac{y^3}{x^3}$

3. $\dfrac{6pq}{4rs} \times \dfrac{8s^2}{3p}$

4. $\dfrac{6ab}{c} \times \dfrac{ad}{2b} \times \dfrac{8cd^2}{4bc}$

5. $\dfrac{ab^2}{bc^2} \div \dfrac{a^2}{bc^3}$

6. $\dfrac{3xy^2}{pq} \div \dfrac{6y^3}{p^2}$

7. $\dfrac{xy^2}{a^2b} \div \dfrac{x^2y}{ab^2}$

8. $\dfrac{m^2n^3}{p^2q} \div \dfrac{m^3n^4}{p^3q^2}$

Addition and Subtraction of Algebraic Fractions

The procedure for adding and subtracting algebraic fractions is the same as for arithmetic fractions:

(1) Find the LCM of the denominators.

(2) Express each fraction with this common denominator.

(3) Add or subtract the numerators of these equivalent fractions.

Example 16

Express $\dfrac{a}{3} + \dfrac{a}{4} + \dfrac{a}{5}$ as a single fraction.

The LCM of 3, 4 and 5 is 60.

$$\dfrac{a}{3} + \dfrac{a}{4} + \dfrac{a}{5} = \dfrac{20a}{60} + \dfrac{15a}{60} + \dfrac{12a}{60}$$

$$= \dfrac{20a + 15a + 12a}{60}$$

$$= \dfrac{47a}{60}$$

Example 17

Simplify $\dfrac{x}{2} - \dfrac{x-1}{3}$

The LCM of 2 and 3 is 6.

$$\dfrac{x}{2} - \dfrac{x-1}{3} = \dfrac{3x}{6} - \dfrac{2(x-1)}{6}$$

$$= \dfrac{3x - 2(x-1)}{6}$$

$$= \dfrac{3x - 2x + 2}{6}$$

$$= \dfrac{x+2}{6}$$

Exercise 13.10

Simplify each of the following by expressing them as single fractions in their lowest terms:

1. $\dfrac{a}{4} + \dfrac{a}{5}$

2. $\dfrac{p}{2} - \dfrac{q}{3}$

3. $\dfrac{3a}{4} - \dfrac{a}{5}$

4. $\dfrac{x}{2} + \dfrac{x}{3} + \dfrac{x}{4}$

5. $\dfrac{3}{2x} + \dfrac{2}{3x}$

6. $\dfrac{m}{x} - \dfrac{3m}{2x} + \dfrac{5m}{3x}$

7. $\dfrac{4x}{3y} - \dfrac{2x}{5y}$

8. $1 + \dfrac{x+2}{3}$

9. $\dfrac{x}{3} + \dfrac{2x-1}{4}$

10. $5m - \dfrac{m-2}{2}$

11. $\dfrac{x-3}{4} - \dfrac{x}{3}$

12. $\dfrac{4}{x} - \dfrac{5}{2x} + \dfrac{3}{4x}$

13. $\dfrac{x}{2} - \dfrac{x-3}{3}$

14. $\dfrac{2a+3b}{3} - \dfrac{a-2b}{2}$

15. $\dfrac{5-x}{5} - \dfrac{x-3}{2}$

Miscellaneous Exercise 13

Section A

1. What is the value of $x^2 + x + 6$ when $x = 4$?

2. Express $\dfrac{a - b}{a + b}$ as a fraction in its lowest terms given that $a = 5$ and $b = 3$.

3. If $a = 0.5$ and $b = 37$, what is the value of $\dfrac{b}{a}$?

4. If $p = 3xy$, what is the value of p when $x = 2$ and $y = 5$?

5. An approximate way of converting degrees Celsius into degrees Fahrenheit is $F = 2C + 30$, where F is the temperature in degrees Fahrenheit and C is the temperature in degrees Celsius. Convert a temperature of $12°C$ into degrees Fahrenheit.

6. If $a = 3$ and $b = 6$, find the value of $5a^2b$.

7. Work out the value of $a + 2b$ when $a = 1$ and $b = \frac{1}{2}$.

Section B

1. Make into single terms:
   (a) $8a - 2a$
   (b) $8a \times 2a$
   (c) $8a \div 2a$

2. Remove the brackets and simplify:
   (a) $3(x + y) + 2(x + 2y)$
   (b) $3(x + y) - (x - 2y)$
   (c) $(2x - y) - (2x + y)$

3. Find the value of $9a^2 + ab - c^2$ when $a = 1$, $b = 2$ and $c = -3$.

4. Simplify the expressions
   (a) $5a \times 2b \times 3a$
   (b) $8x - y - 3x - 5y$

5. Given that $x = 6$, $y = 4$ and $z = -2$, work out the values of the following expressions:
   (a) $x - y + z$ (b) $x - (y - z)$ (c) $\dfrac{x^2}{3y^2}$

6. If $x = 3$ and $y = -5$, find the value of
   (a) $(x + y)^2$ (b) $x^2 + 3xy + y^2$

7. Express as a single fraction in its simplest form $\dfrac{2x}{3} - \dfrac{x}{6}$

8. Express $\dfrac{x}{2} - \dfrac{x - 4}{3}$ as a single fraction, simplifying your answer as far as possible.

9. Simplify $\dfrac{3a^2b^3}{2cd} \times \dfrac{c^3d^2}{ab}$

10. Simplify $\dfrac{6ab}{5cd} \div \dfrac{4a^2}{7bd}$

Multi-Choice Questions 13

1. What is the value of $x^2 + x - 6$ when $x = 3$?
   A 0   B 3   C 6   D 18

2. In its simplest form $3(2p - q) - 2(3p - 2q)$ is
   A $12p + q$ B $12p + 3q$
   C $-7q$ D q

3. What is the value of $\dfrac{a - b}{a + b}$ when $a = 5$ and $b = -3$?
   A -4 B -1 C $\frac{1}{4}$ D 4

4. If $a = 0.2$ and $b = 48$, what is the value of $\dfrac{b}{a}$?
   A 0.24   B 2.4   C 24   D 240

5. Which of the following is the same as $pq + qr$?

   A $q(p + r)$ B $(pq)(qr)$

   C $p(q + r)$ D $r(p + q)$

6. Which of the following gives the factors of $2x^2 + 4x$?

   A $2x(x + 4)$ B $x(x + 4)$

   C $6x^3$ D $2x(x + 2)$

7. Given that $y = 2x^2 - 3x - 7$, what is the value of y when $x = 3$?

   A 16   B 9   C 8   D 2

8. $\dfrac{2}{x} + \dfrac{3}{2x}$ expressed as a single fraction is

   A $\dfrac{5}{3x}$ B $\dfrac{7}{2x}$ C $\dfrac{5}{2x^2}$ D $\dfrac{7}{2x^2}$

Mental Test 13

Try to answer the following questions without writing anything down except the answer.

1. Make into a single term $8x + 3x$.

2. Make into a single term $a + 3a$.

3. Subtract $5b$ from $8b$.

4. Make into a single term $5p - p$.

5. Simplify $2a \times 3b$.

6. Make into a single term $p \times p$.

7. Simplify $3m \times 4m$.

8. Divide $5a^2$ by a.

9. Divide $4a^3$ by a^2.

10. Make into a single term $6p - 2p$.

11. Remove the brackets from $3(p + q)$.

12. Remove the brackets from $-(x - y)$.

13. Remove the brackets from $-2(m + n)$.

14. Simplify $\dfrac{a}{b} \times \dfrac{p}{q}$.

15. Make into a single term $5x - 3x - 4x$.

Indices

Introduction

We have seen that a quick way to write $2 \times 2 \times 2$ is 2^3. The figure 3 which shows the number of twos to be multiplied together is called the **index** (plural: indices). 2^3 is said to be 2 raised to the power of 3. The number 2 is called the **base**.

a^4 is a raised to the power of 4; a is the base.

Multiplication

$$a^2 \times a^3 = a \times a \times a \times a \times a = a^5$$

We can see that in the case of multiplication, it is quicker to **add** the indices. Thus

$$a^2 \times a^3 = a^{2+3} = a^5$$

Example 1

(a) $3^4 \times 3^5 = 3^{4+5} = 3^9$

(b) $p^2 \times p^4 \times p^6 = p^{2+4+6} = p^{12}$

(c) $5y \times 2y^2 \times 3y^4 = 5 \times 2 \times 3 \times y^{1+2+4}$
$$= 30y^7$$

Exercise 14.1

Make each of the following into a single term:

1. $5^2 \times 5^4$
2. $a^3 \times a^4$
3. $p^2 \times p^3 \times p^4$
4. $y^3 \times y^5 \times y^7 \times y^8$
5. $2 \times 2^3 \times 2^4$
6. $2a \times 3a^2 \times 4a^3$
7. $3y^2 \times y \times 5y^4$
8. $6p^4 \times p^2 \times 3p$

Division

$$a^5 \div a^2 = \frac{a^5}{a^2}$$
$$= \frac{a \times a \times a \times \cancel{a} \times \cancel{a}}{\cancel{a} \times \cancel{a}}$$
$$= a^3$$

We can see that in the case of division, it is quicker to **subtract** the indices. Thus

$$a^5 \div a^2 = a^{5-2}$$
$$= a^3$$

Example 2

(a) $p^9 \div p^4 = p^{9-4} = p^5$

(b) $\dfrac{y^3 \times y^5 \times y^4}{y^2 \times y^7} = \dfrac{y^{3+5+4}}{y^{2+7}}$
$$= \frac{y^{12}}{y^9}$$
$$= y^{12-9}$$
$$= y^3$$

Exercise 14.2

Simplify each of the following:

1. $3^6 \div 3^4$
2. $p^7 \div p^5$
3. $d^5 \div d^4$
4. $(q^5 \times q^3) \div q^4$
5. $\dfrac{m^2 \times m^7}{m^5 \times m^3}$
6. $\dfrac{t^4 \times t^9}{t^3 \times t^5}$
7. $\dfrac{am^3}{am^2}$
8. $\dfrac{3p^2 \times 8p^4}{6p^3}$

Raising the Power of a Quantity

$$(a^3)^2 = a^3 \times a^3$$
$$= a^{3+3}$$
$$= a^6$$

We can see that when raising the power of a quantity it is quicker to *multiply* the indices so that

$$(a^3)^2 = a^{3 \times 2}$$
$$= a^6$$

Example 3

(a) $(p^4)^3 = p^{4 \times 3} = p^{12}$

(b) $(2m^3)^5 = 2^{1 \times 5} \times m^{3 \times 5} = 2^5 \times m^{15}$

(c) $\left(\dfrac{x^2}{y^3}\right)^4 = \dfrac{x^{2 \times 4}}{y^{3 \times 4}} = \dfrac{x^8}{y^{12}}$

Exercise 14.3

Simplify each of the following:

1. $(y^2)^4$
2. $(7^2)^3$
3. $(5c^2)^3$
4. $(5ab^2c^3)^5$
5. $\left(\dfrac{3a^2}{2b^3}\right)^4$
6. $(m^3)^5$
7. $(3b)^4$
8. $(2x^2y)^3$
9. $(\frac{3}{4})^4$
10. $(\frac{1}{2})^5$

Negative Indices

$$a^3 \div a^5 = \frac{a^3}{a^5}$$
$$= \frac{\cancel{a} \times \cancel{a} \times \cancel{a}}{\cancel{a} \times \cancel{a} \times \cancel{a} \times a \times a}$$
$$= \frac{1}{a \times a}$$
$$= \frac{1}{a^2}$$

Using the rule for division,

$$a^3 \div a^5 = a^{3-5}$$
$$= a^{-2}$$

so that

$$\frac{1}{a^2} = a^{-2}$$

The reciprocal of a number is $1/\text{number}$. Hence the reciprocal of 5 is $1/5$ and the reciprocal of a^2 is $1/a^2$.

Therefore a negative index indicates the **reciprocal** of a quantity.

Example 4

(a) $5^{-4} = \dfrac{1}{5^4}$

(b) Express 5^{-1} as a fraction.

$$5^{-1} = \frac{1}{5^1} = \frac{1}{5}$$

(c) $7p^{-3} = 7 \times p^{-3} = \dfrac{7}{1} \times \dfrac{1}{p^3} = \dfrac{7}{p^3}$

Zero Index

$$a^3 \div a^3 = \frac{a^3}{a^3}$$
$$= \frac{\cancel{a} \times \cancel{a} \times \cancel{a}}{\cancel{a} \times \cancel{a} \times \cancel{a}}$$
$$= 1$$

Using the rule for division,

$$a^3 \div a^3 = a^{3-3}$$
$$= a^0$$

Hence $a^0 = 1$.

No matter what numbers or symbols we use we find that any quantity raised to the power of 0 equals 1. Thus

$$p^0 = 1, \quad 963^0 = 1 \quad \text{and} \quad 83^0 = 1$$

Example 5

Find the value of (a) 7×3^0, (b) $3x^0$.

(a) $7 \times 3^0 = 7 \times 1 = 7$

(b) $3x^0 = 3 \times x^0 = 3 \times 1 = 3$

Exercise 14.4

Write each of the following fractions with a negative index:

1. $\dfrac{1}{5}$ 2. $\dfrac{1}{3^2}$ 3. $\dfrac{1}{a^3}$

4. $\dfrac{1}{b^5}$ 5. $\dfrac{2}{x^3}$ 6. $\dfrac{5}{x}$

7. $\dfrac{1}{m^6}$ 8. $\dfrac{8}{y^7}$

Write each of the following as a fraction:

9. 3^{-1} 10. 5^{-2} 11. a^{-3}

12. m^{-5} 13. $3b^{-1}$ 14. $7q^{-3}$

Find the value of each of the following, stating the answer as a fraction or a whole number:

15. 10^{-1} 16. 3^{-2} 17. 2^{-4}

18. 5^{-2} 19. 3^{-3} 20. 10^{-3}

21. b^0 22. $3q^0$ 23. $\left(\frac{1}{5}\right)^0$

24. $\left(\dfrac{x}{y}\right)^0$

Place Value

The number 2478 is 2 thousand 4 hundred 7 tens and 8 units. Now 1 thousand is 10^3, 1 hundred is 10^2, 1 ten is 10^1 and 1 unit is 10^0. We can show our number in a table:

10^3	10^2	10^1	10^0
2	4	7	8

This may be written in expanded notation as:

$$2 \times 10^3 + 4 \times 10^2 + 7 \times 10^1 + 8 \times 10^0$$

Example 6

Write the number 8 632 917 in expanded notation.

The number 8 632 917 is 8 millions 6 hundred-thousands 3 ten-thousands 2 thousands 9 hundreds 1 ten and 7 units, i.e.

$$8\,632\,917 = 8 \times 1\,000\,000 + 6 \times 100\,000$$
$$+ 3 \times 10\,000 + 2 \times 1000$$
$$+ 9 \times 100 + 1 \times 10$$
$$+ 7 \times 1$$

Now,

$$1\,000\,000 = 10^6 \qquad 100\,000 = 10^5$$
$$10\,000 = 10^4 \qquad 1000 = 10^3$$
$$100 = 10^2 \qquad 10 = 10^1$$
$$1 = 10^0$$

Hence

$$8\,632\,917 = 8 \times 10^6 + 6 \times 10^5 + 3 \times 10^4$$
$$+ 2 \times 10^3 + 9 \times 10^2$$
$$+ 1 \times 10^1 + 7 \times 10^0$$

Exercise 14.5

Copy the following table:

10^6	10^5	10^4	10^3	10^2	10^1	10^0

By writing figures in the appropriate columns use your table to represent the following numbers:

1. 936   2. 7254

3. 8619   4. 18 342

5. 278 948   6. 563 740

7. 8 621 975   8. 7300

9. 83 507   10. 900 007

Write in expanded notation the numbers given in the table below:

	10^6	10^5	10^4	10^3	10^2	10^1	10^0
11.					7	5	3
12.		3	8	4	6	8	5
13.		1	7	3	2	7	8
14.		2	3	0	0	0	5
15.	3	9	0	7	6	0	0

Decimal Numbers

We can do the same thing with decimal numbers but in this case we use negative indices:

$$10^{-1} = \frac{1}{10} \qquad 10^{-2} = \frac{1}{100}$$
$$= 0.1 \qquad\qquad = 0.01$$
$$10^{-3} = \frac{1}{1000}$$
$$= 0.001 \quad \text{and so on}$$

We can show the number 0.3465 in a table:

10^{-1}	10^{-2}	10^{-3}	10^{-4}
3	4	6	5

This may be written in expanded notation as

$$3 \times 10^{-1} + 4 \times 10^{-2} + 6 \times 10^{-3} + 5 \times 10^{-4}$$

Now consider the number 9346.825. Using a table as before we have

10^3	10^2	10^1	10^0	10^{-1}	10^{-2}	10^{-3}
9	3	4	6	8	2	5

This may be written in expanded notation as

$$9 \times 10^3 + 3 \times 10^2 + 4 \times 10^1 + 6 \times 10^0$$
$$+ 8 \times 10^{-1} + 2 \times 10^{-2} + 5 \times 10^{-3}$$

We see that the decimal point separates the positive powers of 10 from the negative powers of 10.

Exercise 14.6

Copy the following table:

10^4	10^3	10^2	10^1	10^0	10^{-1}	10^{-2}	10^{-3}	10^{-4}

By writing figures in the various columns, use your table to represent the following numbers:

1. 0.52
2. 0.375
3. 0.9134
4. 56 327.58
5. 96.3
6. 187.25
7. 3059.273
8. 0.0635
9. 7300.2763
10. 83 507.0038

11. Write each of the above numbers in expanded notation.

Write in decimal notation the numbers given in the table below:

	10^6	10^5	10^4	10^3	10^2	10^1	10^0	10^{-1}	10^{-2}	10^{-3}
12.				2	7	5	3			
13.			8	4	6	0	5			
14.				8	0	0	0	7	3	
15.	9	0	7	6	0	8	3	2	6	
16.					6	2	7	5	4	

Large Numbers in Standard Form

Any number can be expressed as a value between 1 and 10 multiplied by a power of 10. For instance

$$8372 = 8.372 \times 10^3$$

A number expressed in this way is said to be written in **standard form** or in **scientific notation**.

This way of writing numbers is very useful particularly if the numbers are very large or very small.

Note that $100 = 10^2$, $1000 = 10^3$ and $1\,000\,000 = 10^6$.

The power of 10 is found by counting the number of zeros to the left of the decimal point.

Example 7

(a) $563 = 5.63 \times 100 = 5.63 \times 10^2$

(b) $75\,532 = 7.5532 \times 10\,000$
$= 7.5532 \times 10^4$

(c) $25\,000\,000 = 2.5 \times 10\,000\,000$
$= 2.5 \times 10^7$

Exercise 14.7

Write the following numbers in standard form:

1. 827
2. 1734
3. 17 632
4. 893 762
5. 8 036 000

Write the following as ordinary numbers (i.e. not in standard form):

6. 3.2×10^3
7. 5×10^2
8. 1.87×10^6
9. 4.32×10^4
10. 8.762×10^5

Small Numbers in Standard Form

Now $0.1 = 10^{-1}$, $0.01 = 10^{-2}$ and $0.001 = 10^{-3}$.

The negative power of 10 is found by adding 1 to the number of zeros following the decimal point.

Example 8

(a) $0.043 = 4.3 \div 100 = 4.3 \div 10^2 = \dfrac{4.3}{10^2}$
$= 4.3 \times 10^{-2}$

(b) $0.0036 = 3.6 \div 1000 = 3.6 \div 10^3$
$= \dfrac{3.6}{10^3} = 3.6 \times 10^{-3}$

Exercise 14.8

Write the following numbers in standard form:

1. 0.3
2. 0.05
3. 0.0056
4. 0.000 007
5. 0.000 067

Write the following as ordinary numbers (i.e. not in standard form):

6. 2×10^{-3}
7. 5.67×10^{-1}
8. 3.2×10^{-2}
9. 5×10^{-6}
10. 4.1×10^{-4}

Adding and Subtracting Numbers in Standard Form

(1) If the numbers to be added or subtracted have the same power of 10, then the numbers may be added or subtracted directly.

Example 9

(a) $1.859 \times 10^2 + 2.387 \times 10^2 + 9.163 \times 10^2$
$= (1.859 + 2.387 + 9.163) \times 10^2$
$= 13.409 \times 10^2$
$= 1.3409 \times 10^3$

(b) $8.768 \times 10^{-3} - 4.381 \times 10^{-3}$
$= (8.768 - 4.381) \times 10^{-3}$
$= 4.387 \times 10^{-3}$

(2) If the numbers have different powers of 10, first convert them to decimal form and then add or subtract them.

Example 10

$$3.478 \times 10^3 + 4.826 \times 10^2 = 3478 + 482.6$$
$$= 3960.6$$
$$= 3.9606 \times 10^3$$

Multiplying and Dividing Numbers in Standard Form

By using the laws of indices, numbers expressed in standard form can easily be multiplied or divided.

Example 11

(a) $(8.463 \times 10^2) \times (1.768 \times 10^3)$

$$= (8.463 \times 1.768) \times (10^2 \times 10^3)$$
$$= 14.96 \times 10^5 \text{ (to 4 sig. fig.)}$$
$$= 1.496 \times 10^6$$

(b) $(3.258 \times 10^2) \div (7.197 \times 10^4)$

$$= (3.258 \div 7.197) \times (10^2 \div 10^4)$$
$$= 0.4527 \times 10^{-2} \text{ (to 4 sig. fig.)}$$
$$= 4.527 \times 10^{-3}$$

Exercise 14.9

State which of the following numbers is the larger and by how much:

1. 5.8×10^2 and 2.1×10^3

2. 9.4×10^3 and 3.95×10^4

3. 8.58×10^4 and 9.87×10^3

4. 2.1×10^{-2} and 5.4×10^{-3}

5. 8.73×10^{-3} and 1.26×10^{-1}

State the answers to the following in standard form:

6. $3.582 \times 10^3 + 8.907 \times 10^3$

7. $7.81 \times 10^{-2} + 1.88 \times 10^{-2}$ $+ 8.89 \times 10^{-2}$

8. $1.809 \times 10^2 - 1.705 \times 10^2$

9. $8.89 \times 10^{-3} - 8.85 \times 10^{-3}$

10. $1.78 \times 10^2 + 2.58 \times 10^3$

11. $5.987 \times 10^3 + 8.91 \times 10^2 + 7.635 \times 10^4$

12. $8.902 \times 10^{-2} - 7.652 \times 10^{-3}$

13. $1.832 \times 10^{-1} - 9.998 \times 10^{-2}$

14. $7.58 \times 10^2 \times 6 \times 10^3$ (to 3 significant figures)

15. $6 \times 10^{-1} \times 2.58 \times 10^{-2}$ (to 2 significant figures)

16. $5 \times 10^3 \times 2.11 \times 10^4 \times 4 \times 10^2$ (to 3 significant figures)

17. $(2.68 \times 10^2) \div (8 \times 10^3)$ (to 2 significant figures)

18. $(1.78 \times 10^{-1}) \div (3 \times 10^{-2})$ (to 3 significant figures)

Miscellaneous Exercise 14

Section B

1. Rewrite the number 1 200 000 in standard form.

2. Find the value of $43 \times 3.16 \times 25$, giving your answer in standard form.

3. Write 5.01×10^{-3} as a decimal number not in standard form.

4. Write 0.000 893 in standard form.

5. If $p = 3 \times 10^3$ and $q = 2 \times 10^2$, work out the value of

   (a) $p + q$ (b) pq

   (c) $p - q$ (d) $p \div q$

   expressing the answer in standard form.

6. Find the value of

   (a) 3^{-2} (b) 3^0 (c) 10^{-1}

7. Which is greater, 8.73×10^2 or 1.2×10^3, and by how much?

8. Calculate $10^{20} \div 10^5$, stating the answer as a power of 10.

9. If $3^x = 81$, what is the value of x?

10. Find the value of $2^4 \times 2^2$.

11. Write 8×10^{-2} as a decimal number.

12. Find the value of n in each of the following:
    (a) $12^5 \times 12^n = 12^9$
    (b) $2^5 \div 2^4 = 2^n$
    (c) $37\,500 = 3.75 \times 10^n$

13. The radius, in centimetres, of the nucleus of an atom is
    $$0.000\,000\,000\,000\,3.$$
    Write this number in standard form.

14. Red light has a wavelength of 6.7×10^{-5} cm. Write this number in regular decimal notation.

15. Express 35.7×10^{-3} in the form $A \times 10^n$ where A is a number between 1 and 10 and n is an integer.

16. Evaluate (a) $(\frac{1}{2})^3$, (b) $(3.5)^0$.

Multi-Choice Questions 14

1. 3×10^3 equals
   A 10   B 90   C 300   D 3000

2. $10^2 + 10^5$ is
   A 70   B 1100
   C 100\,100   D 10^7

3. In standard form 43\,000 is
   A 0.43×10^5 B 4.3×10^4
   C 43×1000 D 430×100

4. $\dfrac{2^3 \times 2^4}{2^2}$ is equal to
   A 8   B 16   C 32   D 64

5. When written in standard form, 0.0063 is
   A 63×10^{-4} B 6.3×10^3
   C 6.3×10^{-3} D 0.63×10^{-2}

6. 4.05×10^4 is equivalent to
   A 4\,050\,000   B 405\,000
   C 400\,005   D 40\,500

7. $(3a^2)^3$ is equal to
   A $3a^6$ B $9a^6$ C $27a^3$ D $27a^6$

8. 1.73×10^{-3} is equal to
   A 0.001\,73   B 0.0173
   C 0.519   D 17.3

9. Find the value of $2a^3$ when $a = -2$.
   A 16   B -12 C -16 D -64

10. If $x = 5$ and $y = -3$ then $x^2 - y^3$ is equal to
    A -2 B 16   C 34   D 52

Mental Test 14

Try to answer the following questions without writing anything down except the answer.

1. Find the value of 2^3.

2. Write 3^{-1} as a fraction.

3. Remove the brackets from $(3a^2)^2$.

4. Simplify $a^2 \times a^4$.

5. Write $a^5 \div a^3$ as a single term.

6. Simplify $a^2 \times a^3 \times a^5$.

7. Remove the brackets from $\left(\dfrac{a^2}{b}\right)^3$.

8. Find the value of $2^2 \times 2^3$.

9. Write 8000 in standard form.

10. Write 0.003 in standard form.

Linear Equations, Formulae and Inequalities

Introduction

Fig. 15.1 shows a pair of scales which are in balance. That is, each scale pan contains exactly the same number of kilograms.

Therefore $\qquad x + 2 = 7$

This is an example of an equation.

Fig. 15.1

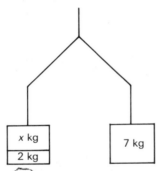

To solve the equation we have to find a value for x so that the scales remain in balance. We may keep the scales in balance by adding or subtracting the same amount from each pan. If we take 2 kg from the left-hand pan then we are left with x kg in this pan. To maintain balance we must take 2 kg from the right-hand pan, leaving 5 kg. That is

$$x + 2 - 2 = 7 - 2$$
$$x = 5$$

Let us take a second example as shown in Fig. 15.2. In the left-hand pan we have 4 packets of exactly the same weight, whilst in the right-hand pan there is 8 kg. The problem is to find the weight of each packet.

Fig. 15.2

If we let each packet have a weight of x kilograms, there is $4x$ kg in the left-hand pan and hence we have the equation

$$4x = 8$$

We can maintain the balance of the scales by dividing or multiplying the quantities in each of the scales by the same amount. In our equation, if we divide each side by 4 we get

$$\frac{4x}{4} = \frac{8}{4}$$

Cancelling the fours on the left-hand side we get

$$x = 2$$

and therefore each packet has a weight of 2 kg.

From these two examples we can say:

(1) An equation expresses balance between two sets of quantities. That is, the equals sign means that one side of the equation is equal to the other side.

(2) If one side of the equation is changed in any way, the equation will still be true so long as the same operation is carried out on the other side. We can add the same amount to each side of the equation without destroying the balance. Also, we can multiply or divide each side of the equation by the same amount.

Example 1

(a) If $5x = 10$, by dividing both sides by 5,
$$x = 2$$

(b) If $x + 3 = 7$, by subtracting 3 from each side, $x = 4$

(c) If $x - 6 = 2$, by adding 6 to each side,
$$x = 8$$

(d) If $\dfrac{x}{2} = 3$, by multiplying each side by 2
$$x = 6$$

Exercise 15.1

Solve the following equations for x:

1. $3x = 6$	2. $7x = 14$
3. $8x = 16$	4. $2x = 12$
5. $x + 3 = 9$	6. $x + 3 = 7$
7. $x + 7 = 11$	8. $x + 5 = 8$
9. $x - 5 = 7$	10. $x - 3 = 6$
11. $x - 2 = 8$	12. $x - 4 = 8$
13. $\dfrac{x}{2} = 4$	14. $\dfrac{x}{3} = 5$
15. $\dfrac{x}{7} = 3$	16. $\dfrac{x}{4} = 6$
17. $\dfrac{x}{5} = 2$	

Solving Harder Equations

Equations Requiring Addition, Subtraction and Division

Example 2

Solve for x the equation $x - 4 = 8$

Adding 4 to each side
$$x - 4 + 4 = 8 + 4$$
$$x = 12$$

The method shown in Example 2 is time-consuming. Provided you understand the method, you can use the quicker method which follows:

The operation of adding 4 to each side can be accomplished by taking 4 to the RHS (right-hand side) and changing its sign.

$$x - 4 = 8$$
$$x = 8 + 4$$
$$x = 12$$

The solution of an equation is that value of the unknown which, when substituted into the equation, makes the LHS equal to the RHS.

Check: LHS $= 12 - 4 = 8 =$ RHS

Since LHS $=$ RHS the solution is correct.

Example 3

Solve $4x - 1 = 7$
$$4x = 7 + 1$$
$$4x = 8$$

Dividing both sides by 4,
$$x = 2$$

Check: Substituting for x in the original equation:

$$LHS = 4 \times 2 - 1$$
$$= 8 - 1$$
$$= 7$$
$$= RHS$$

Since the LHS equals the RHS, the solution is correct. In many cases the given equation contains terms in the unknown and/or constant terms on both sides of the equation. We then take all the terms containing the unknown quantity to one side of the equation and all the other terms to the other side.

Example 4

Solve $5x + 8 = 3x - 2$

$$5x - 3x = -2 - 8$$
$$2x = -10$$
$$x = -5$$

Check:

$$LHS = 5 \times (-5) + 8$$
$$= -25 + 8$$
$$= -17$$
$$RHS = 3 \times (-5) - 2$$
$$= -15 - 2$$
$$= -17$$

Since the LHS equals the RHS the solution is correct.

Example 5

Solve $9 - 4x = 3x - 12$

Gather the terms containing x on the RHS to give a positive term in x. (Taking the terms in x to the LHS would give a negative term in x.)

$$9 - 4x = 3x - 12$$
$$9 + 12 = 3x + 4x$$
$$21 = 7x$$
$$x = 3$$

When an equation contains brackets we remove these first and then solve the equation by the methods previously shown.

Example 6

Solve $3(4x - 5) = 5 - 2(x + 3)$

$$12x - 15 = 5 - 2x - 6$$
$$12x + 2x = 5 + 15 - 6$$
$$14x = 14$$
$$x = 1$$

Exercise 15.2

Solve the following equations:

1. $x - 3 = 5$
2. $2x - 1 = 7$
3. $5x - 8 = 2$
4. $3x + 8 = 5$
5. $2x + 7 = 3$
6. $3x - 4 = x + 6$
7. $3x - 8 = 5x - 20$
8. $23 - x = x + 11$
9. $3(x + 1) = 9$
10. $2(x - 3) - (x - 2) = 5$
11. $5(m + 2) - 3(m - 5) = 29$
12. $4(3p + 2) + 6 = 3(2p - 5) + 5$

Equations Requiring Multiplication and Division

Example 7

Solve $\dfrac{3x}{2} = 12$

Multiplying both sides by 2 gives

$$3x = 24$$

Dividing both sides by 3 gives

$$x = 8$$

When an equation contains fractions, the first step is to get rid of the denominators by multiplying each term in the equation by the LCM of the denominators. The resulting equation is then solved by the methods previously shown.

Example 8

Solve $3p - \dfrac{p}{3} = 8$

Multiplying each term by 3 gives

$$3 \times 3p - \dfrac{p}{3} \times \dfrac{3}{1} = 8 \times 3$$

$$9p - p = 24$$

$$8p = 24$$

$$p = \dfrac{24}{8}$$

$$p = 3$$

Not all equations give a solution which is an integer.

Example 9

Solve $\dfrac{c}{5} = \dfrac{3}{4}$

The LCM of 5 and 4 is 20.

Multiplying each term by 20 gives

$$4c = 15$$

Dividing both sides by 4 gives

$$c = \dfrac{15}{4}$$

Exercise 15.3

Solve each of the following equations:

1. $\dfrac{x}{5} = 3$ 2. $\dfrac{x}{3} = 7$

3. $2x = 14$ 4. $\dfrac{2x}{3} = 8$

5. $\dfrac{4x}{5} = 12$ 6. $\dfrac{x}{3} = \dfrac{2}{5}$

7. $\dfrac{m}{3} + \dfrac{m}{5} = 2$ 8. $\dfrac{x}{5} - \dfrac{x}{3} = 2$

9. $\dfrac{x}{3} + \dfrac{x}{4} + \dfrac{x}{5} = \dfrac{5}{6}$ 10. $3x + \dfrac{3}{4} = 2 + \dfrac{2x}{3}$

11. $\dfrac{3x}{2} + 1 = \dfrac{3}{4} + 2x$

12. $\dfrac{x+3}{4} = 2 + \dfrac{(x-3)}{5}$

Formulae

A formula is an equation which shows the relationship between two or more quantities.

The statement $E = IR$ is a formula for E in terms of I and R. The value of E is found by substituting the given values of I and R.

Example 10

If $v = u + at$, find the value of v when $u = 20$, $a = 3$ and $t = 5$.

Substituting the given values of u, a and t we have

$$v = 20 + (3 \times 5)$$

$$= 20 + 15$$

$$= 35$$

Exercise 15.4

1. If $V = Ah$, find the value of V when $A = 8$ and $h = 4$.

2. The formula $K = Wa + b$ is used in engineering technology. Find K when $W = 25$, $a = 3$ and $b = 5$.

3. $S = 90(n - 4)$ is a formula used in geometry. Find S when $n = 8$.

4. If $P = RT/V$, find the value of P when $R = 56$, $T = 18$, and $V = 7$.

5. A formula used in physics is $E = mv^2/2g$. Find the value of E when $m = 220$, $v = 8$ and $g = 10$.

6. If $A = \frac{1}{2}BH$, find A when $B = 6$ and $H = 7$.

7. If $y = 3t/c$, find y when $t = 12$ and $c = 6$.

Formulae and Equations

Suppose that we are given the formula $M = P/Q$ and we have to find the value of Q, given values for M and P. We can do this by substituting the given values and solving the resulting equation for Q.

Example 11

Find T from the formula $D = \dfrac{T + 2}{P}$, given $D = 5$ and $P = 3$.

Substituting the given values we have

$$5 = \frac{T + 2}{3}$$

Multiplying both sides by 3 gives

$$15 = T + 2$$
$$T = 15 - 2$$
$$T = 13$$

Exercise 15.5

1. Find n from the formula $P = 1/n$ when $P = 2$.

2. Find R from the formula $E = IR$ when $E = 20$ and $I = 4$.

3. Find B from the formula $A = BH$ when $A = 12$ and $H = 4$.

4. Find c from the formula $H = abc$ when $H = 40$, $a = 2$ and $b = 5$.

5. Find P from the formula $I = PRT/100$ when $I = 20$, $R = 5$ and $T = 4$.

6. Find D from the formula $C = \pi D$ when $\pi = 3.142$ and $C = 27$.

7. Find r from the formula $A = \pi rl$ when $\pi = 3.142$, $A = 96$ and $l = 12$.

8. Find W from the formula $K = Wa + b$ when $K = 30$, $a = 4$ and $b = 6$.

Transposition of Formulae

Consider again the formula $M = P/Q$. M is called the **subject** of the formula. We may be given several corresponding values of M and Q, and we want to find the corresponding values of P. We could, of course, find these by the method shown in Example 11 but a lot of time and effort would be spent in solving the resulting equations.

Much of this time and effort would be saved if we could express the formula with P as the subject, because we would then only have to substitute the given values of M and Q in the rearranged formula.

The process of rearranging a formula so that one of the other symbols becomes the subject is called **transposing the formula.** The rules used in transposing formulae are exactly the same as those used in solving linear equations.

Example 12

(a) Transpose the formula $F = ma$ to make a the subject. Dividing both sides of the formula by m gives

$$\frac{F}{m} = \frac{\not{m}a}{\not{m}}$$

$$\frac{F}{m} = a$$

It is usual to write the subject on the LHS of the formula so we write

$$a = \frac{F}{m}$$

(b) Transpose $x = 3y + 5$ for y.

Subtract 5 from each side, giving

$$x - 5 = 3y$$

Divide both sides by 3 giving

$$\frac{x - 5}{3} = y$$

$$y = \frac{x - 5}{3}$$

Exercise 15.6

Transpose the following formulae:

1. $PV = c$ for V
2. $I = PRT$ for R
3. $E = mgh$ for h
4. $x = a/y$ for a
5. $I = E/R$ for R
6. $P = RT/V$ for T
7. $S = ts/T$ for t
8. $v = 3 + p$ for p
9. $p = 3 - q$ for q
10. $v = u + at$ for a
11. $n = p + 3r$ for r
12. $a = b - cx$ for x
13. $y = mx + c$ for m
14. $T = (D - d)/L$ for D
15. $P = S(C - F)/L$ for L

Inequalities

An **inequality** is a statement that one quantity is greater (or less) than a second quantity.

If we want to say that n is greater than 5 we write

$$n > 5$$

If we want to say that x is less than 7 we write

$$x < 7$$

Thus the symbol $>$ means 'greater than' and the symbol $<$ means 'less than'. Note carefully that the arrow always points to the smaller quantity.

A statement using these signs is called an **inequality**.

Exercise 15.7

Using the symbols $>$, $<$ and $=$, fill in the gap between the following pairs of quantities:

1. 2   5
2. $\frac{1}{2}$ $\frac{1}{3}$
3. 10 mm   1 cm
4. £1   80p
5. 9   1
6. $\frac{2}{3}$ $\frac{4}{9}$
7. -3 -7
8. 7   9
9. x^2 $x^2 - 3$
10. 2   4

Two more symbols are used in dealing with inequalities. They are

$\geqslant$ which means 'equal to or greater than'

and

$\leqslant$ which means 'equal to or less than'

$x \geqslant 2$ means that x is equal to or greater than 2

$n \leqslant 5$ means that n is equal to or less than 5

Solution of Simple Inequalities

The solutions of inequalities may be shown by means of a number line. If $x < 3$, all the possible values of x are shown in Fig. 15.3. The empty circle at the end of the arrowed line shows that $x = 3$ is not included.

Fig. 15.3

Figure 15.4 shows the solutions for $x \geqslant -4$. Since $x = -4$ is included, the arrowed line ends in a solid circle.

Fig. 15.4

Exercise 15.8

Use number lines to show the solutions for the following inequalities:

1. $x \leqslant 5$ 2. $x > 7$ 3. $x \geqslant 3$

4. $x > -2$ 5. $x < -6$

Example 13

If x has to be a whole number, find the solution for $x \leqslant 3$ and $x > 1$.

Since x has to be a whole number, the inequalities are represented by distinct points (Fig. 15.5) not continuous lines. We see that the solution must equal 2 or 3, because the points representing the independent solutions for each inequality overlap. We say that the solution set is $\{2, 3\}$.

Fig. 15.5

Inequalities may be combined into one statement. For instance, if we have $4 < x$ and $x < 7$, the two statements may be combined to give $4 < x < 7$, which means that x must lie between 4 and 7.

Example 14

If x is an integer, find the solution sets for
(a) $-3 < x < 2$,   (b) $-3 \leqslant x \leqslant 2$.

(a) Drawing the number line (Fig. 15.6) we see that the value of x must be the integers from -2 to 1. That is, the solution set is $\{-2, -1, 0, 1\}$.

Fig. 15.6

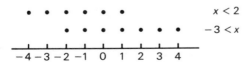

(b) Drawing the number line (Fig. 15.7) we see that the value of x must be the integers from -3 to 2. Hence the solution set is $\{-3, -2, -1, 0, 1, 2\}$.

Fig. 15.7

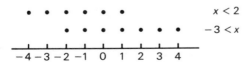

Note that when integers are being discussed we use distinct points instead of continuous lines.

Exercise 15.9

Find the solution sets for the following pairs of inequalities, where x is an integer:

1. $4 < x < 9$ 2. $4 \leqslant x \leqslant 9$

3. $-2 \leqslant x < 3$ 4. $5 \leqslant x \leqslant 8$

5. $-4 < x < 5$

Miscellaneous Exercise 15

Section A

1. If $v = u + at$, find the value of v when $u = 40$, $t = 10$ and $a = 3$.

2. The area of a trapezium is found by using the formula $A = \frac{1}{2}(a + b)h$. Find the value of A when $a = 3$, $b = 5$ and $h = 7$.

3. Find the value of C in the formula $C = 2\pi r$ when $\pi = 3.14$ and $r = 10$.

4. If $A = \pi r^2$, find the value of A when $\pi = 3.14$ and $r = 5$.

5. If $V = \frac{1}{3}\pi r^2 h$ find V when $\pi = 3.14$, $r = 10$ and $h = 6$.

6. If x is a positive integer such that $10 < x < 100$, state the least positive value of x.

7. Use one of the symbols $>$ $=$ $<$ in each of the following to make true statements:

   (a) $5^2 - 3^2$ 4^2 (b) $9^2 - 3^2$ 3^2

   (c) $\frac{5}{9} - \frac{1}{3}$ $\frac{2}{3}$ (d) $\sqrt{14.4}$ 1.2

   (e) 0.5^2 0.5^3

Section B

1. Solve for x, $2(x - 1) - 4(x + 1) = 0$.

2. Solve the equations
   (a) $3x - 8 = 7$
   (b) $-(x - 5) = 2$

3. Solve the equation $2(4x - 3) = 18$.

4. Solve the equations
   (a) $2x + 4 = 5$
   (b) $(8/x) + 1 = 3$

5. Solve the equations
   (a) $2(x - 3) + 5 = 3$
   (b) $(3x - 2)/2 = 17$

6. Consider the formula $S = 180(n - 2)$. Find the value of S when $n = 8$.

7. If $P = 2(a + b)$, find the value of a when $P = 10$ and $b = 3$.

8. Transpose the formula $v = u + at$ to make t the subject.

9. Use the formula $P = q + (n - 1)p$ to find the value of P when $q = 6$, $n = 5$ and $p = 4$.

10. Transpose the formula $T = 20 + \frac{1}{3}N$ to make N the subject.

Multi-Choice Questions 15

1. The solution of the equation $2x - 9 = 25$ is

   A $x = \frac{1}{2}$ B $x = 5$

   C $x = 8$ D $x = 17$

2. Consider the equation $y = 3x - 2$. The value of x when $y = 7$ is

   A $\frac{5}{3}$ B 3   C 6   D 27

3. If $x - 10 = 22$ then x is equal to

   A 2.2   B 12   C 22   D 32

4. If $\frac{x}{2} = 7$ then the value of x is

   A $3\frac{1}{2}$ B 5   C 9   D 14

5. If $4(2x - 1) = 60$ what is the value of x?

   A $\frac{1}{2}$ B 7   C 8   D 15

6. If $\frac{5x}{6} + \frac{3x}{4} = \frac{19}{12}$ then the value of x equals

   A -8 B -2 C -1 D 1

7. What is the value of c when $3c + 8 = 5c - 16$?

   A 1   B 3   C 4   D 12

8. If $v = u - at$ then t is equal to

   A $\dfrac{v - u}{a}$ B $\dfrac{u - v}{a}$

   C $\dfrac{v}{au}$ D $\dfrac{u}{av}$

Mental Test 15

Try to answer the following questions without writing anything down except the answer.

1. If $3x = 9$, what is the value of x?

2. If $\frac{x}{2} = 4$, what is the value of x?

3. If $y + 5 = 8$, what is the value of y?

4. Solve the equation $x - 4 = 2$.

5. If $F = rs$, work out the value of F when $r = 2$ and $s = 5$.

6. Transpose the formula $MN = p$ for N.

7. Transpose the formula $x = a/y$ to make a the subject.

8. Solve the equation $8x - 5x = 12$.

Simultaneous and Quadratic Equations

Look at the equation

$$3x + 6 = 12$$

This equation is easy to solve and the solution is

$$x = 2.$$

Now look at the equation

$$3x + 2y = 12$$

We cannot solve this equation because there are two unknown quantities, x and y. We need another equation with x and y in which the values of x and y are the same as in the first equation. Thus we may have

$$3x + 2y = 12 \qquad (1)$$

$$5x - 2y = 4 \qquad (2)$$

Because x and y have the same value in both equations, they are called **simultaneous equations**.

Elimination Method in Solving Simultaneous Equations

Since we are familiar with solving simple equations in one variable, simultaneous equations may be solved by getting rid of one of the unknown quantities, that is, by eliminating one of the unknown quantities, thereby forming an equation in one variable.

Example 1

Solve the equations

$$3x + 2y = 12 \qquad (1)$$

$$5x - 2y = 4 \qquad (2)$$

Since we have the same coefficient of y in both equations, we can get rid of the y term by adding the two equations:

$$(3x + 2y) + (5x - 2y) = 12 + 4$$

$$3x + 2y + 5x - 2y = 16$$

$$8x = 16$$

$$x = 2$$

To find the value of y we substitute for x in either of the original equations. Thus, substituting for x in equation (1) gives

$$(3 \times 2) + 2y = 12$$

$$6 + 2y = 12$$

$$2y = 6$$

$$y = 3$$

To check the values of x and y, substitute them in equation (2). There is no point in using equation (1) because this was used to find the value of y.

$$\text{LHS} = (5 \times 2) - (2 \times 3)$$

$$= 10 - 6$$

$$= 4$$

$$= \text{RHS}$$

Since the LHS and the RHS are the same, the solutions are correct.

Example 2

Solve the equations

$$4x + 5y = 30 \tag{1}$$
$$4x + 3y = 26 \tag{2}$$

Because we have the same coefficient of x in both equations, we can get rid of x by subtracting equation (2) from equation (1):

$$(4x + 5y) - (4x + 3y) = 30 - 26$$
$$4x + 5y - 4x - 3y = 4$$
$$2y = 4$$
$$y = 2$$

To find the value of x, substitute for y in equation (1):

$$4x + (5 \times 2) = 30$$
$$4x + 10 = 30$$
$$4x = 20$$
$$x = 5$$

Thus the solutions are $x = 5$ and $y = 2$.

Checking the solutions in equation (2):

$$\text{LHS} = (4 \times 5) + (3 \times 2)$$
$$= 20 + 6$$
$$= 26 = \text{RHS}$$

Because the LHS equals the RHS, the solutions are correct.

Exercise 16.1

Solve the following pairs of equations for x and y and check their solutions:

1. $x + y = 8$
   $x - y = 2$

2. $x + 3y = 10$
   $2x - 3y = 2$

3. $3x + 4y = 25$
   $3x + 2y = 17$

4. $5x - 2y = 1$
   $4x - 2y = 0$

5. $6x + 3y = 15$
   $6x - 5y = 7$

6. $4x - 3y = 14$
   $2x - 3y = 4$

7. $x + 2y = 10$
   $5x + 2y = 26$

8. $3x + 2y = 20$
   $3x + 4y = 34$

Example 3

Solve the equations

$$3x + 5y = 26 \tag{1}$$
$$2x + 3y = 16 \tag{2}$$

If we multiply equation (1) by 2 and equation (2) by 3, we shall have the same coefficient of x in each equation. We can then eliminate x by subtracting equation (2) from equation (1).

Alternatively, if we multiply equation (1) by 3 and equation (2) by 5, we shall have the same coefficient of y in each equation. We can then eliminate y by subtracting equation (2) from equation (1).

You have to decide which of the variables x or y you are going to eliminate. We will eliminate x.

Multiplying equation (1) by 2 and equation (2) by 3:

$$6x + 10y = 52 \tag{3}$$
$$6x + 9y = 48 \tag{4}$$

Subtracting equation (4) from equation (3):

$$(6x + 10y) - (6x + 9y) = 52 - 48$$
$$6x + 10y - 6x - 9y = 4$$
$$y = 4$$

Substituting for y in equation (1):

$$3x + (5 \times 4) = 26$$
$$3x + 20 = 26$$
$$3x = 6$$
$$x = 2$$

The solutions are $x = 2$ and $y = 4$.

Checking in equation (2):

$$\text{LHS} = (2 \times 2) + (3 \times 4)$$
$$= 4 + 12$$
$$= 16$$
$$= \text{RHS}$$

Since the LHS equals the RHS, the solutions are correct.

Example 4

Solve the equations

$$5x - 4y = 13 \qquad (1)$$
$$7x - 2y = 29 \qquad (2)$$

This time we can eliminate y by multiplying equation (2) by 2:

$$14x - 4y = 58 \qquad (3)$$

Subtracting equation (1) from equation (3):

$$(14x - 4y) - (5x - 4y) = 58 - 13$$
$$14x - 4y - 5x + 4y = 45$$
$$9x = 45$$
$$x = 5$$

Substituting for x in equation (1):

$$(5 \times 5) - 4y = 13$$
$$25 - 4y = 13$$
$$-4y = -12$$
$$y = 3$$

The solutions are $x = 5$ and $y = 3$.

Checking in equation (2):

$$LHS = (7 \times 5) - (2 \times 3)$$
$$= 35 - 6$$
$$= 29$$
$$= RHS$$

Hence the solutions are correct.

Exercise 16.2

Solve the following pairs of equations for x and y and check their solutions:

1. $x + 3y = 11$
       $2x + 5y = 19$

2. $3x + 2y = 21$
       $2x - y = 7$

3. $5x - 4y = 7$
       $7x - 2y = 17$

4. $3x + 5y = 31$
       $2x + 3y = 19$

5. $5x + 2y = 9$
       $3x + 5y = 13$

6. $2x + 5y = 27$
       $3x + 2y = 13$

7. $3x + 2y = 13$
       $2x + 3y = 7$

8. $7x - 4y = 8$
       $3x - 5y = -13$

The Product of Two Binomial Expressions

A binomial expression consists of *two terms*. Thus $3x + 5$, $a + b$, $2x + 3y$ and $4p - q$ are all binomial expressions.

To find the product of $(a + b)(c + d)$ consider the diagram (Fig. 16.1).

Fig. 16.1

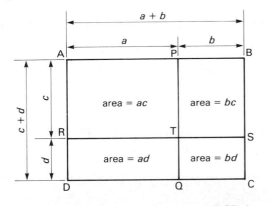

In Fig. 16.1 the rectangular area ABCD is made up as follows:

$$ABCD = APTR + RTQD + PBST + TSQC$$

i.e. $(a + b)(c + d) = ac + ad + bc + bd$

It will be noticed that the expression on the right-hand side is obtained by multiplying each term in the one bracket by each term in the *other* bracket. The process is illustrated below where each pair of terms connected by a line are multiplied together.

$$(a + b)(c + d) = ac + ad + bc + bd$$

Example 5

(a) $(x + 3)(x + 2) = x \times x + 3 \times x + x \times 2$
$$+ 3 \times 2$$
$$= x^2 + 5x + 6$$

(b) $(x - 4)(x + 3) = x \times x - 4 \times x + x \times 3$
$$- 4 \times 3$$
$$= x^2 - x - 12$$

(c) $(x - 2)(x - 5) = x \times x - 2 \times x - x \times 5$
$$- 2 \times (-5)$$
$$= x^2 - 7x + 10$$

Exercise 16.3

Remove the brackets from each of the following:

1. $(x + 1)(x + 3)$ 2. $(x + 2)(x + 4)$

3. $(x + 5)(x + 1)$ 4. $(x + 3)(x + 6)$

5. $(x - 1)(x - 2)$ 6. $(x - 3)(x - 4)$

7. $(x - 2)(x - 6)$ 8. $(x - 4)(x - 5)$

9. $(x + 1)(x - 3)$ 10. $(x + 2)(x - 3)$

11. $(x - 5)(x + 2)$ 12. $(x - 4)(x + 6)$

$x^2 + 6x - 4x - 24$

Factors of Quadratic Expressions

$x^2 + 2x - 24$

A quadratic expression is an expression in which the highest power of the symbol used is 2. Thus

$$x^2 - 5x + 7 \quad \text{and} \quad x^2 + 3x - 8$$

are both quadratic expressions.

Now $(x + 3)(x + 2) = x^2 + 5x + 6$

Note that the factors of 6 are 3 and 2. The coefficient of x is 5 which equals the factors of 6 added together (i.e. $3 + 2 = 5$). This example gives a clue as to how we factorise quadratic expressions.

Example 6

(a) Factorise $x^2 + 7x + 12$.

We see that $12 = 4 \times 3$
and that $7 = 4 + 3$.

Hence $x^2 + 7x + 12 = (x + 4)(x + 3)$

(b) Factorise $x^2 - 2x - 3$.

We see that $-3 = (-3) \times 1$
and that $-3 + 1 = -2$.

Hence $x^2 - 2x - 3 = (x - 3)(x + 1)$

Exercise 16.4

Factorise the following quadratic expressions:

1. $x^2 + 5x + 6$ 2. $x^2 + 6x + 8$

3. $x^2 - 7x + 10$ 4. $x^2 - 11x + 30$

5. $x^2 - x - 2$ 6. $x^2 - 2x - 15$

7. $x^2 + 2x - 8$ 8. $x^2 + x - 12$

Now consider

$$(x + 2)^2 = (x + 2)(x + 2)$$
$$= x^2 + 4x + 4$$

The square of a binomial expression therefore consists of:

$$(\text{first term})^2 + 2 \times (\text{first term})$$
$$\times (\text{second term}) + (\text{last term})^2$$

Example 7

(a) Factorise $x^2 + 6x + 9$.

$$x^2 = (x)^2; \quad 9 = 3^2; \quad 2 \times x \times 3 = 6x$$
$$\therefore \qquad x^2 + 6x + 9 = (x + 3)^2$$

(b) Factorise $x^2 - 4x + 4$.

$$x^2 = (x)^2; \quad 4 = (-2)^2;$$
$$2 \times x \times (-2) = -4x$$
$$\therefore \qquad x^2 - 4x + 4 = (x - 2)^2$$

Next consider

$$(x + 3)(x - 3) = x \times x + 3 \times x - 3 \times x$$
$$+ 3 \times (-3)$$
$$= x^2 - 9$$

Expressions like $x^2 - 9$ are called the **difference of two squares**. Their factors are the sum and difference of the square roots of each of the squares.

Example 8

Factorise $x^2 - 16$.

$$x^2 = (x)^2 \quad \text{and} \quad \sqrt{x^2} = x$$
$$16 = 4^2 \quad \text{and} \quad \sqrt{4^2} = 4$$
$$\therefore \qquad x^2 - 16 = (x + 4)(x - 4)$$

Exercise 16.5

Factorise the following:

1. $x^2 + 2x + 1$ 2. $x^2 - 6x + 9$
3. $x^2 + 8x + 16$ 4. $x^2 - 2x + 1$
5. $x^2 - 10x + 25$ 6. $x^2 - 1$
7. $x^2 - 9$ 8. $x^2 - 25$
9. $x^2 - 64$ 10. $x^2 - 81$

Quadratic Equations

An equation of the type $ax^2 + bx + c = 0$ is called a **quadratic equation**. The constants a, b and c can have any numerical values. Thus:

$$x^2 - 36 = 0$$

in which $a = 1$, $b = 0$ and $c = -36$.

$$5x^2 + 7x + 8 = 0$$

in which $a = 5$, $b = 7$ and $c = 8$.

$$2.5x^2 - 3.1x - 2 = 0$$

in which $a = 2.5$, $b = -3.1$ and $c = -2$.

These are all examples of quadratic equations. A quadratic equation may contain only the square of the unknown quantity, as in the first of the above equations, or it may contain both the square and the first power as in the remaining two equations.

Solution by Factors

If the product of two factors is zero, then one factor or the other factor must be zero or they may both be zero. Thus if $ab = 0$, then either $a = 0$ or $b = 0$ or both $a = 0$ and $b = 0$.

We make use of this fact in solving quadratic equations.

Example 9

(a) Solve the equation $(x + 4)(x - 3) = 0$.
Since the product of the two factors $(x + 4)$ and $(x - 3)$ is zero then either

$$x + 4 = 0 \quad \text{giving} \quad x = -4$$

or $\qquad x - 3 = 0 \quad \text{giving} \quad x = 3$

The solutions (sometimes called the **roots of the equation**) are therefore $x = -4$ or $x = 3$.

(b) Solve the equation $x^2 - 5x + 6 = 0$.
Factorising,

$$(x - 3)(x - 2) = 0$$

either $x - 3 = 0 \quad \text{giving} \quad x = 3$

or $\qquad x - 2 = 0 \quad \text{giving} \quad x = 2$

The solutions are $x = 3$ or $x = 2$.

(c) Solve the equation $x^2 - 9 = 0$
Factorising,

$$(x + 3)(x - 3) = 0$$

either $(x + 3) = 0 \quad \text{giving} \quad x = -3$

or $\qquad (x - 3) = 0 \quad \text{giving} \quad x = 3$

We say the roots are ± 3 meaning that the solutions are $x = +3$ or $x = -3$. The double sign $\pm$ is read as 'plus or minus'.

(d) Solve the equation $x^2 + 8x - 16 = 0$.

Factorising,

$$(x-4)^2 = 0 \quad \text{or} \quad (x-4)(x-4) = 0$$

Therefore the two solutions are the same and

$$x - 4 = 0 \quad \text{giving} \quad x = 4$$

We say the roots are identical and equal 4.

Exercise 16.6

Solve the following quadratic equations:

1. $(x - 5)(x - 2) = 0$

2. $(x + 3)(x + 6) = 0$

3. $(x + 2)(x - 3) = 0$

4. $x^2 + 6x + 8 = 0$

5. $x^2 - 5x + 6 = 0$

6. $x^2 + 3x - 10 = 0$

7. $x^2 - 3x - 18 = 0$

8. $(x + 2)^2 = 0$

9. $(x - 5)^2 = 0$

10. $x^2 - 6x + 9 = 0$

11. $x^2 + 10x + 25 = 0$

12. $x^2 - 2x + 1 = 0$

13. $x^2 - 1 = 0$

14. $(x - 4)(x + 4) = 0$

15. $x^2 - 25 = 0$

16. $x^2 - 36 = 0$

17. $x^2 - 64 = 0$

18. $x^2 - 49 = 0$

Miscellaneous Exercise 16

Section B

1. Write without brackets:
   (a) $(x + 5)(x + 2)$
   (b) $(x - 6)(x - 3)$
   (c) $(x + 3)(x - 7)$

2. Factorise each of the following:
   (a) $x^2 - 121$
   (b) $x^2 + 14x + 49$

3. Factorise each of the following:
   (a) $x^2 - 10x + 25$
   (b) $x^2 + 7x + 12$

4. Solve the quadratic equation $x^2 - 100 = 0$.

5. Solve the equation $x^2 - 8x + 16 = 0$.

6. Find the solutions of the quadratic equation $x^2 - 7x + 10 = 0$.

7. Solve the equation $x^2 + 5x + 6 = 0$.

8. Find the two possible values of x for which $x^2 - 11x + 18 = 0$.

Solve the following pairs of simultaneous equations:

9. $4x - y = 5$
   $2x + y = -2$

10. $3x - 2y = 5$
    $x - 2y = -1$

11. $4x - 3y = 15$
    $2x + 3y = 3$

12. If $p - q = 8$ and $2p - q = 20$, calculate the value of p.

13. Given the two equations

$$3x + 2y = 9$$
$$2x + 3y = 16$$

(a) by adding the two equations find the value of $x + y$,

(b) by subtracting the two equations find the value of $x - y$,

(c) hence, or otherwise, solve the equations for x and y.

14. Solve the following equations for x and y:

$$2x - 3y = 17$$
$$2x + 3y = 11$$

Multi-Choice Questions 16

1. The solution of the simultaneous equations $2x - y = 2$ and $x + y = -5$ is

   A $x = 1, y = -4$
   B $x = -1, y = -4$
   C $x = -1, y = 4$
   D $x = 1, y = 4$

2. $3x - y = 2$ (1)
   $3x - 2y = 1$ (2)

   The result of subtracting equation (2) from equation (1) is

   A $-3y = 1$ B $6x - 3y = 1$
   C $y = 1$ D $6x + y = 1$

3. If $x + y = 5$ and at the same time $2x - y = 7$, find the value of x.

   A 1   B 2   C 3   D 4

4. Which pair of values of x and y is the solution of $4x + y = 23$ and $3x - y = 12$?

   A $x = 6, y = 6$ B $x = 4, y = 7$
   C $x = 5, y = 3$ D $x = 11, y = 21$

5. What is the value of x in these simultaneous equations?

   $$x + y = 7; \quad x - y = 1$$

   A 3   B 4   C 5   D 6

6. The solution of the equation $(x - 2)(x + 3) = 0$ is

   A 0   B -3
   C 2   D 2 or -3

7. The solution of $x^2 - 11x + 30 = 0$ is

   A 0   B -11 or 30
   C -5 or -6 D 5 or 6

Angles and Straight Lines

Some Definitions

(1) A **point** has position but no size, although to show it on paper it obviously must be given some size.

(2) If two points A and B are chosen (Fig. 17.1) then only one straight line can contain them.

Fig. 17.1

A ●   B ●

(3) If the two points chosen are the end points of the line then AB is called a line segment (Fig. 17.2).

Fig. 17.2

A ●————————————● B

Angles

When two lines meet at a point they form an angle. The size of the angle depends upon the amount of opening between the two lines. It does not depend upon the lengths of the lines forming the angle. In Fig. 17.3 the angle *A* is larger than the angle *B* despite the fact that the arms are shorter.

Fig. 17.3

Measurement of Angles

An angle may be looked upon as the amount of rotation or turning. In Fig. 17.4 the line OA has been turned about O until it takes up the position OB. The angle through which the line OA has turned is the amount of opening between the lines OA and OB.

Fig. 17.4

If the line OA is rotated further until it returns to its starting position it will have completed one revolution. Therefore we can measure an angle as a fraction of a revolution.

Fig. 17.5 shows a circle divided into 36 equal parts. The first division is split into 10 equal divisions so that each small division is $\frac{1}{360}$ of a complete revolution. This small division is called a **degree**. Therefore

$$360 \text{ degrees } = 1 \text{ revolution}$$

which is written for brevity as

$$360° = 1 \text{ rev}$$

Fig. 17.5

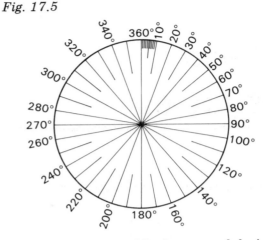

Most angles are stated in degrees and decimals of a degree. A typical angle might be $36.7°$ or $206.4°$.

Example 1

(a) Find the angle, in degrees, corresponding to one-third of a revolution.

$$1 \text{ rev} = 360°$$
$$\tfrac{1}{3} \text{ rev} = \tfrac{1}{3} \times 360°$$
$$= 120°$$

(b) Find the angle, in degrees, corresponding to 0.7 of a revolution.

$$0.7 \text{ rev} = 0.7 \times 360°$$
$$= 252°$$

Example 2

(a) Add together $46.3°$ and $36.9°$.

$$\begin{array}{r} 46.3° \\ 36.9° + \\ \hline 83.2° \end{array}$$

(b) Subtract $38.75°$ from $65.62°$.

$$\begin{array}{r} 65.62° \\ 38.75° - \\ \hline 26.87° \end{array}$$

Exercise 17.1

Find the angle in degrees corresponding to each of the following:

1. $\tfrac{2}{3}$ of a revolution.

2. $\tfrac{3}{8}$ of a revolution.

3. 0.2 of a revolution.

4. 0.9 of a revolution.

5. 0.15 of a revolution.

Add the following angles:

6. $18.9°$ and $27.6°$.

7. $39.2°$ and $17.8°$.

8. $43.2°$, $54.5°$ and $72.6°$.

Subtract the following angles:

9. $18.3°$ from $54.6°$.

10. $37.9°$ from $46.1°$.

Types of Angles

An **acute angle** is an angle of less than $90°$ (Fig. 17.6).

Fig. 17.6

Acute angle

A **right angle** is an angle equal to $90°$ or $\tfrac{1}{4}$ of a revolution (Fig. 17.7). Note carefully how a right angle is marked.

Fig. 17.7

Right angle

An **obtuse angle** is an angle between 90°
and 180° (Fig. 17.8).

Fig. 17.8

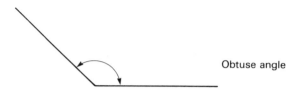

Obtuse angle

A **reflex angle** is an angle greater than 180°
(Fig. 17.9).

Fig. 17.9

Reflex angle

Complementary angles are angles whose sum
is 90°. Thus 18° and 72° are complementary
angles because 18° + 72° = 90°.

Supplementary angles are angles whose sum
is 180°. Thus 103° and 77° are supplementary angles because 103° + 77° = 180°.

Exercise 17.2

1. Look at each of the angles in
   Fig. 17.10. State which are acute,
   which are obtuse and which are reflex.

Fig. 17.10

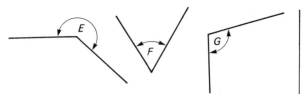

2. State the angle, in degrees, corresponding to each of the following:
   (a) $\frac{2}{3}$ of a right angle
   (b) $\frac{1}{4}$ of a right angle
   (c) 2 right angles
   (d) $1\frac{1}{2}$ right angles
   (e) 0.6 of a right angle
   (f) 0.35 of a right angle.

3. In each of the diagrams of Fig. 17.11,
   the angles at the centre of each one are
   of equal size. Work out the number of
   degrees in each of the angles.

Fig. 17.11

(a)　　　　　　　　(b)

(c)　　　　　　　　(d)

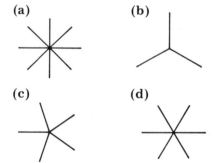

4. Find the number of degrees in each of
   the angles marked x in Fig. 17.12.

Fig. 17.12

(a)　　　　　(b)　　　　　(c)

(d)　　　　　(e)

5. **(a)** Two angles are complementary. One of them is 57°. What is the other?

(b) Two angles are complementary. If one is 34°, find the other.

(c) Two angles are supplementary. One is 27°, what is the other?

(d) Angles A and B are supplementary. If $A = 98°$, what is the size of B?

Properties of Angles and Straight Lines

(1) The total angle on a straight line is 180°, i.e. two right angles. The angles A and B in Fig. 17.13 are called **adjacent angles** on a straight line and their sum is 180°.

Fig. 17.13

$$\angle A + \angle B = 180°$$

(2) When two straight lines intersect, the **vertically opposite angles** are equal (Fig. 17.14).

Fig. 17.14

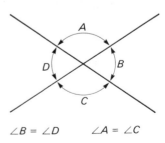

$$\angle B = \angle D \qquad \angle A = \angle C$$

Parallel Lines

Two lines in a plane that have no points in common, no matter how far they are produced, are called **parallel lines**.

(1) When two parallel lines are cut by a transversal (Fig. 17.15), the **corresponding angles** are equal. That is

$$a = l \quad b = m \quad c = p \quad d = q$$

Fig. 17.15

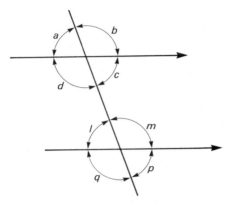

(2) The **alternate angles** are also equal. That is

$$d = m \quad \text{and} \quad c = l$$

(3) The **interior angles** are supplementary. That is

$$d + l = 180° \quad \text{and} \quad c + m = 180°$$

Example 3

In Fig. 17.16, AB and CD are two parallel straight lines. Find the size of the angles marked a, b, c, d and e.

Fig. 17.16

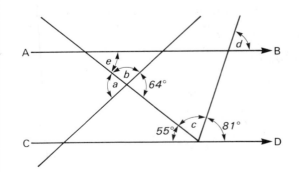

$a = 64°$ (vertically opposite angles)

$b = 180° - 64°$

$= 116°$ (sum of the angles on a straight line equals $180°$)

$c = 180° - 55° - 81°$

$= 44°$ (sum of the angles on a straight line $= 180°$)

$d = 81°$ (AB $\parallel$ CD, corresponding angles)

$e = 55°$ (AB $\parallel$ CD, alternate angles)

(Note the symbol $\parallel$ which means 'is parallel to'.)

Exercise 17.3

1. Find the size of the angle a in Fig. 17.17.

Fig. 17.17

2. In Fig. 17.18, find the size of the angles marked x and y.

Fig. 17.18

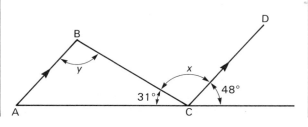

3. In Fig. 17.19, AB and CD are parallel. Calculate the size of the angles marked m, n, p and q.

Fig. 17.19

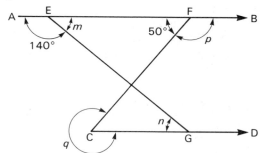

4. In Fig. 17.20, AB and CD are parallel lines crossed by the line ST. Find the size of the angle marked y.

Fig. 17.20

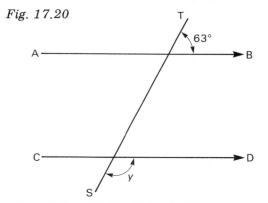

5. ABC and DEF (Fig. 17.21) are two parallel straight lines with BD, EC and BF transversals. CE and BF intersect at $90°$. Find the size of the angles marked p, q, r and s.

Fig. 17.21

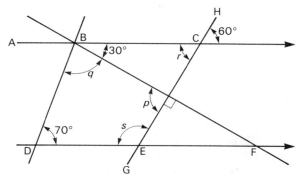

6. In Fig. 17.22, AB is parallel to CD. Find the size of the angles marked x and y.

Fig. 17.22

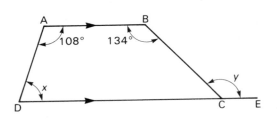

7. In Fig. 17.23, calculate the angle marked a.

Fig. 17.23

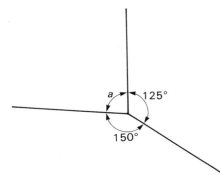

8. In Fig. 17.24, ABCDE and FGH are two parallel straight lines with BKG, FKCJ and GDJ transversals. Find the size of the angles marked u, v, w, x, y and z.

Fig. 17.24

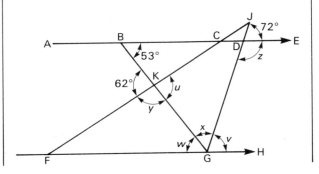

Multi-Choice Questions 17

1. The size of the angle x in Fig. 17.25 is
   A 40° B 50° C 130° D 140°

Fig. 17.25

2. What is the size of the angle a in Fig. 17.26?
   A 60° B 90° C 100° D 110°

Fig. 17.26

3. In Fig. 17.27, PQ and RS are parallel and the point T lies on RS. Find the size of the angle marked X.
   A 20° B 40° C 60° D 70°

Fig. 17.27

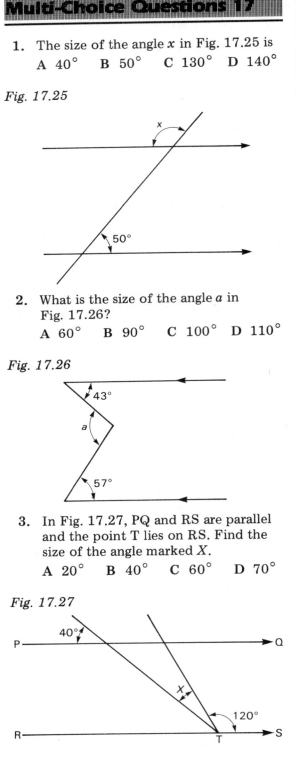

4. The size of the angle marked *a* in Fig. 17.28 is

    A 73°   B 107°   C 117°   D 146°

Fig. 17.28

5. The size of the angle marked *b* in Fig. 17.29 is

    A 60°   B 110°   C 120°   D 130°

Fig. 17.29

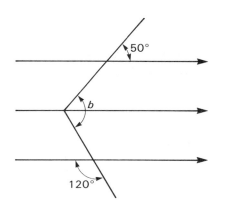

In Fig. 17.30, the lines AB and CD are parallel and the point E lies on CD. Use this diagram to answer questions 6, 7 and 8.

Fig. 17.30

6. The size of angle *a* is

    A 25°   B 50°   C 75°   D 80°

7. The size of angle *b* is

    A 80°   B 100°   C 105°   D 150°

8. The size of angle *c* is

    A 25°   B 50°   C 80°   D 135°

9. In Fig. 17.31, the value of *x* is

    A 45°   B $67\frac{1}{2}°$ C 90°   D 135°

Fig. 17.31

10. The angle shown in Fig. 17.32 is

    A acute   B right

    C reflex   D obtuse

Fig. 17.32

Mental Test 17

Try to answer the following questions without writing anything down except the answer.

1. Express a degree as a fraction of a complete revolution.

2. Find the angle in degrees corresponding to one-sixth of a revolution.

3. Find the angle in degrees corresponding to 0.3 of a revolution.

4. What is the angle in Fig. 17.33 called?

Fig. 17.33

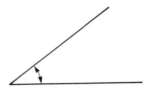

5. Two angles are complementary. One is 60°. What is the size of the other?

6. Two angles are supplementary. One is 130°. What is the size of the other?

7. Find the number of degrees in each of the angles marked x in Fig. 17.34.

Fig. 17.34

(a)   **(b)**   **(c)**

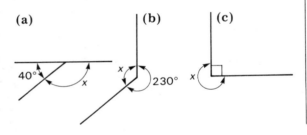

8. In Fig. 17.35, write down the size of the angles A and B.

Fig. 17.35

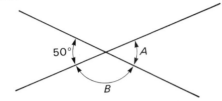

9. In Fig. 17.36, write down the size of the angles marked b, c, d, e, f, g and h.

Fig. 17.36

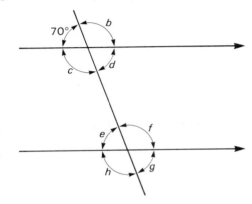

10. Add 35.3° and 44.7°.

Symmetry

Lines of Symmetry

If you fold a piece of paper so that one half of the paper covers the other half exactly, then the fold is called a **line of symmetry**.

The shape shown in Fig. 18.1 is symmetrical only about the line AA'. The shape is said to have one line of symmetry.

Fig. 18.1

The shape shown in Fig. 18.2 is symmetrical about the lines XX' and YY'. The shape therefore has two lines of symmetry.

Fig. 18.2

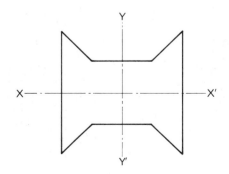

It is possible for a shape to have several lines of symmetry. For instance, the square shown in Fig. 18.3 has four lines of symmetry AA', BB', CC' and DD'.

Fig. 18.3

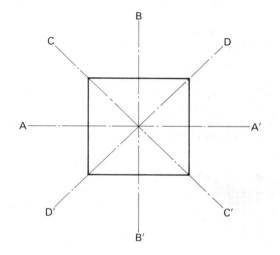

Some shapes have no lines of symmetry. The parallelogram shown in Fig. 18.4 is an example.

Fig. 18.4

From the examples given above it can be seen that a line of symmetry may be horizontal, vertical or oblique. Symmetry is also discussed on page 343.

Exercise 18.1

1. Each of the shapes shown in Fig. 18.5 has one line of symmetry. Copy the shapes on squared paper and then draw the line of symmetry.

2. Fig. 18.6 shows a number of half-shapes with the line of symmetry indicated in chain dot. Draw the complete symmetrical shape using squared paper.

Fig. 18.6

Fig. 18.5

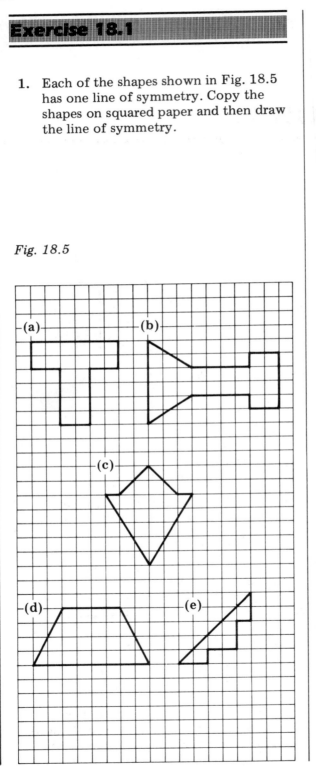

3. Each of the shapes shown in Fig. 18.7 has two lines of symmetry. Using squared paper copy each of these shapes and then draw the two lines of symmetry.

Fig. 18.7

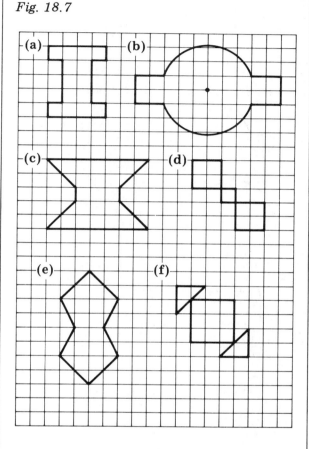

4. Each of the capital letters M, K, T, C, X and H has one or more lines of symmetry. Sketch the letters and draw all the lines of symmetry for each of the letters.

Rotational Symmetry

Fig. 18.8 shows a square ABCD whose diagonals intersect at O. If we rotate the square through 90°, 180° and 270° the square does not appear to have moved (unless we label the corners A, B, C and D, when the change is apparent). Because

there are four positions where it appears not to have moved, we say that the square has **rotational symmetry of order four.**

Fig. 18.8

Original position and rotated through 360°

Rotated through 90°

Rotated through 180°

Rotated through 270°

Point Symmetry

Rectangles and parallelograms appear to be in the same position when rotated through 180°. They are said to have **point symmetry.**

The parallelogram (Fig. 18.9) does not appear to have moved after being rotated through 180° and hence it has point symmetry, but the trapezium (Fig. 18.10) appears upside down after a rotation of 180°. A trapezium therefore has no point symmetry.

Fig. 18.9

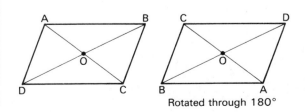

Rotated through 180°

Fig. 18.10

Rotated through 180°

Exercise 18.2

For each of the shapes shown in Fig. 18.11 write down

(a) the number of axes of symmetry

(b) the order of rotational symmetry

(c) whether the shape has point symmetry.

Fig. 18.11

1. Rhombus

2. Regular pentagon

3. Ellipse

4. Regular octagon

5. Star

6. Letter X

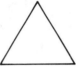

7. Letter E

8. Equilateral triangle

19 Triangles

Types of Triangles

(1) An **acute-angled triangle** (Fig. 19.1) has each of its angles less than 90°.

Fig. 19.1   Acute-angled triangle

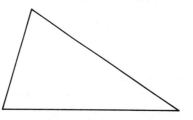

(2) A **right-angled triangle** has one of its angles equal to 90° (Fig. 19.2). The side opposite to the right angle is the longest side and it is called the **hypotenuse**.

Fig. 19.2   Right-angled triangle

(3) An **obtuse-angled triangle** (Fig. 19.3) has one angle greater than 90°.

Fig. 19.3   Obtuse-angled triangle

∠A greater than 90°

(4) A **scalene triangle** has all three sides of different length and all three angles of different size.

(5) An **isosceles triangle** has two sides and two angles equal (Fig. 19.4). The equal sides lie opposite to the equal angles.

Fig. 19.4   Isosceles triangle

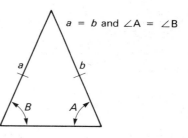

$a = b$ and $\angle A = \angle B$

(6) An **equilateral triangle** has all its sides and all its angles equal (Fig. 19.5). Each angle of the triangle is 60°.

Fig. 19.5   Equilateral triangle

Exercise 19.1

Look at the triangles shown in Fig. 19.6 and decide:

1. Which are equilateral triangles.
2. Which are obtuse-angled triangles.
3. Which are scalene triangles.
4. Which are isosceles triangles.
5. Which are right-angled triangles.

Fig. 19.6

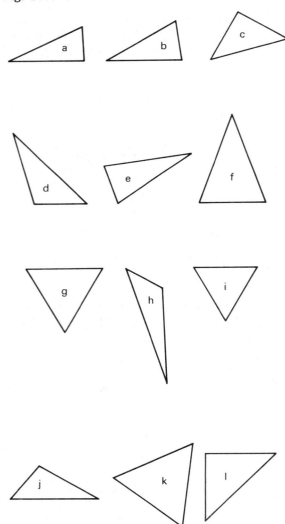

Angle Properties of Triangles

(1) The sum of the angles of a triangle is 180° (Fig. 19.7).

Fig. 19.7

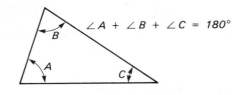

Example 1

Two of the angles of a triangle are 35° and 72°. What is the size of the third angle?

$$\text{Third angle} = 180° - (35° + 72°)$$
$$= 180° - 107°$$
$$= 73°$$

(2) When the side of a triangle is produced, the exterior angle so formed is equal to the sum of the opposite interior angles (Fig. 19.8).

Fig. 19.8

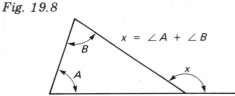

Example 2

In Fig. 19.9, find the size of the angle marked y.

Fig. 19.9

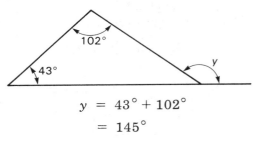

$$y = 43° + 102°$$
$$= 145°$$

Exercise 19.2

1. In each of the triangles listed in the table below, the size of two angles is given. Find by calculation the size of the third angle.

Triangle	Angles		
A	28°	67°	
B	37°		82°
C	80°	80°	
D		90°	28°
E	104°		63°
F		60°	30°

2. In Fig. 19.10, *x* is the exterior angle and *A* and *B* are the opposite interior angles. For each of the triangles listed below, find the size of the angle *x*.

Fig. 19.10

Triangle	∠A	∠B
R	49°	54°
S	39°	87°
T	68°	94°
U	107°	35°
V	27°	54°

3. Using Fig. 19.10, find the missing angle for each of the triangles listed below.

Triangle	∠A	∠B	*x*
G	39°	73°	
H	41°		108°
I	65°		89°
J	54°		165°
K	78°	90°	
L		86°	97°

4. Find each of the angles marked *x* and *y* in Fig. 19.11.

Fig. 19.11

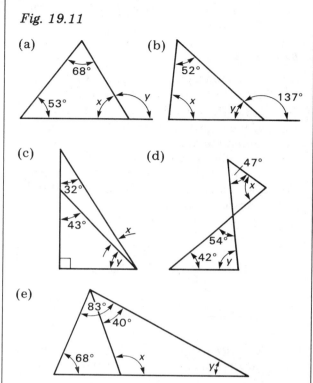

Standard Notation for a Triangle

Fig. 19.12 shows the standard notation for a triangle. The three vertices are labelled A, B and C. The three angles are called by the same letters as the vertices. Then the side *a* lies opposite to the angle *A*, the side *b* lies opposite to the angle *B* and the side *c* lies opposite to the angle *C*.

Fig. 19.12

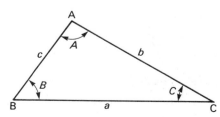

Constructing Triangles

(1) *To construct a triangle given the lengths of the three sides.*

Example 3

Construct the triangle ABC given that $a = 6$ cm, $b = 3$ cm and $c = 4$ cm.

Draw BC = 6 cm (Fig. 19.13), then with centre B and radius 4 cm using compasses draw an arc. With centre C and radius 3 cm, draw a circular arc to cut the first arc at A. Join A and B and also A and C, then ABC is the required triangle.

Fig. 19.13

(2) *To construct a triangle given two sides and the angle included between the two sides.*

Example 4

Construct the triangle ABC given that $b = 5$ cm, $c = 6$ cm and $A = 65°$.

Draw AB = 6 cm (Fig. 19.14). Using a protractor draw AX such that $\angle BAX = 65°$. Along AX̀ mark off AC = 5 cm, and join B and C. Then ABC is the required triangle.

Fig. 19.14

(3) *To construct a triangle given the lengths of two of the sides and an angle which is not the angle included between the two sides.*

Example 5

Construct the triangle ABC given that $a = 5$ cm, $b = 6$ cm and $B = 70°$.

Draw BC = 5 cm (Fig. 19.15). Using a protractor draw BX such that $\angle CBX = 70°$. With centre C and a radius of 6 cm, use compasses to draw an arc to cut BX at A. Join CA, then ABC is the required triangle.

Fig. 19.15

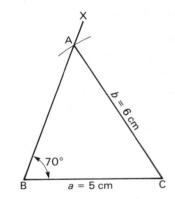

(4) *To construct a triangle given two angles and the length of the side opposite to the third angle of the triangle.*

Example 6

Construct the triangle ABC given that $a = 4.7$ cm, $B = 54°$ and $C = 46°$.

Draw BC = 4.7 cm (Fig. 19.16). At B draw a line BX making an angle of 54° with BC. At C draw a line CY making an angle of 46° with BC. The point of intersection of BX and CY is A, and ABC is the required triangle.

Fig. 19.16

Fig. 19.18

Construct each of the triangles ABC listed in the table below. Then measure the lengths of the sides and the sizes of the angles not given. In the table the lengths of *a*, *b* and *c* are stated in centimetres.

Triangle	$\angle A$	$\angle B$	$\angle C$	a	b	c
1	75°	34°		10		
2	19°		105°			11
3	116°		18°	8.5		
4	50°			7		9
5				5	8	7
6	43°			7	9	
7	62°				9	7
8		84°		5	6	
9				7	6	10
10			85°		7	9
11		60°	70°	8		
12	110°	40°				7

The Right-Angled Triangle

In any right-angled triangle the hypotenuse is the longest side and it always lies opposite to the right angle. Thus in Fig. 19.17 the side AC is the hypotenuse because it lies opposite to the right angle at B.

Fig. 19.17

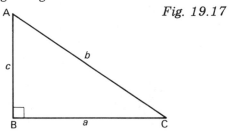

Fig. 19.18 shows a triangle ABC which is right angled at B. The side AB is 3 cm long and the side BC is 4 cm long. By accurately constructing the triangle the side AC is found to be 5 cm long. By constructing squares on each of the three sides we find that:

the area of the square on AC is

$$5 \times 5 = 5^2$$

$$= AC^2$$

$$= 25 \text{ cm}^2$$

the area of the square on BC is

$$4 \times 4 = 4^2$$

$$= BC^2$$

$$= 16 \text{ cm}^2$$

the area of the square on AB is

$$3 \times 3 = 3^2$$

$$= AB^2$$

$$= 9 \text{ cm}^2$$

Therefore

the area of the square on AC

$$= \text{ the area of the square on BC}$$

$$+ \text{ the area of the square on AB}$$

That is

the area of the square on the hypotenuse

= the sum of the squares on the other

two sides

No matter how many right-angled triangles we draw we will always find this statement to be true. This statement is known as Pythagoras' theorem. For Fig. 19.18,

$$AC^2 = BC^2 + AB^2 \quad \text{or} \quad b^2 = a^2 + c^2$$

Example 7

In $\triangle ABC$, $A = 90°$, $b = 3$ cm and $c = 4$ cm. Find a.

The triangle is sketched in Fig. 19.19 and we see that a is the hypotenuse because it lies opposite to the right angle at A. Hence

$$a^2 = b^2 + c^2$$
$$a^2 = 3^2 + 4^2$$
$$= 9 + 16$$
$$a = \sqrt{25}$$
$$= 5 \text{ cm}$$

Fig. 19.19

It is worth remembering that triangles with sides of $3:4:5$, $5:12:13$ and $7:24:25$ are right-angled triangles. Multiples of these lengths also give right-angled triangles. Hence triangles with sides 10, 24, 26 and 14, 48, 50 are right-angled.

Example 8

In $\triangle ABC$, $A = 90°$, $a = 6.4$ cm and $b = 5.2$ cm. Find the length of the side c.

The triangle is sketched in Fig. 19.20 and we see that a is the hypotenuse. Hence

$$a^2 = b^2 + c^2$$
$$c^2 = a^2 - b^2$$
$$= 6.4^2 - 5.2^2$$
$$= 40.96 - 27.04$$
$$= 13.92$$
$$c = \sqrt{13.92}$$
$$= 3.731 \text{ cm}$$

Fig. 19.20

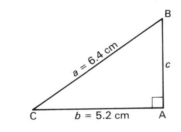

Exercise 19.4

1. Fig. 19.21 shows several right-angled triangles. For each triangle, name the hypotenuse.

Fig. 19.21

(a)   (b)   (c)

(d)   (e)

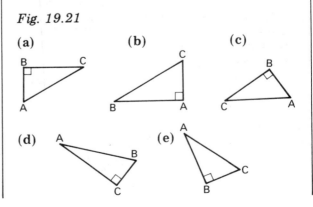

2. Fig. 19.22 shows a number of right-angled triangles. For each triangle, calculate the side marked x.

Fig. 19.22

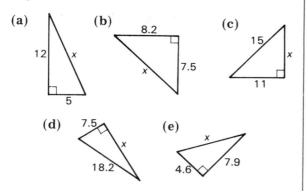

3. For each of the triangles ABC listed in the table below, the lengths, in centimetres, of two of the sides are given. Sketch each of the triangles and then work out the length of the third side.

Triangle	$\angle A$	$\angle B$	$\angle C$	a	b	c
(a)	$90°$				4	5
(b)			$90°$	8	11	
(c)		$90°$			15	8
(d)			$90°$	9		17
(e)	$90°$			6	3	

Properties of the Isosceles Triangle

It will be remembered that an isosceles triangle (Fig. 19.23) has two sides and two angles equal. The equal angles lie opposite to the equal sides.

Fig. 19.23

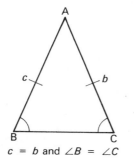

$c = b$ and $\angle B = \angle C$

Example 9

In Fig. 19.24, find the size of the angles A and C.

Fig. 19.24

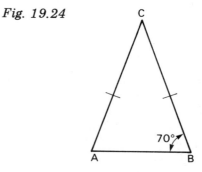

Since $AC = BC$, ABC is isosceles. Hence

$$A = B$$
$$= 70°$$

Since the sum of the angles of a triangle equals $180°$,

$$C = 180° - (70° + 70°)$$
$$= 180° - 140°$$
$$= 40°$$

Looking at Fig. 19.25, it is clear that the isosceles triangle ABC is symmetrical about the line AD. Hence the line AD bisects the apex angle A (i.e. divides the angle A into two equal angles). AD also bisects the base BC of the triangle (i.e. BD = DC) and finally AD is perpendicular to BC (i.e. it makes an angle of $90°$ with BC). Note that AD is the perpendicular height or altitude of the triangle ABC.

Fig. 19.25

Example 10

△ABC (Fig. 19.26) is isosceles with altitude AD = 10.2 cm and base BC = 8.4 cm. Calculate the length of the equal sides.

Fig. 19.26

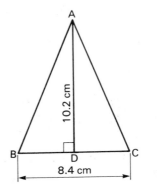

Since AD is the perpendicular dropped from the apex A,

$$BD = CD$$
$$= \tfrac{1}{2}BC$$
$$= \tfrac{1}{2} \times 8.4 \text{ cm}$$
$$= 4.2 \text{ cm}$$

In △ABD, AB is the hypotenuse and by Pythagoras' theorem,

$$AB^2 = BD^2 + AD^2$$
$$= 4.2^2 + 10.2^2$$
$$= 17.64 + 104.04$$
$$= 121.68$$
$$AB = \sqrt{121.68}$$
$$= 11.03 \text{ cm}$$

Hence the two equal sides AB and AC are 11.03 cm long.

Properties of the Equilateral Triangle

It will be recalled that an equilateral triangle has all its sides equal in length and all its angles equal in size. As can be seen from

Fig. 19.27, the equilateral triangle ABC has three axes of symmetry AD, BE and CF. Hence AD is perpendicular to BC, BE is perpendicular to AC and CF is perpendicular to AB. Since ∠BAC = ∠ACB = ∠ABC = 60°, each of the angles BAD, CAD, ACF, BCF, ABE and EBC equals 30°.

Fig. 19.27

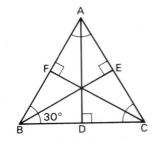

Example 11

△ABC (Fig. 19.28) is equilateral with sides 8 cm long. Calculate the altitude BD of this triangle. Since BD is the perpendicular dropped from the vertex B,

$$AD = CD$$
$$= \tfrac{1}{2}AC$$
$$= \tfrac{1}{2} \times 8 \text{ cm}$$
$$= 4 \text{ cm}$$

Fig. 19.28

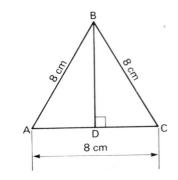

In △ADB, AB is the hypotenuse and by Pythagoras' theorem,

$$AB^2 = AD^2 + BD^2$$
$$BD^2 = AB^2 - AD^2$$
$$= 8^2 - 4^2$$
$$= 64 - 16$$
$$= 48$$
$$BD = \sqrt{48}$$
$$= 6.928 \text{ cm to 4 sig. fig.}$$

Hence the altitude of ABC is approx. 6.9 cm.

Exercise 19.5

1. Find the altitude AD of △ABC (Fig. 19.29).

Fig. 19.29

2. Find the side marked *a* in Fig. 19.30.

Fig. 19.30

3. Find the length of the base, BC, of △ABC in Fig. 19.31.

Fig. 19.31

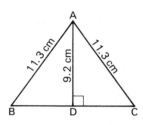

4. Find the angles marked *A* and *B* in Fig. 19.32.

Fig. 19.32

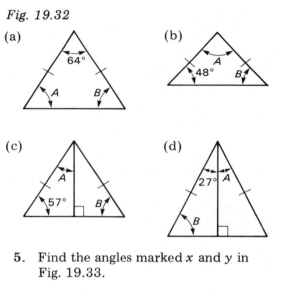

(a)   (b)

(c)   (d)

5. Find the angles marked *x* and *y* in Fig. 19.33.

Fig. 19.33

(a)   (b)

6. An equilateral triangle has sides 12 cm long. Calculate its altitude.

7. In △ABC, *a* = 12 cm, *b* = 12 cm and *c* = 10 cm. Calculate its altitude.

8. What is the vertical height of an equilateral triangle whose sides are each 9 cm long?

Miscellaneous Exercise 19

Section A

1. In Fig. 19.34 calculate the length of PQ.

Fig. 19.34

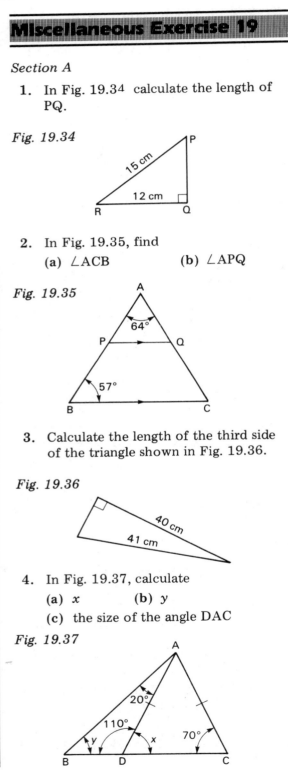

2. In Fig. 19.35, find
    (a) $\angle$ACB   (b) $\angle$APQ

Fig. 19.35

3. Calculate the length of the third side of the triangle shown in Fig. 19.36.

Fig. 19.36

4. In Fig. 19.37, calculate
    (a) x (b) y
    (c) the size of the angle DAC

Fig. 19.37

5. In Fig. 19.38 find
    (a) x (b) y (c) z
    (d) w (e) t

Fig. 19.38

6. A rectangle is 9 cm long and 6 cm wide. Find the length of its diagonal.

7. Find the height of the isosceles triangle shown in Fig. 19.39.

Fig. 19.39

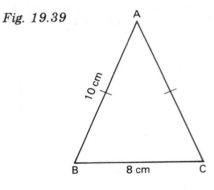

Section B

1. In Fig. 19.40, calculate the length of PQ.

Fig. 19.40

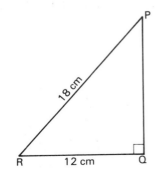

2. In Fig. 19.41, calculate the length of AB.

Fig. 19.41

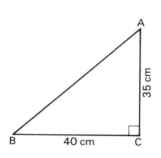

3. In Fig. 19.42, find
   (a) the size of the angle marked x
   (b) the size of the angle marked y
   (c) the size of $\angle ABC$
   (d) What kind of triangle is ABD?

Fig. 19.42

4. Fig. 19.43 shows a template. Calculate the length marked x which is needed to check the template.

Fig. 19.43

5. Fig. 19.44 shows a triangular piece of land. The owner wishes to fence the plot. What length of fencing does he need?

Fig. 19.44

6. A piece of wire 12 cm long can be bent in different ways to form different triangles. The sides of the triangle must always be a whole number of centimetres. Write down the lengths of the three sides if the triangle is
   (a) equilateral   (b) isosceles
   (c) right-angled

7. In Fig. 19.45, find the size of the angles marked x, y and z. The line AC bisects $\angle BAD$.

Fig. 19.45

8. In Fig. 19.46, write down the size of the angles marked p, q, r and s.

Fig. 19.46

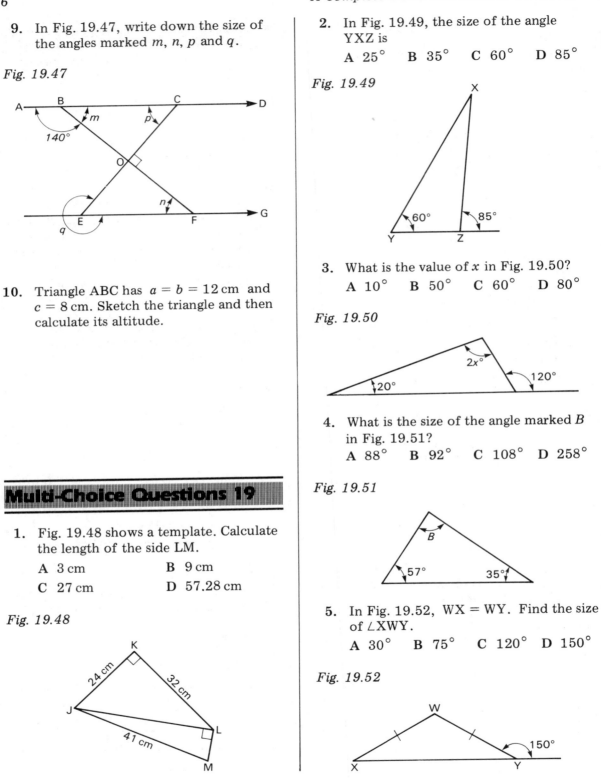

9. In Fig. 19.47, write down the size of the angles marked m, n, p and q.

Fig. 19.47

10. Triangle ABC has $a = b = 12$ cm and $c = 8$ cm. Sketch the triangle and then calculate its altitude.

Multi-Choice Questions 19

1. Fig. 19.48 shows a template. Calculate the length of the side LM.

   A 3 cm   B 9 cm
   C 27 cm   D 57.28 cm

Fig. 19.48

2. In Fig. 19.49, the size of the angle YXZ is
   A 25°   B 35°   C 60°   D 85°

Fig. 19.49

3. What is the value of x in Fig. 19.50?
   A 10°   B 50°   C 60°   D 80°

Fig. 19.50

4. What is the size of the angle marked B in Fig. 19.51?
   A 88°   B 92°   C 108°   D 258°

Fig. 19.51

5. In Fig. 19.52, WX = WY. Find the size of ∠XWY.
   A 30°   B 75°   C 120°   D 150°

Fig. 19.52

6. In Fig. 19.53, what is the size of ∠BAC?

   A 30°   B 45°   C 60°   D 90°

Fig. 19.53

7. In Fig. 19.53, what is the length of AC?

   A 4 cm   B 5.65 cm

   C 6 cm   D 8 cm

8. In Fig. 19.54, find the size of the angle QRS.

   A 60°   B 50°   C 40°   D 30°

Fig. 19.54

Try to answer the following questions without writing anything down except the answer.

1. Two angles of a triangle are 40° and 60°. What is the size of the third angle?

2. An isosceles triangle has an apex angle of 20°. What is the size of each of the other two angles?

3. In Fig. 19.55, what is the size of the angle marked x?

Fig. 19.55

4. In Fig. 19.55, what is the size of the angle marked y?

5. Using Fig. 19.56, find the missing angle for each of the triangles listed below.

Triangle	∠A	∠B	x
(a)	20°	40°	?
(b)	100°	20°	?
(c)	40°	?	100°
(d)	?	80°	150°

Fig. 19.56

6. Using Fig. 19.57, find the length of the missing side for each of the triangles listed below.

Triangle	a	b	c
(a)	6	8	?
(b)	?	12	20
(c)	?	5	13
(d)	15	36	?
(e)	7	?	25
(f)	21	72	?

Fig. 19.57

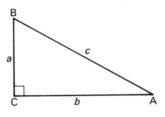

7. In Fig. 19.58, $\triangle ABC$ is isosceles. Find the missing angles for each of the triangles listed below.

Triangle	A	B	C	x
(a)	70°	?	?	?
(b)	?	?	60°	?
(c)	?	30°	?	?
(d)	?	?	?	100°
(e)	?	?	80°	?
(f)	?	?	?	120°
(g)	45°	?	?	?

Fig. 19.58

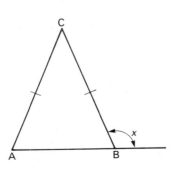

8. What is the triangle shown in Fig. 19.59 called?

Fig. 19.59

9. Fig. 19.60 shows an isosceles triangle. Find the size of the angles x and y.

Fig. 19.60

10. Two of the angles of a triangle are 70° and 80°. What is the size of the third angle?

11. In Fig. 19.61, find the size of the angle marked x.

Fig. 19.61

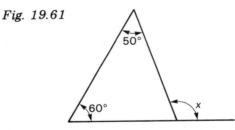

12. In Fig. 19.62 find the size of the angle marked a.

Fig. 19.62

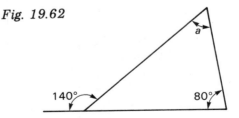

13. What is the longest side of a right-angled triangle called?

14. Two sides of a right-angled triangle are 5 and 12 cm long. What is the length of the third size?

Congruent Triangles

Triangles which are exactly the same shape and size are said to be **congruent.**

Two triangles are congruent if the six elements of one triangle (i.e. three sides and three angles) are equal to the six elements of the second triangle. In Fig. 20.1

$$AC = XZ \qquad AB = XY \qquad BC = YZ$$
$$\angle B = \angle Y \qquad \angle C = \angle Z \qquad \angle A = \angle X$$

and the two triangles also have a scale factor of 1.

Fig. 20.1

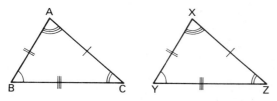

The notation for congruent triangles is $\triangle ABC \equiv \triangle XYZ$. To prove that two triangles are congruent it is not necessary to prove all six equalities. Any of the following conditions is sufficient to prove that two triangles are congruent:

(1) One side and two angles in one triangle are equal to one side and two similarly located angles in the second triangle (Fig. 20.2a).

(2) Two sides and the angle between them in one triangle are equal to two sides and the angle between them in the second triangle (Fig. 20.2b).

(3) If the triangles are right-angled and the hypotenuse and one side in one triangle are equal to the hypotenuse and one side in the second triangle (Fig. 20.2c).

(4) Three sides in one triangle are equal to three sides in the second triangle (Fig. 20.2d).

Fig. 20.2

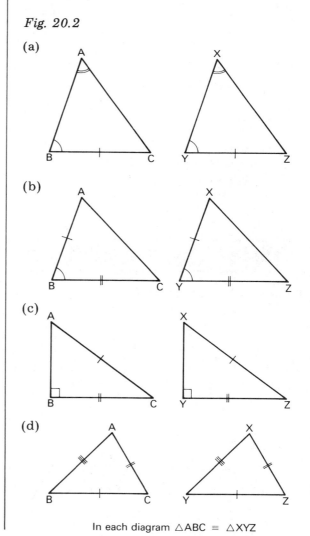

(a)

(b)

(c)

(d)

In each diagram $\triangle ABC \equiv \triangle XYZ$

Note carefully that the following do not prove congruency (Fig. 20.3):

(1) Three angles in one triangle equal to three angles in the second triangle.

(2) Two sides and an angle which does not lie between these sides in one triangle equal to two sides and a similarly located angle in the second triangle.

Fig. 20.3

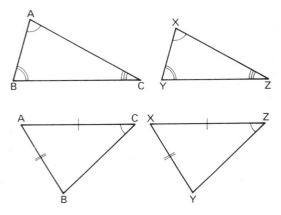

In neither diagram is △ABC necessarily congruent to △XYZ.

Exercise 20.1

State which of the following pairs of triangles are definitely congruent.

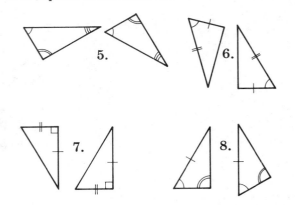

Example 1

(a) In Fig. 20.4, find the length of AB.

Fig. 20.4

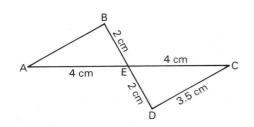

In triangles AEB and CED,

$$BE = ED$$
$$= 2\,cm \quad \text{(given)}$$
$$AE = EC$$
$$= 4\,cm \quad \text{(given)}$$

∠AEB = ∠CED (vertically opposite angles)

We now have two sides and the angle between them in △AEB equal to two sides and the angle between them in △CED. Hence △AEB ≡ △CED.

∴   AB = CD
$$= 3.5\,cm$$

(b) In Fig. 20.5. AB is parallel to CD and AB = CD = 4 cm. If AD = 5 cm, find the length of BC.

Fig. 20.5

In triangles ABC and CAD

$$\angle BAC = \angle ACD$$

(AB∥CD, alternate angles)

AC is common

$$AB = CD$$

$$= 4\,cm \quad \text{(given)}$$

We now have two sides and the included angle in △ABC equal to two sides and the included angle in △CAD. Hence △ABC ≡ △CAD and therefore

$$BC = AD$$

$$= 5\,cm$$

Exercise 20.2

1. In Fig. 20.6, find the lengths CD and EC.

Fig. 20.6

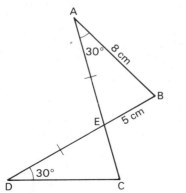

2. In Fig. 20.7, name all of the triangles which are congruent. G, H and J are the mid-points of DF, EF and DE respectively.

Fig. 20.7

3. In Fig. 20.8, PQ = RS and PQ is parallel to RS. If TS = 8 cm and TR = 6 cm, find the lengths of TP and TQ.

Fig. 20.8

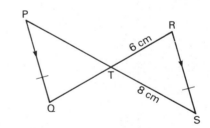

4. Fig. 20.9 shows two right-angled triangles ABC and XYZ. Find the length of BC.

Fig. 20.9

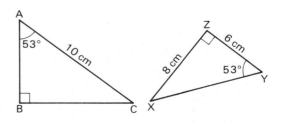

5. Fig. 20.10 shows two isosceles triangles LMP and RST. Find the lengths of ST, LX and RY. Find also the size of the angles MLP and SRY.

Fig. 20.10

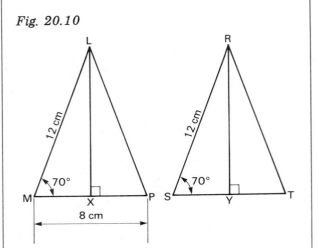

6. In Fig. 20.11, ABCD is a quadrilateral whose diagonals meet at E. Find the length of CD.

Fig. 20.11

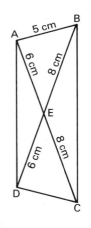

7. In Fig. 20.12, BC and AD are perpendicular to each other. Find the lengths of AE and CD.

Fig. 20.12

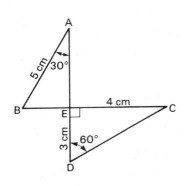

8. In Fig. 20.13, DE is parallel to BC, EF is parallel to AB and AE = EC. Find (a) the size of ∠ADE, (b) the size of ∠FEC, (c) the length of EF.

Fig. 20.13

Quadrilaterals

Introduction

A **quadrilateral** is any plane figure bounded by four straight lines.

If we draw a straight line from one corner to the opposite corner of a quadrilateral (Fig. 21.1) the quadrilateral is divided into two triangles. We have seen in Chapter 19 that the sum of the angles of a triangle is $180°$. Hence *the sum of the angles of a quadrilateral is 360°*.

Fig. 21.1

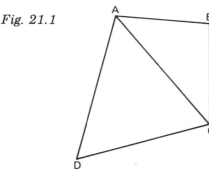

Example 1

Fig. 21.2 shows a quadrilateral. Find the size of the angle marked x.

Fig. 21.2

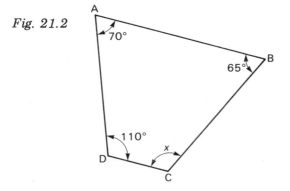

$$x + 70° + 65° + 110° = 360°$$
$$x + 245° = 360°$$
$$x = 115°$$

The Parallelogram

The **parallelogram** has both pairs of opposite sides parallel (Fig. 21.3a).

As shown in Chapter 18, a parallelogram has point symmetry. That is, if we rotate the parallelogram about O (Fig. 21.3b) through $180°$ it does not appear to have moved. However, we see that the vertices A, B, C and D become C′, D′, A′ and B′ respectively.

Fig. 21.3

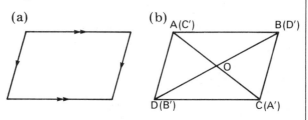

(a)   (b)

From Fig. 21.3b we see that:

(1) AB = CD and AD = BC. That is, the sides which are opposite to each other are equal in length.

(2) ∠ABC = ∠ADC and ∠DAB = ∠BCD. That is, the angles which are opposite to each other are equal in size.

(3) AO = OC and BO = OD. That is, the diagonals bisect each other.

(4) Triangles ACD and ABC are congruent, as are triangles BCD and ABD.

Example 2

Fig. 21.4 shows a parallelogram. Find the size of angles marked w, x, y and z.

Fig. 21.4

In $\triangle ACD$,

$w + 92° + 54° = 180°$ (sum of the
$\qquad\qquad\qquad\qquad$ angles of a triangle)

$w + 146° = 180°$

$w = 180° - 146°$

$= 34°$

In the parallelogram ABCD,

$x = w$

$= 34°$ (opposite angles of a
$\qquad\qquad$ parallelogram are
$\qquad\qquad$ equal)

$y = 54°$ (AB $\parallel$ CD, alternate
$\qquad\qquad$ angles)

$z + 54° = 92° + y$ (opposite
$\qquad\qquad\qquad\qquad$ angles of
$\qquad\qquad\qquad\qquad$ parallelogram
$\qquad\qquad\qquad\qquad$ are equal)

$z = 92° + 54° - 54°$

$= 92°$

Alternatively

$z = 92°$ (BC $\parallel$ AD, alternate angles)

The Rhombus

The **rhombus** is a parallelogram with all its sides equal in length. Hence it possesses all the properties of a parallelogram. Whereas the parallelogram has no axes of symmetry, the rhombus is symmetrical about each of its diagonals. This leads to two further properties (Fig. 21.5):

(1) Triangles ABD and BCD are congruent. Hence $\angle ABD = \angle DBC$ and $\angle ADB = \angle BDC$. That is, the diagonal bisects the angles through which it passes. Likewise the diagonal AC also bisects the angles through which it passes.

(2) From Fig. 21.5 it can be seen that the triangles AOB, BOC, COD and AOD are all congruent. Hence the four angles at O are all equal to 90°. This means that the diagonals of a rhombus bisect at right angles.

Fig. 21.5

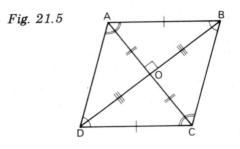

Example 3

ABCD (Fig. 21.6) is a rhombus whose sides are 12 cm long. The diagonal BD is 11.3 cm long. If $\angle ABD = 70°$, find the size of angles BDC, ADC and BAD and calculate the length of the diagonal AC.

Fig. 21.6

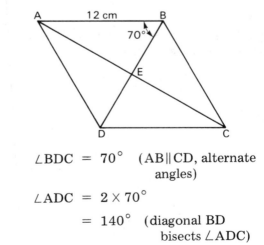

$\angle BDC = 70°$ (AB $\parallel$ CD, alternate
$\qquad\qquad\qquad$ angles)

$\angle ADC = 2 \times 70°$

$= 140°$ (diagonal BD
$\qquad\qquad$ bisects $\angle ADC$)

$$\angle BAD + \angle ABC = 180°$$
$$\angle ABC = 2 \times \angle ABD$$
$$= 2 \times 70°$$
$$= 140°$$
$$\angle BAD = 180° - 140°$$
$$= 40°$$

In $\triangle ABE$,

$$\angle AEB = 90° \text{ and } AE = 5.65 \text{ cm}$$

(diagonals of a rhombus bisect at right angles)

Also AB = 12 cm (given). Using Pythagoras' theorem,

$$AE^2 = AB^2 - BE^2$$
$$= 12^2 - 5.65^2$$
$$= 144 - 31.9$$
$$= 112.1$$
$$AE = \sqrt{112.1}$$
$$\simeq 10.59 \text{ cm}$$
$$AC = 2 \times AE$$
$$\simeq 2 \times 10.59$$
$$= 21.18 \text{ cm}$$

The Rectangle

A **rectangle** is a parallelogram with each of its vertex angles equal to 90°. Hence it possesses all the properties of a parallelogram.

Looking at Fig. 21.7 we see that triangles ACD and BCD are congruent. Hence the diagonals AC and BD are equal in length.

Fig. 21.7

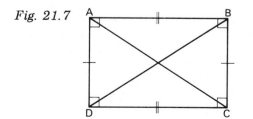

Example 4

A rectangle has sides 8 cm and 12 cm in length.

Calculate the length of its diagonals.

Fig. 21.8

In $\triangle ABD$ (Fig. 21.8), by Pythagoras' theorem,

$$BD^2 = AB^2 + AD^2$$
$$= 12^2 + 8^2$$
$$= 144 + 64$$
$$= 208$$
$$BD = \sqrt{208}$$
$$\simeq 14.4$$

Hence the length of the diagonals is 14.4 cm.

The Square

A **square** is a rectangle with all its sides equal in length. It has all the properties of a parallelogram, rhombus and rectangle. A square is symmetrical about the four axes shown in Fig. 21.9.

Fig. 21.9

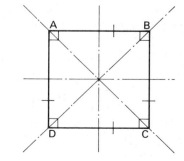

Example 5

A square has diagonals 8 cm long. Find the length of the sides of the square.

Fig. 21.10

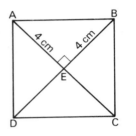

Since the diagonals bisect at right angles (Fig. 21.10), in $\triangle AEB$ by Pythagoras' theorem,

$$AB^2 = AE^2 + BE^2$$
$$= 4^2 + 4^2$$
$$= 16 + 16$$
$$= 32$$
$$AB = \sqrt{32}$$
$$\simeq 5.7$$

Hence the sides of the square are 5.7 cm long.

The Trapezium

A **trapezium** is a quadrilateral with one pair of sides parallel (Fig. 21.11). Generally speaking a trapezium has no axes of symmetry nor does it possess point symmetry.

Fig. 21.11

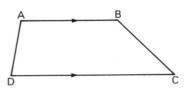

An **isosceles trapezium** has its two non-parallel sides equal in length (Fig. 21.12) and it has the one axis of symmetry shown in the diagram.

Fig. 21.12

In the isosceles trapezium shown in Fig. 21.12,
$\angle BCD = \angle ADC$ and $\angle DAB = \angle CBA$.
Also $\angle CDA + \angle DAB = 180°$
and $\angle DCB + \angle ABC = 180°$.

Example 6

(a) ABCD (Fig. 21.13) is a trapezium with AB parallel to CD. Given the angles shown in the diagram, calculate the angle marked p.

Fig. 21.13

$$\angle DAB + \angle ABC + \angle BCD + \angle CDA$$
$$= 360° \quad \text{(sum of the angles of a quadrilateral)}$$
$$DAB + 108° + 51° + 75° = 360°$$
$$\angle DAB = 126°$$
$$\angle BAC = 126° - 84°$$
$$= 42°$$

Since $AB \| CD$,
$$p = \angle BAC \quad \text{(alternate angles)}$$
$$= 42°$$

(b) In Fig. 21.14, ABCD is an isosceles trapezium with AD = BC.
Find the size of the angles marked x and y.

Fig. 21.14

Since ABCD is an isosceles trapezium,

$$y = 45°$$

Also,

$$x + y = 180°$$
$$x + 45° = 180°$$
$$x = 135°$$

The Kite

The **kite** is a quadrilateral having two pairs of adjacent sides equal in length (Fig. 21.15). It is symmetrical about the line BD, and hence this line bisects the angles ADC and ABC. The angles DAB and DCB are also equal and the diagonals AC and BD intersect at right angles.

Fig. 21.15

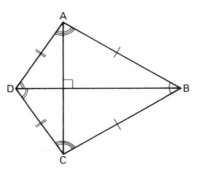

Example 7

Fig. 21.16 shows a kite ABCD. Given that ∠DAC = 15° and ∠CBD = 24°, find the size of the angles ACD, ADC and BAD.

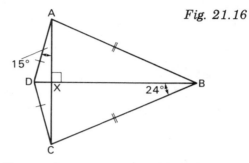

Fig. 21.16

Since AD = DC, △ADC is isosceles and therefore

$$∠DAC = ∠ACD$$
$$= 15°$$
$$∠ADC = 180° - (15° + 15°)$$
$$= 180° - 30°$$
$$= 150°$$

Since ∠CBD = 24°, ∠ABD = 24°

$$∠BAX = 180° - (90° + 24°)$$
$$= 180° - 114°$$
$$= 76°$$
$$∠BAD = 76° + 15°$$
$$= 91°$$

Exercise 21.1

1. Fig. 21.17 shows a kite. Find the angles marked x, y and z.

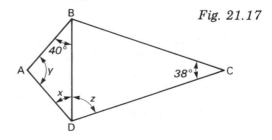

Fig. 21.17

2. In Fig. 21.18, ABCD is a quadrilateral. Find the angles marked x and y.

Fig. 21.18

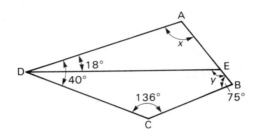

3. Fig. 21.19 shows a parallelogram. Write down the sizes of the angles marked a and b.

Fig. 21.19

4. For the rhombus ABCD (Fig. 21.20), find the sizes of the angles DAC, BCD, ABC and DOC.

Fig. 21.20

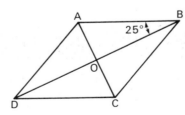

5. In a rhombus ABCD, the diagonal BD is 20 cm long and the diagonal AC is 12 cm long. Calculate the length of the sides of the rhombus.

6. A square has sides which are 8 cm long. Calculate the length of its diagonals.

7. The rectangle ABCD (Fig. 21.21) has its diagonal AC = 15 cm and its side AB = 11 cm. Calculate the length of the side BC.

Fig. 21.21

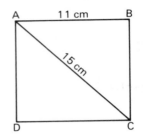

8. Fig. 21.22 shows an iron rectangular gate made from thin tubing. What length of tubing is needed to make it?

Fig. 21.22

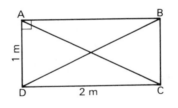

9. In Fig. 21.23, ABCD is a rhombus. BD = 10 cm, AC = 6 cm and XE is a straight line parallel to BC.

(a) What is the size of the angle AXD?

(b) What is the length of AX?

(c) What is the name of the figure EXCB?

Fig. 21.23

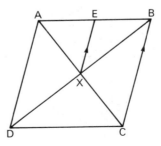

10. In Fig. 21.24, ABCD is a parallelogram.

   (a) What is the name of the quadrilateral BCDX?

   (b) Calculate the size of the angles XDC and DXB.

Fig. 21.24

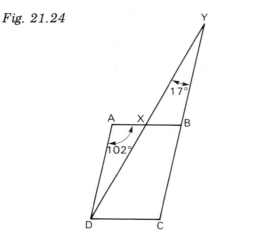

11. Fig. 21.25 shows a parallelogram ABCD. Find

   (a) The size of the angle marked x.

   (b) The size of the angle marked y.

   (c) The size of $\angle$ACB.

   (d) What type of triangle is ACD?

Fig. 21.25

12. In a rhombus ABCD (Fig. 21.26), the diagonal AC = 24 cm and the diagonal BD = 10 cm. Find AO, BO and AB.

Fig. 21.26

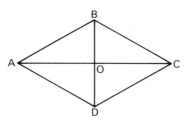

13. Name and draw two quadrilaterals which have one axis of symmetry.

14. Name a quadrilateral which has two pairs of equal sides and two axes of symmetry.

15. In Fig. 21.27, find the size of the angles marked a and p.

Fig. 21.27

16. In Fig. 21.28, ABCD is a quadrilateral. ABC is bisected by BE. Find the size of the angles ABE and BED.

Fig. 21.28

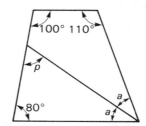

17. In Fig. 21.29, ABCD is a quadrilateral. $\angle$BAD = 60° and the other three angles of the quadrilateral are equal. EC is drawn parallel to AB. Find the size of the angles ABC and ECD.

Fig. 21.29

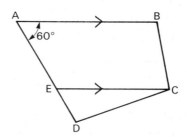

18. In Fig. 21.30, ABCD is a rhombus. Find the angle y if $x = 29°$.

Fig. 21.30

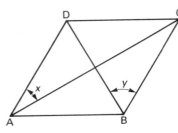

Multi-Choice Questions 21

1. The opposite angles of a parallelogram are
   A supplementary   B always equal
   C complementary   D never equal

2. What is the value of x in Fig. 21.31.

Fig. 21.31

   A 75°   B 95°   C 105°   D 115°

3. In the trapezium shown in Fig. 21.32, the angles marked x and y are
   A 124°, 65°   B 130°, 56°
   C 115°, 56°   D 115°, 62°

Fig. 21.32

4. A quadrilateral has one pair of sides parallel. It is a
   A rhombus   B parallelogram
   C rectangle   D trapezium

5. A quadrilateral has diagonals which intersect at right angles but does not bisect them. It is a
   A rhombus   B square
   C kite   D rectangle

6. In Fig. 21.33, x is equal to
   A $a + b + c$
   B $360° - (a + b + c)$
   C $a + b + c + 180°$
   D $360° - a + b + c$

Fig. 21.33

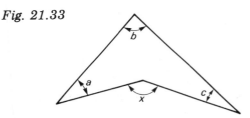

7. A quadrilateral has three angles of 70°, 80° and 90°. The size of the fourth angle is
   A 60°   B 90°   C 100°   D 120°

8. In Fig. 21.34, the size of the angle marked x is
   A 105°   B 110°   C 140°   D 145°

Fig. 21.34

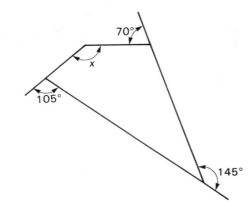

9. In Fig. 21.35, triangle DEC is equilateral and ABCD is a square. The size of ∠DEA is

   A 15°   B 20°   C 30°   D 45°

Fig. 21.35

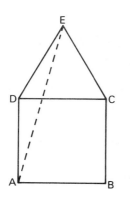

10. Three angles of a quadrilateral are equal. The fourth angle is twice the size of the other three. What is the size of the largest angle?

    A 72°   B 120°   C 144°   D 180°

Mental Test 21

Try to answer the following questions without writing anything down except the answer.

1. Fig. 21.36 shows a quadrilateral. Find the angle marked *x*.

Fig. 21.36

2. Find the size of the angle marked *y* in the parallelogram shown in Fig. 21.37.

Fig. 21.37

3. In the rhombus ABCD (Fig. 21.38) write down the size of the angles BEA and ABC.

Fig. 21.38

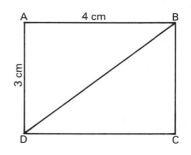

4. A rectangle ABCD (Fig. 21.39) has AB = 4 cm and AD = 3 cm. What is the length of the diagonal BD?

Fig. 21.39

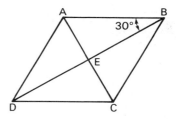

5. How many axes of symmetry has a square?

6. Draw an isosceles trapezium and on it show the axis (or axes) of symmetry.

7. In the kite ABCD (Fig. 21.40) find the size of the angles ACD and ACB.

8. Fig. 21.41 shows a square ABCD surmounted by an equilateral triangle ABE. What is the size of the angle EAD?

Fig. 21.40

Fig. 21.41

Polygons

Types of Polygons

A **polygon** is any plane figure bounded only by straight lines. Thus a triangle is a polygon with three sides, and a quadrilateral is a polygon with four sides.

A **convex polygon** (Fig. 22.1) has no interior angle greater than 180°.

Fig. 22.1

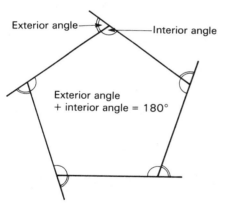

Exterior angle — Interior angle

Exterior angle + interior angle = 180°

A **re-entrant polygon** (Fig. 22.2) has at least one angle greater than 180°.

Fig. 22.2

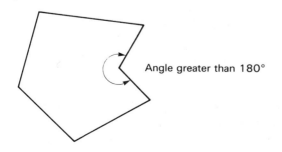

Angle greater than 180°

A **regular polygon** has all its sides and angles equal.

Sum of the Interior Angles of a Polygon

To find the sum of the interior angles of a polygon, choose any point such as O (Fig. 22.3) which lies within the polygon. Next draw lines from O to the vertices A, B, C ... to form triangles. The number of triangles is the same as the number of sides of the polygon: if the polygon has n sides then there are n triangles. Since the sum of the angles of each triangle is 180°, the sum of the angles of the triangles equals $180n$ degrees or $2n$ right angles. The sum of the angles round O equals 360° or 4 right angles. Hence *the sum of the interior angles of a polygon is $2n - 4$ right angles.*

Fig. 22.3

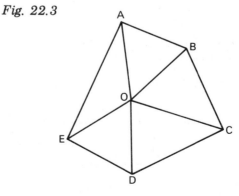

It makes no difference whether the polygon is convex or re-entrant, the sum of the interior angles is $(2n - 4)$ right angles.

Names of Polygons

Name	Number of sides
Pentagon	5
Hexagon	6
Heptagon	7
Octagon	8
Nonagon	9
Decagon	10
Hendecagon	11
Dodecagon	12

Example 1

(a) Find the sum of the interior angles of a heptagon (seven-sided polygon).

Here we have the number of sides
$$= n = 7$$

Sum of the
interior angles $= (2n - 4)$ right angles

$= (2 \times 7 - 4)$ right angles

$= 10$ right angles

$= 10 \times 90$ degrees

$= 900$ degrees

(b) Find the size of each interior angle of a regular pentagon.

Since a pentagon has 5 sides, $n = 5$.

Sum of the
interior angles $= (2n - 4)$ right angles

$= (2 \times 5 - 4)$ right angles

$= 6$ right angles

$= 540$ degrees

Size of each interior angle $= 540° \div 5$

$= 108°$

Sum of the Exterior Angles of a Convex Polygon

Fig. 22.4

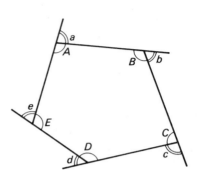

In Fig. 22.4, the interior angles of the polygon are A, B, C, D and E. The exterior angles are a, b, c, d and e. Now

$$A + a = 180°$$
$$= 2 \text{ right angles}$$
$$B + b = 180°$$
$$= 2 \text{ right angles}$$

and so on. Therefore:

Sum of the interior angles + Sum of the exterior angles $= 2n$ right angles

Sum of the
interior angles $= (2n - 4)$ right angles

Sum of the
exterior angles $= [2n - (2n - 4)]$ right angles

$= 4$ right angles

$= 360°$

No matter how many sides the convex polygon has, the sum of its exterior angles is 360°.

Example 2

(a) Each interior angle of a regular polygon is 150°. How many sides has it?

$$\text{Each exterior angle} = 180° - 150°$$
$$= 30°$$

Since the sum of the exterior angles is 360°,

$$\text{Number of sides} = 360 \div 30$$
$$= 12$$

(b) A regular polygon has 15 sides. What is the size of each interior angle?

$$\text{Each exterior angle} = 360° \div 15$$
$$= 24°$$

$$\text{Each interior angle} = 180° - 24°$$
$$= 156°$$

Symmetry of Regular Polygons

Regular polygons have the number of axes of symmetry equal to the number of sides of the polygon. Thus a pentagon (Fig. 22.5) has five axes of symmetry and an octagon (Fig. 22.6) has eight axes of symmetry.

Fig. 22.5

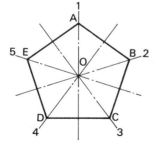

In Fig. 22.5, each of the triangles AOB, BOC, COD, DOE and EOA are congruent. The angle AOB = 360° ÷ 5 = 72°. Therefore ∠ABO + ∠OBC = 180° − 72° = 108°. Therefore each interior angle of a pentagon is 108°. This method gives us a third way of finding the size of the interior angle of a regular polygon.

Fig. 22.6

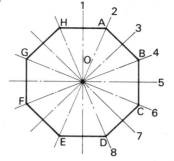

In Fig. 22.6 we note that there are 8 congruent triangles. Therefore the size of the interior angle of a regular octagon is $180° - (360° \div 8) = 180° - 45° = 135°$.

Exercise 22.1

1. Find the sum, in degrees, of the interior angles of polygons having

   (a) 6 sides   **(b)** 9 sides   **(c)** 11 sides

2. A regular polygon has 8 sides. Find the size of

   (a) each exterior angle

   (b) each interior angle

3. Draw a regular hexagon and show on it all the axes of symmetry. Hence find the size of each interior angle of the hexagon.

4. Draw a regular seven-sided polygon and on it show all the axes of symmetry. Hence find the size of each interior angle of the polygon.

5. A regular polygon has 12 sides. Find the size of each interior angle.

6. A pentagon (five-sided polygon) has interior angles of 104°, 108° and 112°. If the remaining two angles are equal, what is their size?

7. Each interior angle of a regular polygon is 120°. How many sides has it?

8. An irregular hexagon has interior angles of 110°, 90°, 130°, 140° and 135°. Find the size of its sixth angle.

9. A regular polygon has an interior angle of 160°. Calculate **(a)** the size of each exterior angle, **(b)** the number of sides possessed by the polygon.

10. Fig. 22.7 shows a regular pentagon ABCDE inscribed in a circle of centre O. Calculate the size of angles AOB and ABC.

Fig. 22.7

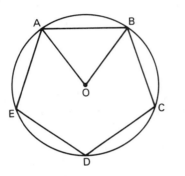

11. The exterior angle of a regular polygon is 24°. How many sides has the polygon?

12. UVWXYZ is a regular hexagon (Fig. 22.8). What is the size of the angle UVO? O is the centre of the circumscribing circle of the polygon.

Fig. 22.8

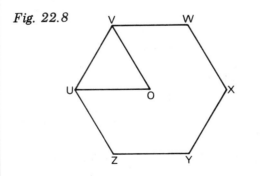

13. Calculate the size of the angles marked x and y for the regular polygon shown in Fig. 22.9. O is the centre of the circumscribing circle of the polygon.

Fig. 22.9

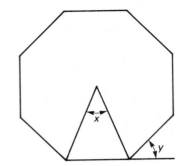

14. AB, BC and CD are three adjacent sides of a regular polygon with ten sides (Fig. 22.10). Calculate **(a)** the size of the exterior angle at B, **(b)** the size of angle ACD.

Fig. 22.10

15. ABCDE is a regular pentagon and ABX is an equilateral triangle (Fig. 22.11). Calculate the angle AEX.

Fig. 22.11

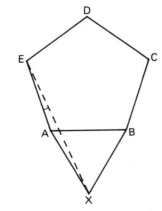

16. A polygon has n sides. Two of its angles are right angles and each of the remaining angles is 144°. Calculate n.

17. In the irregular pentagon ABCDE (Fig. 22.12) the interior angles at D and E are equal. The sides AB and DC when produced meet at right angles at F. Calculate the angles BCD and AED.

Fig. 22.12

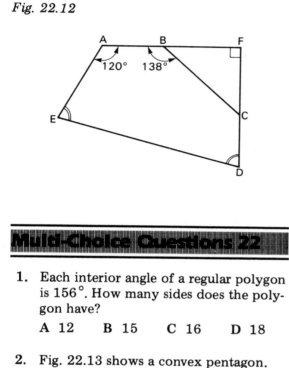

Multi-Choice Questions 22

1. Each interior angle of a regular polygon is 156°. How many sides does the polygon have?

   A 12   B 15   C 16   D 18

2. Fig. 22.13 shows a convex pentagon. What is the value of x?

   A 15°   B 20°   C 30°   D 40°

Fig. 22.13

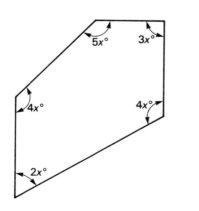

3. A regular polygon has each interior angle equal to 108°. How many sides does it have?

   A 4   B 5   C 6   D 7

4. The sum of the interior angles of an octagon is

   A 360°   B 1080°
   C 1260°   D 1440°

5. A regular polygon, centre O, can be subdivided into isosceles triangles identical to POQ (Fig. 22.14). How many sides does the polygon have?

   A 9   B 10   C 12   D 15

Fig. 22.14

6. Five interior angles of an irregular hexagon are equal and the sixth angle is three times the size of one of the other five. What is the size of the equal angles?

   A 60°   B 90°   C 120°   D 144°

7. A regular polygon has each of its interior angles equal to 162°. How many sides does the polygon have?

   A 8   B 10   C 20   D 24

8. For an eight-sided regular polygon, what is the ratio of the exterior angle to the interior angle?

   A 1:1   B 1:3   C 1:4   D 1:5

9. A regular polygon has each interior angle equal to $108°$. How many sides has it?

    A 7   B 6   C 5   D 4

10. A regular polygon has each exterior angle equal to $60°$. How many sides has it?

    A 5   B 6   C 7   D 9

Mental Test 22

Try to answer the following questions without writing anything down except the answer.

1. The exterior angle of a polygon is $40°$. What is the size of its interior angle?

2. A regular polygon has exterior angles of $60°$. How many sides has the polygon?

3. A regular polygon has interior angles of $135°$. How many sides has the polygon?

4. A regular polygon has 10 sides. Work out the size of its exterior angles.

5. A regular polygon has each interior angle equal to $120°$. How many sides does it have?

6. What is the name given to the polygon in question 5?

7. A regular polygon has 8 sides. How many axes of symmetry does it possess?

The Circle

Figure 23.1 shows the main parts of a circle.

Fig. 23.1

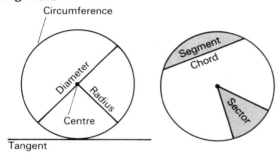

Angles in Circles

(1) *The angle which an arc of a circle subtends at the centre is twice the angle which the arc subtends at the circumference.*

Thus in Fig. 23.2, $\angle AOB = 2 \times \angle APB$.

Fig. 23.2

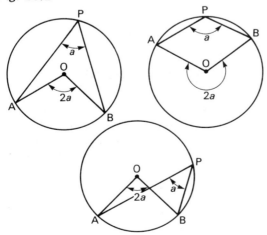

Example 1

In Fig. 23.3, O is the centre of the circle and $\angle ACB = 40°$. Find the size of $\angle AOB$.

Because $\angle AOB$ is the angle subtended by the arc AB at the centre of the circle and $\angle ACB$ is the angle subtended by the arc AB at the circumference

$$\angle AOB = 2 \times \angle ACB$$
$$= 2 \times 40°$$
$$= 80°$$

Fig. 23.3

(2) *If a triangle is drawn in a circle with a diameter as the base of the triangle, then the angle opposite that diameter is a right angle.*

Because the diameter AB (Fig. 23.4) forms the angle (180°) subtended by the semicircle ABD, the angle subtended by this arc at the circumference is 90°.

Fig. 23.4

Example 2

In Fig. 23.5, calculate the diameter of the circle whose centre is O.

Fig. 23.5

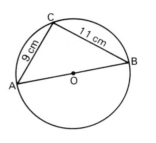

Since AB is a diameter, $\angle ACB = 90°$.

In $\triangle ABC$, using Pythagoras' theorem

$$AB^2 = AC^2 + BC^2$$
$$= 9^2 + 11^2$$
$$= 81 + 121$$
$$= 202$$
$$AB = \sqrt{202}$$
$$= 14.2$$

Hence the diameter of the circle is 14.2 cm.

(3) *Angles in the same segment of a circle are equal.*

In Fig. 23.6, the angles ACB and ADB stand on the same chord AB and so they are angles in the same segment.
Therefore $\angle ACB = \angle ADB$.

Fig. 23.6

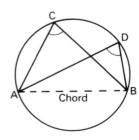

Example 3

In Fig. 23.7, find the size of the angles marked x and y.

Fig. 23.7

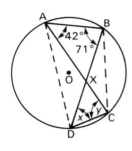

First draw the chords AD and BC.

The angles BAX and XDC are angles in the same segment because they stand on the same chord BC. Hence

$$\angle XDC = \angle BAX$$
$$= 42°$$

i.e. $x = 42°$

The angles ABX and XCD are angles in the same segment because they stand on the same chord AD. Hence

$$\angle XCD = \angle ABX$$
$$= 71°$$

i.e. $y = 71°$

Cyclic Quadrilaterals

The opposite angles of any quadrilateral inscribed in a circle are supplementary (i.e. the sum is $180°$). A quadrilateral inscribed in a circle is called a **cyclic quadrilateral**.

Thus the cyclic quadrilateral ABCD (Fig. 23.8) has $\angle A + \angle C = 180°$ and $\angle B + \angle D = 180°$. Also, $\angle CDX = \angle B$ and $\angle BCY = \angle A$ etc.

Fig. 23.8

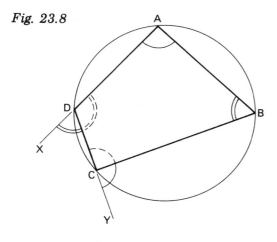

Example 4

ABCD (Fig. 23.9) is a cyclic quadrilateral with $\angle A = 100°$ and $\angle B$ and $\angle D$ equal. Find the angles C and D.

Fig. 23.9

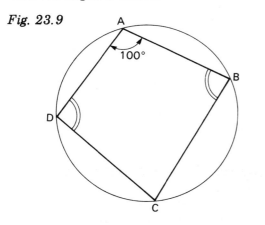

Since ABCD is a cyclic quadrilateral

$$\angle A + \angle C = 180°$$

$$\angle C = 180° - \angle A$$

$$= 180° - 100°$$

$$= 80°$$

Also

$$\angle B + \angle D = 180°$$

$$\angle B = \angle D$$

Therefore $\angle D = 90°$

Exercise 23.1

1. In Fig. 23.10, AC is the diameter of the circle. Find the angles ACB, ADB and BDC.

Fig. 23.10

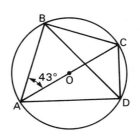

2. In Fig. 23.11, AC is the diameter of the circle. Calculate the length of AC.

Fig. 23.11

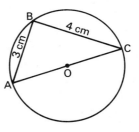

3. A circle centre O (Fig. 23.12) has $\angle AOB = 40°$. Find the size of the angles ABO and OAC.

Fig. 23.12

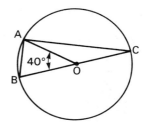

4. In Fig. 23.13, ABC is an isosceles triangle with AB = AC. O is the centre of the circle and BOY is a straight line. Find the angles BAC, BYC, ACY and YBC.

Fig. 23.13

5. In Fig. 23.14, K is the centre of the circle, AC is a diameter and BD is parallel to AC. Find the angles CBA, ACB and CAD.

Fig. 23.14

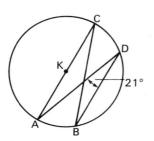

6. In Fig. 23.15, O is the centre of the circle. Find the size of the angles AOB and ADB.

Fig. 23.15

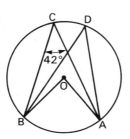

7. In Fig. 23.16, O is the centre of the circle. Find the size of the angles marked *x* and *y*. Also find the size of ∠BDA.

Fig. 23.16

8. In Fig. 23.17, find the angles C and D of the cyclic quadrilateral ABCD. What is the size of ∠BCX?

Fig. 23.17

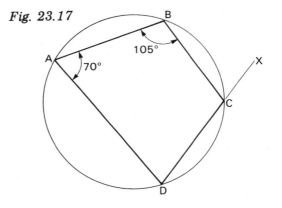

Tangent Properties of a Circle

A **tangent** is a line which touches a circle at one point. This point is called the **point of tangency** (Fig. 23.18).

Fig. 23.18

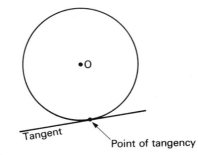

(1) *A tangent to a circle lies at right angles to a radius drawn to the point of tangency.*

Thus in Fig. 23.19, ∠OTS is a right angle.

Fig. 23.19

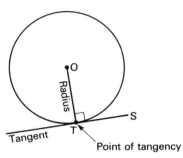

Example 5

In Fig. 23.20, TS is a tangent, T being the point of tangency and O the centre of the circle. If the radius of the circle is 5 cm and OS = 7 cm, find the distance TS.

Fig. 23.20

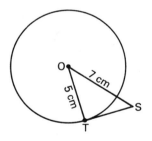

Since OT is a radius drawn to the point of tangency, ∠OTS = 90°.
By Pythagoras' theorem

$$TS^2 = OS^2 - OT^2$$
$$= 7^2 - 5^2$$
$$= 49 - 25$$
$$= 24$$
$$TS = \sqrt{24}$$
$$= 4.90$$

Hence the distance TS is 4.90 cm.

(2) *If, from a point outside a circle, tangents are drawn to the circle, then the two tangents are equal in length.*

Thus in Fig. 23.21, PX = PY. The quadrilateral OXPY is a kite since OX = OY (radii) and it is symmetrical about the line OP. Hence ∠XPY is bisected and ∠KXP = ∠KYP.

Fig. 23.21

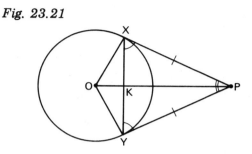

Example 6

In Fig. 23.22, calculate the size of the angles marked *a* and *b*.

Fig. 23.22

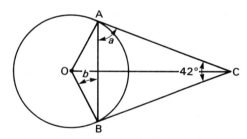

The two tangents AC and BC meet at C. Hence AC = BC and ABC is isosceles. Therefore

$$a = \tfrac{1}{2} \times (180° - 42°)$$
$$= \tfrac{1}{2} \times 138°$$
$$= 69°$$

Since OBC is the angle between a tangent and a radius, it equals 90°. Therefore

$$b = ∠OBC - ∠ABC$$
$$= 90° - 69°$$
$$= 21°$$

(3) *If two circles touch internally or exter-
nally then the line which passes through
their centres also passes through the
point of tangency* (Fig. 23.23).

Fig. 23.23

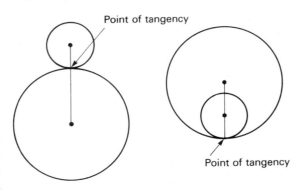

Point of tangency

Point of tangency

Example 7

Three circles are arranged as shown in
Fig. 23.24. Find the distance h.

Fig. 23.24

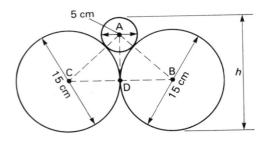

5 cm

15 cm

15 cm

h

Because the circles are tangential to
each other

$$AC = 7.5 + 2.5$$
$$= 10\,\text{cm}$$
$$AB = 7.5 + 2.5$$
$$= 10\,\text{cm}$$
$$BC = 7.5 + 7.5$$
$$= 15\,\text{cm}$$

Therefore triangle ABC is isosceles, and
hence

$$CD = \tfrac{1}{2} \text{ of } 15$$
$$= 7.5\,\text{cm}$$

In $\triangle ACD$, using Pythagoras' theorem
$$AD^2 = AC^2 - CD^2$$
$$= 10^2 - 7.5^2$$
$$= 100 - 56.25$$
$$= 43.75$$
$$AD = \sqrt{43.75}$$
$$= 6.61$$
$$h = 7.5 + 6.61 + 2.5$$
$$= 16.61\,\text{cm}$$

Exercise 23.2

1. In Fig. 23.25, calculate the distance OP
   given that OA is a radius and PA and
   PB are tangents meeting at P.

Fig. 23.25

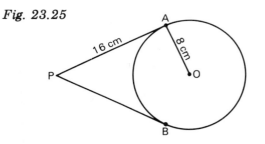

16 cm

8 cm

P

A

O

B

2. In Fig. 23.26, OA is a radius and AP
   and BP are tangents meeting at P. Find
   the size of the angles marked a and b.

Fig. 23.26

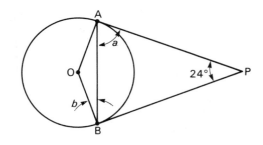

A

a

O

$24°$

P

b

B

3. In Fig. 23.27, OA is the radius of the circle and CA and CB are tangents meeting at C. Find the length of CB.

Fig. 23.27

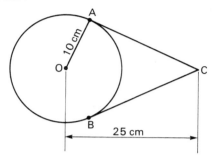

6. Fig. 23.30 shows two circles which are touching. By using Pythagoras' theorem, calculate the distance *x*.

Fig. 23.30

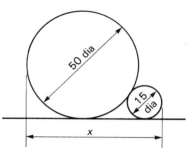

4. In Fig. 23.28, AB and AC are tangents to the circle whose centre is O. AOD is a straight line. Find the size of the angles AOB, OBD and CBD.

Fig. 23.28

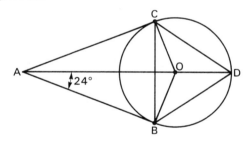

7. In Fig. 23.31, find the distance *h*.

Fig. 23.31

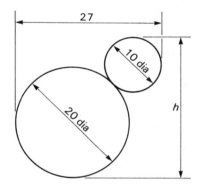

5. In Fig. 23.29, C is the centre of the circle and AP is a tangent to the circle. Calculate the radius of the circle.

Fig. 23.29

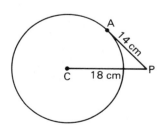

8. Three circles are arranged as shown in Fig. 23.32. Find the distance *h*.

Fig. 23.32

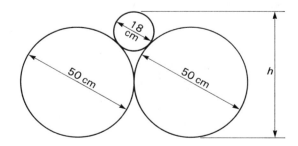

Multi-Choice Questions 23

The circle in Fig. 23.33 has a centre O and COD is a straight line. Use this diagram to answer questions 1, 2 and 3.

Fig. 23.33

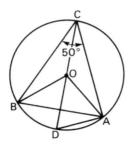

1. If ∠ACB = 50°, then the obtuse angle AOB is

    A 40°   B 50°   C 90°   D 100°

2. ∠DAC is

    A 40°   B 50°   C 90°   D 100°

3. ∠ADC is the same size as the angle

    A ABC   B ACD   C CAB   D CAD

4. In Fig. 23.34, O is the centre of the circle. Calculate the size of the angle marked x.

    A 20°   B 40°   C 50°   D 20°

Fig. 23.34

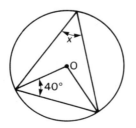

5. In Fig. 23.35, O is the centre of the circle. What is the size of the angle marked x?

    A 40°   B 45°   C 80°   D 160°

Fig. 23.35

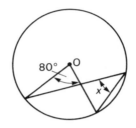

6. In Fig. 23.36, O is the centre of the circle and ∠PSR = 38°. What is the size of ∠OPR?

    A 38°   B 45°   C 52°   D 76°

Fig. 23.36

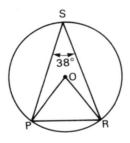

7. In Fig. 23.37, O is the centre of the circle and ROP is a diameter. The angle marked x is

    A 20°   B 40°   C 50°   D 60°

Fig. 23.37

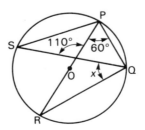

8. In Fig. 23.38, O is the centre of the
   circle and QT and PT are tangents to
   the circle. What is the size of ∠QTP?

   A 90°   **B** 63°   **C** 54°   **D** 27°

Fig. 23.38

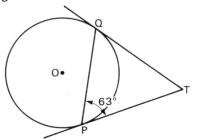

9. In Fig. 23.39, O is the centre of the
   circle and AB = CD. It is necessarily
   true that ABCD is a

   A parallelogram   **B** square
   C rhombus   **D** rectangle

Fig. 23.39

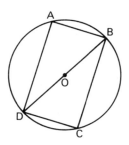

10. In Fig. 23.40, AB is the diameter of
    the circle and angles CBX and ABX
    are equal. Hence the angle CAX is
    equal to

    A 15°   **B** 30°   **C** 45°   **D** 60°

Fig. 23.40

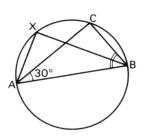

Mental Test 23

1. Find the angle marked *a* in Fig. 23.41.
   O is the centre of the circle.

Fig. 23.41

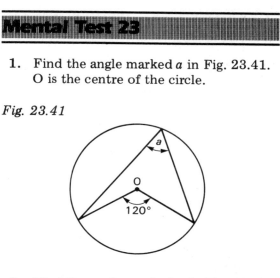

2. Find the angle marked *x* in Fig. 23.42,
   O being the centre of the circle.

Fig. 23.42

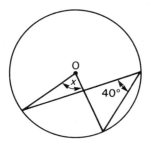

3. Calculate the diameter of the circle
   shown in Fig. 23.43.

Fig. 23.43

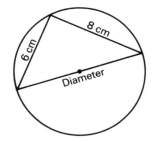

4. What is the size of the angle marked *p* in Fig. 23.44?

Fig. 23.44

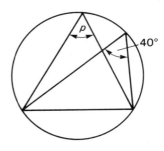

5. In Fig. 23.45, what is the size of ∠ABO, O being the centre of the circle?

Fig. 23.45

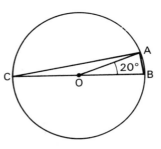

6. In Fig. 23.45, what is the size of ∠CAO?

7. In Fig. 23.46, ABC is an equilateral triangle. What is the size of ∠BYC?

Fig. 23.46

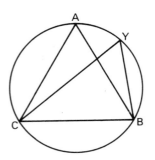

8. In Fig. 23.47, TS is a tangent and OT is a radius. Find the length of OS.

Fig. 23.47

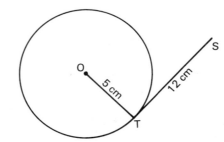

9. In Fig. 23.48, find the angle marked *x*. O is the centre of the circle and AT and BT are tangents.

Fig. 23.48

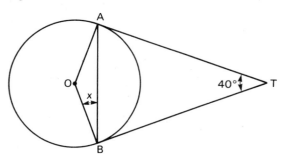

10. In Fig. 23.49, the three circles touch. Find the lengths AB, AC and BC if A, B and C are the centres of the circles.

Fig. 23.49

Geometrical Constructions

(1) *To divide a line AB into two equal parts:*

With A and B as centres and a radius greater than $\frac{1}{2}$AB, draw circular arcs which intersect at X and Y (Fig. 24.1). Join XY. The line XY divides AB into two equal parts and is also perpendicular to AB.

Fig. 24.1

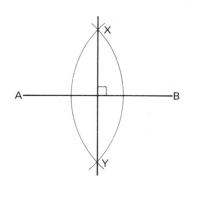

(2) *To draw a perpendicular from a given point A on a straight line:*

With centre A and any radius, draw a circle to cut the straight line at points P and Q (Fig. 24.2). With centres P and Q and a radius greater than AP (or AQ), draw circular arcs to intersect at X and Y. Join XY. This line will pass through A and is perpendicular to the given line.

Fig. 24.2

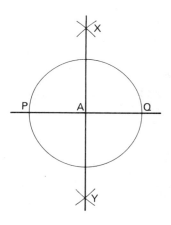

(3) *To draw a perpendicular from a point A at the end of a line:*

From any point O outside the line and with radius OA, draw a circle to cut the line at B (Fig. 24.3). Draw the diameter BC and join AC. AC is perpendicular to the straight line (because the angle in a semicircle is 90°).

Fig. 24.3

(4) *To draw the perpendicular to a line AB from a given point P which is not on the line:*

With P as the centre, draw a circular arc to cut AB at points C and D. With C and D as centres and a radius greater than $\frac{1}{2}$CD, draw circular arcs to intersect at E. Join PE. The line PE is the required perpendicular (Fig. 24.4).

Fig. 24.4

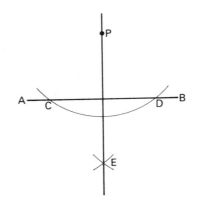

(5) *To construct an angle of 60°:*

Draw a line AB. With A as centre, and any radius, draw a circular arc to cut AB at D. With D as centre and the *same* radius, draw a second arc to cut the first arc at C. Join AC. The angle CAD is then 60° (Fig. 24.5).

Fig. 24.5

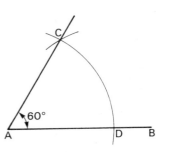

(6) *To bisect a given angle BAC:*

With centre A and any radius, draw an arc to cut AB at D and AC at E. With centres D and E and a radius greater than $\frac{1}{2}$DE, draw arcs to intersect at F. Join AF, then AF bisects ∠BAC (Fig. 24.6). Note that by bisecting an angle of 60°, an angle of 30° is obtained. An angle of 45° is obtained by bisecting a right angle.

Fig. 24.6

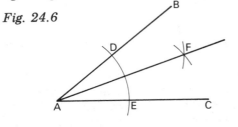

(7) *To construct an angle equal to a given angle BAC:*

With centre A and any radius, draw an arc to cut AB at D and AC at E. Draw a line XY. With centre X and the same radius, draw an arc to cut XY at W. With centre W and radius equal to DE draw an arc to cut the first arc at V. Join VX, then ∠VXW = ∠BAC (Fig. 24.7).

Fig. 24.7

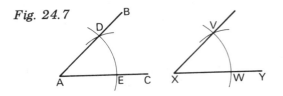

(8) *To draw through point P a line parallel to a given line AB:*

Mark off any two points X and Y on AB. With centre P and radius XY, draw an arc. With centre Y and radius XP, draw a second arc to cut the first arc at Q. Join PQ, then PQ is parallel to AB (Fig. 24.8).

Fig. 24.8

(9) *To divide a straight line AB into a number of equal parts:*

Suppose that AB has to be divided into four equal parts. Draw AC at an angle to AB. Mark off on AC four equal parts AP, PQ, QR, RS of any convenient length. Join SB. Draw RV (using construction 8), QW and PX each parallel to SB. Then AX = XW = WV = VB (Fig. 24.9).

Fig. 24.9

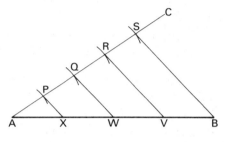

(10) *To draw the circumscribed circle of a given triangle ABC:*

Construct the perpendicular bisectors of the sides AB and AC (using construction 1) so that they intersect at O. With centre O and radius AO, draw a circle which is the required circumscribed circle (Fig. 24.10).

Fig. 24.10

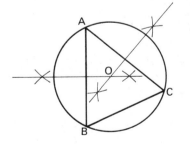

(11) *To draw the inscribed circle of a given triangle ABC:*

Construct the internal bisectors of ∠B and ∠C (using construction 6) to intersect at O. Construct the perpendicular ON from O to

BC (using construction 4). With centre O, and radius ON, draw the inscribed circle of the triangle ABC (Fig. 24.11).

Fig. 24.11

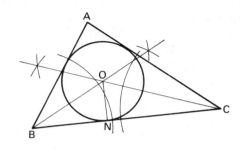

(12) *To construct a common tangent to two given circles:*

The two given circles have centres X and Y and radii x and y, respectively (Fig. 24.12). With centre X, draw a circle whose radius is $(x - y)$. Bisect XY at Z. With radius XZ and centre Z, draw an arc to cut the previously drawn circle at M. Join XM and produce to P at the circumference of the circle. Draw YQ parallel to XP (using construction 8), Q being at the circumference of the circle. Join PQ, which is the required tangent.

Fig. 24.12

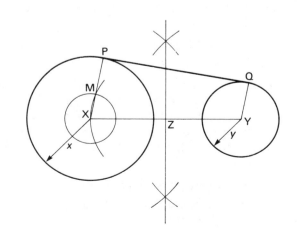

(13) *To construct a pair of tangents from an external point to a given circle:*

It is required to draw a pair of tangents from the point P to the circle, centre O (Fig. 24.13). Join OP. Bisect OP at X. With OX as radius and centre X draw a circular arc to cut the given circle at points A and B. Join PA and PB, which are the required pair of tangents.

Fig. 24.13

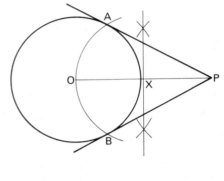

Exercise 24.1

In this exercise, only a rule and compasses should be used:

1. Draw a line PQ 8 cm long and construct its perpendicular bisector.

2. Draw a line XY 9 cm long and mark off a point A such that AX = 4 cm. Through the point A draw a line perpendicular to XY.

3. Draw a line PQ 7 cm long, and through P draw a line perpendicular to PQ.

4. Draw a horizontal line AB which is 9 cm long. Then mark off a point P such that P is 6 cm from A and 5 cm from B. Now construct a line perpendicular to AB which also passes through point P.

5. (a) Construct an angle of 60°.

   (b) By bisecting your 60° angle, construct an angle of 30°.

6. Draw a line AB 8 cm long and construct a line perpendicular to AB which passes through the point A. By bisecting the right angle so formed, construct an angle of 45°.

7. Draw a line 9 cm long and divide it into 5 equal parts without measuring each division.

8. (a) Construct the triangle ABC given that AB = 7 cm, BC = 6 cm and AC = 5 cm.

   (b) Draw the circumscribed circle for triangle ABC and measure its diameter.

9. (a) Construct triangle XYZ given XY = 5 cm, YZ = 7 cm and ∠XYZ = 50°.

   (b) Draw the inscribed circle for △XYZ and state its diameter.

10. (a) Draw a triangle ABC given that AB = 6.3 cm, AC = 7.7 cm and BC = 8.4 cm.

    (b) Draw the circle which passes through A, B and C. Measure the radius of this circle.

11. Construct the quadrilateral ABCD from the following data: AB = 10 cm, BC = 6 cm, AC = 14 cm, ∠DAC = 60° and ∠DCA = 45°. Measure and record the length DB.

12. Draw a line AB 8 cm long. Now construct an equilateral triangle on the line segment AB.

13. Draw two circles whose centres are 9 cm apart and whose diameters are 5 cm and 7 cm. Draw the common tangent to these circles.

14. Draw a circle, centre O, whose radius is 5 cm. Mark off any point P so that OP = 9 cm. Construct a pair of tangents from P to the circle.

Orthographic Projection

In **orthographic projection** a full view of each of the faces of an object is shown in turn. Usually, but not always, three of these views are enough to show all the details of the object.

There are two versions of orthographic projection, first-angle projection and third-angle projection. Only first-angle projection will be described in this book.

First-Angle Projection

Figure 25.1 shows a pictorial drawing of a wooden block. To represent this in first-angle projection, we look at the front of the block in the direction of arrow A. The view we see is called the **elevation**.

Fig. 25.1

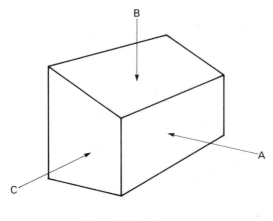

Pictorial drawing

Next we look at the object in the direction of arrow B. This view is called the **plan view.** The plan view is drawn directly underneath the elevation, as shown in Fig. 25.2. Finally we look at the object in the direction of arrow C. The view we see is called the **end elevation** or the **end view.** This view is drawn in line with the elevation and the plan view, as shown in Fig. 25.2.

Fig. 25.2

Elevation
(view on arrow A)

End elevation
or end view
(view on arrow C)

45°

Projection lines

Plan view
(view on arrow B)

Projection Lines

Projection lines are used in orthographic projection to make sure that the views are correctly positioned relative to each other. In Fig. 25.2, the faint projection lines are shown and these lines ensure that the plan view, elevation and end view line up properly.

Hidden Details

Hidden details are shown by means of dotted lines. In drawing the object shown in Fig. 25.3 we cannot see the face AB when looking at the top of the object. Hence this face is represented by dotted lines in the plan view.

Fig. 25.3

Dotted lines
representing
face AB

Two-View Drawings

A two-view drawing is often enough to represent an object fully. Thus the object shown in Fig. 25.4 may be fully represented by an elevation and an end view.

Fig. 25.4

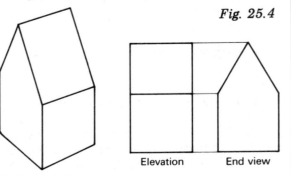

Elevation   End view

Pictorial drawing

Make three-view drawings, in correct projection, of each of the objects shown pictorially below:

1.

2.

3.

4.

5.

6.

7.

8.

9.

10.

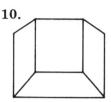

Make two-view drawings of the objects shown below:

11.

12.

13.

14.

15.

Isometric Projection

Isometric projection is a way of representing an object by means of a pictorial drawing. It is often used to make a pictorial drawing from two- and three-view drawings.

Fig. 25.5 shows the **isometric axes** OX, OY and OZ from which an isometric drawing is made.

Fig. 25.5

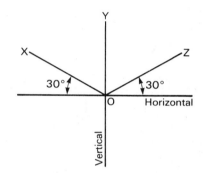

Example 1

Fig. 25.6 shows a two-view drawing of a metal block. Make an isometric drawing of the block.

Fig. 25.6

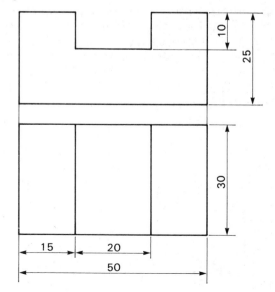

The first step is to draw the isometric axes OX, OY and OZ (Fig. 25.7). Vertical edges of the block are represented by vertical lines in the isometric drawing. For rectangular-type objects, all the lines making up the isometric drawing either lie along the isometric axes or are parallel to them, as shown in Fig. 25.7. Note that when making an isometric drawing, all measurements are made full size.

Fig. 25.7

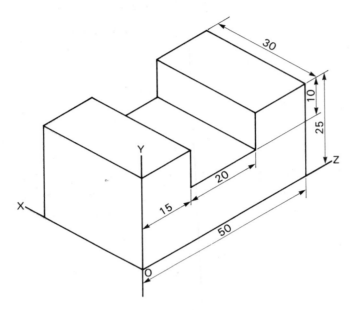

Isometric Representation of Circles

Representing circles in an isometric drawing sometimes causes difficulty. Fig. 25.8 gives the clue necessary to overcome this difficulty. Fig. 25.8(a) shows a true circle inscribed in a square. At the points marked X the circumference of the circle and the sides of the square touch. This must be the same in the isometric view shown in Fig. 25.8(b).

Fig. 25.8

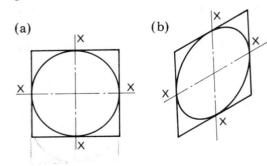

Fig. 25.9 shows the method of making an isometric drawing of a cylinder. The rectangular shape shown by the faint lines is needed to obtain the shape of the circular ends of the cylinder.

Fig. 25.9

Oblique Projection

Oblique projection is a second way of representing objects pictorially. Figure 25.10 shows the axes used. OA is horizontal, OB is vertical whilst OC is drawn at 45° to the horizontal. Measurements made along OA and OB (or parallel to them) are made full-size but measurements made along OC (or parallel to it) are made half-size.
Fig. 25.11 shows a 50 mm cube drawn in oblique projection. The lengths of OA and OB are both made 50 mm but the length OC is 25 mm.

Fig. 25.10

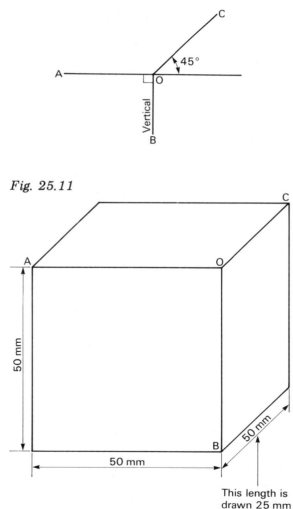

Fig. 25.11

This length is drawn 25 mm

Example 2

Figure 25.12 shows a two-view orthographic drawing of a metal block. Draw an oblique projection of this block.

Fig. 25.12

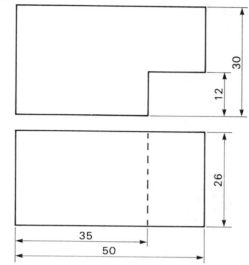

To make the oblique drawing, first draw the three axes OA, OB and OC. The front face of the block in oblique projection is the same as the front elevation of the block. Thus in Fig. 25.13 the front face of the block is drawn using the axes OA and OB. The length OC is made 13 mm (i.e. half of 26 mm).

Fig. 25.13

This length is drawn 13 mm

It is easier to represent circles in oblique projection than in isometric projection. Figure 25.14 is an example. It will be noticed that the circles are drawn as true circles at the front face of the oblique projection.

Fig. 25.14

Exercise 25.2

Make **(a)** isometric drawings, **(b)** oblique drawings of each of the objects shown in orthographic projection below. All dimensions are in millimetres:

1.

2.

3.

4.

5.

6.

7.

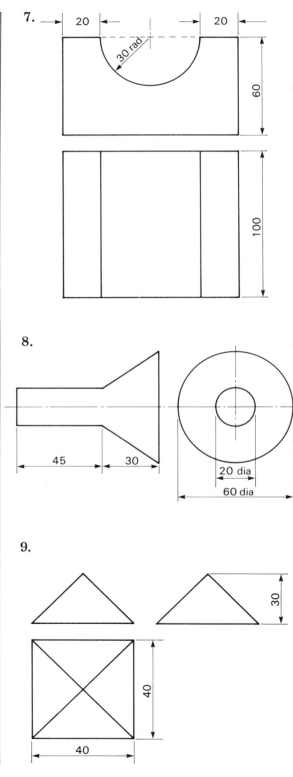

8.

9.

1. Figure 25.15 shows a block of wood. Draw a front elevation, end view and plan view using 1 cm to represent 5 cm.

Fig. 25.15

2. Figure 25.16 illustrates a blackboard cleaner consisting of a piece of felt inserted into a wooden holder. Draw, full-size, the plan, front elevation and end elevation of the cleaner. The three views must be correctly positioned to each other.

Fig. 25.16

3. Make an isometric drawing of the object shown in Fig. 25.17.

Fig. 25.17

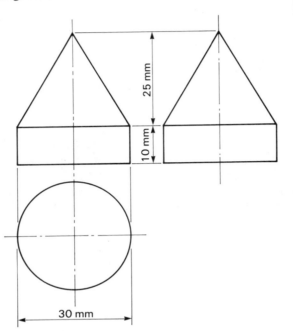

4. A pyramid with a square base stands on a horizontal table (Fig. 25.18). Make a three-view drawing of the pyramid with the views correctly positioned to each other.

Fig. 25.18

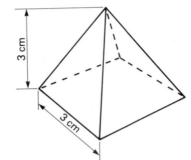

5. Make an isometric drawing of the object shown in Fig. 25.19.

Fig. 25.19

Area of Plane Figures

Types of Plane Figures

A plane figure (or shape) is bounded by lines which are called the **sides** of the figure.

Triangles (Fig. 26.1) have three sides.

Fig. 26.1

Quadrilaterals (Fig. 26.2) have four sides.

Fig. 26.2

Polygons (Fig. 26.3) are plane figures bounded by straight lines. Thus a triangle is a polygon with three sides and a quadrilateral is a polygon with four sides.

Fig. 26.3

A **regular polygon** has all its sides equal in length and all its angles equal in size. An equilateral triangle and a square are examples of regular polygons.

A **pentagon** has five sides, a **hexagon** has six sides and an **octagon** has eight sides. The names of these polygons should be memorised.

Perimeters

The distance round a plane figure is called its **perimeter**.

Example 1

(a) An equilateral triangle has all its sides 5 cm in length. Calculate its perimeter.

$$\text{Perimeter} = (5 + 5 + 5)\,\text{cm}$$
$$= 15\,\text{cm}$$

(b) The longer side of a rectangle is 8 cm in length and its shorter side is 5 cm long. What is the perimeter of the rectangle?

$$\text{Perimeter} = (8 + 5 + 8 + 5)\,\text{cm}$$
$$= 26\,\text{cm}$$

(c) A regular hexagon has each of its sides 7 cm long. Calculate the perimeter of the hexagon. Since a regular hexagon has six equal sides

$$\text{Perimeter} = (6 \times 7)\,\text{cm}$$
$$= 42\,\text{cm}$$

Exercise 26.1

1. Each side of a square is 8 cm long. Find its perimeter.

2. A rhombus has sides 5 cm long. Calculate its perimeter.

3. A regular pentagon has each of its sides 6 cm long. What is its perimeter?

4. Figure 26.4 shows a scalene triangle. Find its perimeter.

Fig. 26.4

5. Fig. 26.5 shows a trapezium. Calculate its perimeter.

Fig. 26.5

6. A rectangle has a long side 8.2 cm in length and a short side 7.5 cm in length. Calculate the perimeter of the rectangle.

7. An isosceles triangle has a base 7.8 cm long and equal sides 8.2 cm long. What is the perimeter of the triangle?

8. A kite has a long side 7.3 cm in length and a short side 3.8 cm long. Find its perimeter.

Units of Area

The amount of surface that a shape possesses is called its **area**.

The area of a shape is measured by counting the number of equal squares contained in the shape.

Example 2

Look at the shapes shown in Fig. 26.6. By counting the number of equal squares contained in each plane figure, find out which figure has the greatest area.

Fig. 26.6

Figure A contains 11 equal squares
Figure B contains 12 equal squares
Figure C contains 13 equal squares
Figure D contains 12 equal squares

Therefore Figure C has the greatest area because it contains the largest number of equal squares.

In practice the squares used for measuring area have sides of 1 metre or 1 centimetre or 1 millimetre. The area inside a square having a side of 1 metre is 1 square metre; 1 square centimetre is the area inside a square with a side of 1 centimetre; 1 square millimetre is the area inside a square having a side of 1 millimetre.

The standard abbreviations for the units of area are:

$$\text{square metre} = \text{m}^2$$

$$\text{square centimetre} = \text{cm}^2$$

$$\text{square millimetre} = \text{mm}^2$$

In the imperial system, areas are measured in square inches (in^2), square feet (ft^2) and square yards (yd^2).

Area of a Rectangle

The rectangle shown in Fig. 26.7 contains $4 \times 2 = 8$ equal squares, each of which has an area of 1 square centimetre. Hence the area of the rectangle is $8\,\text{cm}^2$. What we have done is to multiply the length by the breadth in order to find the area of the rectangle.

Fig. 26.7

This applies to any rectangle, hence

Area of rectangle = Length × Breadth

If we let A = area of the rectangle, l = length of the rectangle and b = breadth of the rectangle then

$$A = lb$$

When using this formula the units of l and b must be the same. That is, they both must be stated in metres, or centimetres, or millimetres, etc.

Example 3

A rectangular carpet measures 6 m by 5 m. Calculate the area of the carpet.

Here $l = 6\,\text{m}$ and $b = 5\,\text{m}$.

$$A = lb$$
$$= (6 \times 5)\,\text{m}^2$$
$$= 30\,\text{m}^2$$

Example 4

A room 9 m long and 7 m wide is to be carpeted so as to leave a surround 50 cm wide as shown in Fig. 26.8. Find the area of the surround.

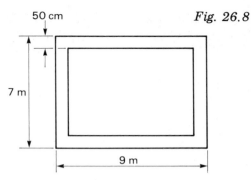

Fig. 26.8

One way of solving this problem is to calculate the area of the room, and from it subtract the area of the carpet.

Area of room $= (7 \times 9)\,\text{m}^2$
$= 63\,\text{m}^2$
Area of carpet $= (6 \times 8)\,\text{m}^2$
$= 48\,\text{m}^2$
Area of surround $= (63 - 48)\,\text{m}^2$
$= 15\,\text{m}^2$

Many shapes can be split up into rectangles. Then by calculating the area of each rectangle the area of the shape can be found by simple addition or subtraction.

Example 5

Find the area of the steel girder shown in Fig. 26.9.

Fig. 26.9

All dimensions in millimetres

The shape can be divided up into three rectangles as shown in the diagram.

Area of shape $=$ Area of rectangle 1

$\qquad$ $+$ Area of rectangle 2

$\qquad$ $+$ Area of rectangle 3

$\qquad$ $= (50 \times 10)\,\text{mm}^2$

$\qquad$ $+ (90 \times 8)\,\text{mm}^2$

$\qquad$ $+ (150 \times 20)\,\text{mm}^2$

$\qquad$ $= 500\,\text{mm}^2 + 720\,\text{mm}^2$

$\qquad$ $+ 3000\,\text{mm}^2$

$\qquad$ $= 4220\,\text{mm}^2$

Example 6

A piece of wood is 85 cm long and 30 cm wide. What is its area in square metres?

> In problems of this kind it is best to express each dimension in metres before attempting to find the area. Thus 85 cm $=$ 0.85 m, and 30 cm $=$ 0.3 m.
>
> Area of wood $= (0.85 \times 0.3)\,\text{m}^2$
>
> $\qquad$ $= 0.255\,\text{m}^2$

Sometimes we are given the area of a rectangle and one dimension (such as the width) and we have to find the other dimension (such as the length). The formula $A = lb$ can be transposed to give

$$l = \frac{A}{b} \qquad \text{or} \qquad b = \frac{A}{l}$$

Note carefully that, when using these transposed formulae, the units for each term in the formula must be of the same kind. Thus if the area is given in square metres, the given dimension (length or width) must be in metres and the remaining dimension will then also be in metres.

Example 7

The area of a rectangle is 63 in². If its width is 7 in, find the length of the rectangle.

We are given that $A = 63\,\text{in}^2$ and $b = 7\,\text{in}$.

$$l = \frac{A}{b}$$

$$= \frac{63}{7}\,\text{in}$$

$$= 9\,\text{in}$$

Hence the length of the rectangle is 9 in.

Area of a Square

Since a square is a rectangle with all sides equal in length

$$\text{Area of square} = \text{Side} \times \text{Side}$$

$$= \text{Side}^2$$

If $A =$ area of the square and $a =$ length of side

$$A = a^2$$

Example 8

(a) A square has sides 8 cm long. Calculate its area.

We are given that $a = 8$ cm, hence

$$A = a^2$$

$$= 8\,\text{cm}^2$$

$$= 64\,\text{cm}^2$$

(b) A square plate has an area of 25 ft². Calculate the length of its sides.

We are given $A = 25\,\text{ft}^2$ and we have to find a.

$$a^2 = A$$

$$a = \sqrt{A}$$

$$= \sqrt{25}$$

$$= 5$$

Hence the plate has sides which are 5 ft long.

Exercise 26.2

Find the areas of the following rectangles:

1. Length 7 cm, width 5 cm.

2. Length 20 mm, width 11 mm.

3. Length 35 ft, width 8 ft.

4. Length 8.3 cm, width 5.2 cm.

5. Length 91 mm, width 20 mm.

6. A piece of wood is 3.7 m long and 30 cm wide. Find its area in square metres.

7. A rectangular metal plate is 120 cm long and 80 cm wide. Calculate its area in square metres.

8. A rectangular floor 5.8 ft long and 4.9 ft wide is to be covered with vinyl. What area of vinyl is needed?

9. What is the total area of the walls of a room 6 m long, 5 m wide and 2 m high?

10. A rectangular lawn is 32 yd long and 23 yd wide. A path 1.5 yd wide is made around the lawn. What is the area of the path?

11. A room 8.3 m long and 6.3 m wide is to be carpeted to leave a surround 60 cm wide round the carpet. Work out **(a)** the area of the room, **(b)** the area of the carpet, **(c)** the area of the surround.

12. Find the areas of the steel sections shown in Fig. 26.10.

Fig. 26.10

(a)

(g)

(h)

13. The area of a rectangle is $72\,\text{ft}^2$. If its length is $12\,\text{ft}$, what is its width?

14. The floor area of a room is $84\,\text{m}^2$. If its width is $7\,\text{m}$, calculate its length.

15. A square metal plate has an area of $36\,\text{cm}^2$. What is the length of its sides?

16. A tile is square and its sides are $8\,\text{cm}$ long. Calculate its area.

17. A room is $5.4\,\text{m}$ long and $4.2\,\text{m}$ wide. It takes 1575 square tiles to cover the floor. Calculate the area of each tile and then find the length of its sides.

Area of a Parallelogram

A parallelogram is in effect a rectangle pushed out of square as shown in Fig. 26.11, where the equivalent rectangle is shown in dotted outline. Hence

  Area of parallelogram = Length of base
                 × Vertical height

If A is the area, b the length of the base and h the vertical height (often called the altitude) then

$$A = bh$$

Fig. 26.11

Example 9

(a) Find the area of a parallelogram whose base is $15\,\text{cm}$ long and whose altitude is $8\,\text{cm}$.

  We are given $b = 15$ and $h = 8$ and we have to find A.

$$A = bh$$
$$= 15 \times 8$$
$$= 120$$

  Hence the area of the parallelogram is $120\,\text{cm}^2$.

(b) A parallelogram has an area of $36\,\text{in}^2$. If its base is $9\,\text{in}$ long, find its altitude.

  Since $A = bh$

$$h = \frac{A}{b}$$

  We are given $A = 36$ and $b = 9$ and we have to find h.

$$h = \frac{36}{9}$$
$$= 4$$

  Therefore the altitude is $4\,\text{in}$.

Exercise 26.3

Find the areas of the following parallelograms:

1. Base $7\,\text{cm}$, altitude $5\,\text{cm}$.

2. Base $15\,\text{mm}$, altitude $12\,\text{mm}$.

3. Base $5\,\text{ft}$, altitude $3\,\text{ft}$.

4. The area of a parallelogram is 64 m². If its base is 16 m long, find its altitude.

5. The area of a parallelogram is 56 in². If its altitude is 7 in, find its base.

6. Figure 26.12 shows a steel section. Calculate its area in square centimetres.

Fig. 26.12

7. Determine the length of the side of a square whose area is equal to that of a parallelogram with a base of 6 m and an altitude of 3 m.

8. The perimeter of a square is 36 cm.

   (a) Calculate the area of the square.

   (b) Find the altitude of a parallelogram with the same area as the square, if its base is 6 cm long.

Area of a Triangle

The diagonal of a parallelogram (Fig. 26.13) splits the parallelogram into two equal triangles. Hence

Area of triangle $= \frac{1}{2} \times$ Base $\times$ Altitude

As a formula the statement becomes

$$A = \tfrac{1}{2}bh$$

where A is the area of the triangle, b the length of the base and h the altitude (or vertical height); see Fig. 26.14.

Fig. 26.13

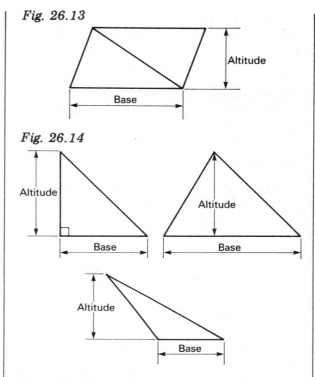

Fig. 26.14

Example 10

(a) A triangle has a base 8 cm long and an altitude of 5 cm. Calculate its area.

$$A = \tfrac{1}{2}bh$$
$$= (\tfrac{1}{2} \times 8 \times 5)\,\text{cm}^2$$
$$= 20\,\text{cm}^2$$

(b) A triangle has an area of 30 in² and a base 6 in long. What is its altitude?

Since $A = \tfrac{1}{2}bh$,

$$h = \frac{2A}{b}$$

We are given $A = 30$ and $b = 6$ and we have to find h.

$$h = \frac{2A}{b}$$
$$= \frac{2 \times 30}{6}$$
$$= 10$$

Hence the altitude of the triangle is 10 in.

When we are given the length of the three sides of a triangle, the area of the triangle may be found by using the formula

$$A = \sqrt{s(s-a)(s-b)(s-c)}$$

where s stands for half the perimeter of the triangle and a, b and c are the lengths of the sides of the triangle.

Example 11

The lengths of the sides of a triangle are 7 cm, 8 cm and 13 cm. Calculate the area of the triangle.

$$s = \frac{a+b+c}{2}$$

$$= \frac{7+8+13}{2}$$

$$= \frac{28}{2}$$

$$= 14$$

$$A = \sqrt{14 \times (14-7) \times (14-8) \times (14-13)}$$

$$= \sqrt{14 \times 7 \times 6 \times 1}$$

$$= \sqrt{588}$$

$$= 24.2$$

Hence the area of the triangle is 24.2 cm².

Exercise 26.4

Calculate the area of the following triangles:

1. Base 6 cm, altitude 3 cm.
2. Base 8 cm, altitude 7 cm.
3. Base 11 in, altitude 16 in.
4. Base 3.2 m, vertical height 4.3 m.
5. Base $4\frac{1}{2}$ cm, altitude $1\frac{1}{3}$ cm.
6. Fig. 26.15 shows several right-angled triangles. Work out the area of each.
7. A triangle has an area of 60 yd² and a base 12 yd long. Calculate its altitude.

Fig. 26.15

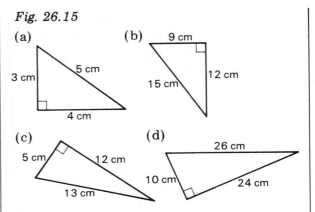

8. A triangle has an area of 80 mm² and a vertical height of 5 mm. Find the base of the triangle.

9. A triangle has sides which are 4 cm, 5 cm and 7 cm long. Calculate its area.

10. The sides of a triangle are 3 ft, 5 ft and 6 ft long. Find the area of the triangle.

11. Find the area of the right-angled triangle shown in Fig. 26.16.

Fig. 26.16

12. Calculate the area of the quadrilateral shown in Fig. 26.17.

Fig. 26.17

13. In Fig. 26.18, ABCD is a rhombus. The length of diagonal BD is 10 cm and the length of diagonal AC is 6 cm. Calculate the area of ABCD.

Fig. 26.18

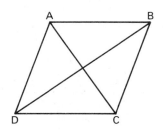

14. A triangle has sides 37 mm, 52 mm and 63 mm long. Calculate the area of the triangle in square centimetres.

Area of a Trapezium

A trapezium is a quadrilateral with one pair of parallel sides (Fig. 26.19).

Area of trapezium = $\frac{1}{2}$ × Sum of parallel sides

× Distance between parallel sides

As a formula this becomes

$$A = \tfrac{1}{2}(a + b)h$$

where A is the area of the trapezium, a and b are the lengths of the parallel sides and h is the distance between the parallel sides.

Fig. 26.19

Example 12

(a) The parallel sides of a trapezium are 12 cm and 16 cm long. The distance between them is 9 cm. Calculate the area of the trapezium.

We are given $a = 12$, $b = 16$ and $h = 9$. We have to find A.

$$\begin{aligned} A &= \tfrac{1}{2}(a + b)h \\ &= \tfrac{1}{2} \times (12 + 16) \times 9 \\ &= \tfrac{1}{2} \times 28 \times 9 \\ &= 126 \end{aligned}$$

Hence the area of the trapezium is 126 cm^2.

(b) The area of a trapezium is 220 cm^2. The parallel sides are 26 cm and 14 cm long. Find the distance between the parallel sides.

We are given $A = 220$, $a = 26$ and $b = 14$. We have to find h.

$$\begin{aligned} 220 &= \tfrac{1}{2} \times (26 + 14) \times h \\ 220 &= \tfrac{1}{2} \times 40 \times h \\ 220 &= 20 \times h \\ h &= \frac{220}{20} \\ &= 11 \end{aligned}$$

Hence the distance between the parallel sides is 11 cm.

Exercise 26.5

1. A trapezium has parallel sides 12 in and 8 in long. If the distance between the parallel sides is 6 in, calculate the area of the trapezium.

2. Find the area of a trapezium whose parallel sides are 7 cm and 9 cm long if the distance between them is 5 cm.

3. The area of a trapezium is 300 cm^2. Its parallel sides are 35 cm and 65 cm long. What is the distance between the parallel sides?

4. Find the area of a trapezium whose parallel sides are 25 in and 35 in, the distance between them being 20 in.

5. Find the area of the trapezium shown in Fig. 26.20.

Fig. 26.20

Similar Shapes

Two shapes are similar if they are equiangular and the ratio of corresponding sides is the same. The parallelograms shown in Fig. 26.21 are similar because they are equiangular and $AD/WZ = CD/YZ = \frac{1}{3}$.

Fig. 26.21

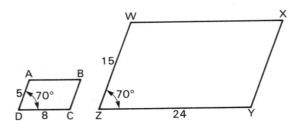

The ratio of the areas of similar shapes is equal to the ratio of the squares on corresponding sides. For the parallelograms of Fig. 26.21, AD and WZ are corresponding sides because they lie opposite to equal angles. Therefore

$$\frac{\text{Area of WXYZ}}{\text{Area of ABCD}} = \frac{\text{WZ}^2}{\text{AD}^2}$$

$$= \frac{15^2}{5^2}$$

$$= \frac{9}{1}$$

Example 13

(a) Find the area of triangle XYZ, given that the area of triangle ABC is 12 cm^2 (Fig. 26.22).

Fig. 26.22

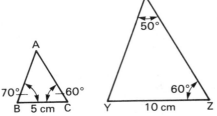

In $\triangle$XYZ, $Y = 70°$ and in $\triangle$ABC, $A = 50°$. Hence triangles XYZ and ABC are equiangular, i.e. similar. YZ and BC are corresponding sides because they lie opposite to the equal angles A and X.

Hence $$\frac{\text{Area of XYZ}}{\text{Area of ABC}} = \frac{\text{YZ}^2}{\text{BC}^2}$$

$$\frac{\text{Area of XYZ}}{12} = \frac{10^2}{5^2}$$

$$= \frac{4}{1}$$

$$\text{Area of XYZ} = 4 \times 12$$

$$= 48$$

Hence the area of triangle XYZ is 48 cm^2.

(b) Two similar polygons have areas of 8 cm^2 and 18 cm^2. The shortest side of the larger polygon is 4 cm long. Calculate the length of the shortest side of the smaller polygon.

Let the length of the shortest side of the smaller polygon be x cm. Then

$$\frac{8}{18} = \frac{x^2}{4^2}$$

$$x^2 = \frac{8 \times 4^2}{18}$$

$$= 7.11$$

$$x = \sqrt{7.11}$$

$$= 2.67$$

Hence the shortest side of the smaller polygon is 2.67 cm long.

Exercise 26.6

1. In Fig. 26.23, triangles ABC and EFG are similar. If the area of triangle ABC is 8 m², calculate the area of triangle EFG.

Fig. 26.23

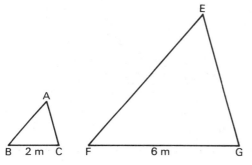

2. In Fig. 26.24, the area of triangle XYZ is 9 m². What is the area of triangle ABC?

Fig. 26.24

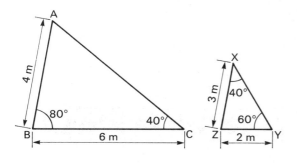

3. A rectangle ABCD has a length of 30 in and an area of 600 in². A similar rectangle WXYZ has a length of 20 in. What is the area of WXYZ?

4. A trapezium ABCD (Fig. 26.25) has an area of 12 300 mm². Calculate the area of the trapezium PQRS.

Fig. 26.25

5. A rhombus ABCD (Fig. 26.26) has an area of 18 m². The rhombus EFGH has an area of 72 m². Find the length of EF.

Fig. 26.26

6. Two similar polygons have areas of 160 in² and 360 in². The shortest side of the larger polygon is 18 in long. Calculate the length of the shortest side of the smaller polygon.

Miscellaneous Exercise 26

Section B

1. A rectangular piece of cardboard 10 cm by 6 cm has equal squares of side 2 cm cut from its corners. The final shape is shown in Fig. 26.27. Calculate (a) the perimeter, (b) the area of the shape.

Fig. 26.27

2. A square has a perimeter of 20 in. What is its area?

3. Fig. 26.28 shows a trapezium ABCD. Calculate its area.

Fig. 26.28

4. Refer to Fig. 26.29. Calculate
   (a) the length of AB
   (b) the area of the triangle AED

Fig. 26.29

5. Find the area of the parallelogram shown in Fig. 26.30.

Fig. 26.30

6. A rectangular table-cloth is 150 cm by 200 cm. What is its area in square metres?

7. In Fig. 26.31, ABCD is a rectangle 12 cm by 8 cm. The shaded portions are cut off. Find the area of DLMNPC.

Fig. 26.31

8. A square has an area of 81 cm². Calculate its perimeter.

9. Express the area of a rectangle 10 in long and 6 in wide as a fraction of the area of a square with sides 8 in long.

10. Find the area of the kite shown in Fig. 26.32.

Fig. 26.32

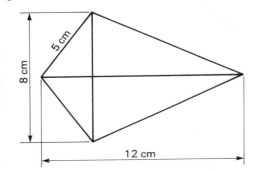

Multi-Choice Questions 26

1. Calculate the perimeter of a square whose area is 16 cm².

   A 4 cm   **B** 8 cm
   C 12 cm   **D** 16 cm

2. Find the cost of a rectangular sheet of glass 0.75 m long and 1.08 m wide when glass costs £5 per square metre.

   A £4.05   **B** £6.75
   C £40.50   **D** £67.50

3. A rectangular lawn is 8 yd long and 4.5 yd wide. What is its area in square yards?

   A 12.5   B 25   C 32   D 36

4. How many square centimetres are equal to one square metre?

   A 10   B 100
   C 1000   D 10 000

5. The perimeter of a square is 36 cm. Its area in square centimetres is

   A 6   B 9   C 36   D 81

6. A parallelogram has a base 10 cm long and a vertical height of 4 cm. Its area in square centimetres is

   A 14   B 20   C 28   D 40

7. A trapezium has parallel sides 5 cm and 9 cm long. The distance between them is 4 cm. What is the area, in square centimetres, of the trapezium?

   A 28   B 36   C 56   D 81

8. Triangle ABC (Fig. 26.33) is right-angled at B. What is its area in square centimetres?

   A 8.5   B 30   C 32.5   D 78

Fig. 26.33

9. Triangle XYZ (Fig. 26.34) has XY and YZ equal in length. The triangle is right-angled at Y. If the area of the triangle is 32 ft^2, what is the length of XY?

   A 4 ft   B 8 ft
   C 16 ft   D 32 ft

Fig. 26.34

10. Find the area of the angle section shown in Fig. 26.35.

    A 16 cm^2 B 20 cm^2
    C 24 cm^2 D 48 cm^2

Fig. 26.35

11. An equilateral triangle ABC has sides 2 cm long. A second equilateral triangle PQR has sides 3 cm long. The ratio area of ABC : area of PQR is

    A $\dfrac{2}{3}$ B $\dfrac{2^2}{3^2}$ C $\dfrac{2^3}{4^3}$ D $\dfrac{6}{9}$

Mental Test 26

1. What is the area of a rectangle whose length is 7 cm and whose width is 3 cm?

2. A square has an area of 25 in^2. What is the length of its sides?

3. A parallelogram has a base 8 cm long and a vertical height of 5 cm. Find its area.

4. A parallelogram has an area of 80 cm^2. Its base is 10 cm long. What is its vertical height?

5. A triangle has a base 8 ft long and an altitude of 4 ft. What is its area?

6. A triangle has an area of 36 cm^2. Its altitude is 9 cm. What is the length of its base?

7. A trapezium has parallel sides 12 cm and 8 cm long. The distance between these sides is 5 cm. What is the area of the trapezium?

8. An isosceles right-angled triangle has equal sides 4 cm long. What is its area?

Mensuration of the Circle

The names of the main parts of a circle are shown in Fig. 27.1.

Fig. 27.1

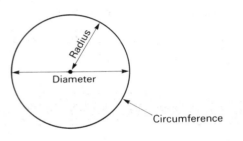

Circumference of a Circle

In a circle the value of

$$\frac{\text{Circumference}}{\text{Diameter}} = 3.141\,59\ldots$$

It has been proved that the exact value cannot be worked out, but for most problems a value of 3.142 is sufficiently accurate when working in decimals. When working in fractions a value of $\frac{22}{7}$ may be used. When making mental calculations, a value of $\frac{25}{8}$ is acceptable.

The value circumference/diameter is so important that it has been given the special symbol π (the Greek letter pi).
We take the value of π as being 3.142 or $\frac{22}{7}$.

Since Circumference/Diameter $= \pi$
it follows that

  Circumference $= \pi \times$ Diameter

or   Circumference $= 2 \times \pi \times$ Radius

or, as a formula

$$C = 2\pi r$$
$$= \pi d$$

where C is the circumference, r the radius and d the diameter.

Example 1

(a) The diameter of a circle is 30 cm. Calculate its circumference.

$$C = \pi d$$
$$= 3.142 \times 30$$
$$= 94.3$$

Hence the circumference is 94.3 cm.

(b) The radius of a circle is 14 in. Calculate its circumference.

$$C = 2\pi r$$
$$= \frac{2}{1} \times \frac{22}{\cancel{7}_1} \times \frac{\cancel{14}^2}{1}$$
$$= 2 \times 22 \times 2$$
$$= 88$$

Hence the circumference is 88 in.

The formulae $C = \pi d$ and $C = 2\pi r$ can be transposed to give

$$d = \frac{C}{\pi} \qquad \text{and} \qquad r = \frac{C}{2\pi}$$

Example 2

Find the radius of a circle whose circumference is 93.8 cm.

$$r = \frac{C}{2\pi}$$

$$= \frac{93.8}{2 \times 3.142}$$

$$= 14.9$$

This calculation is best done using a calculator:

Input	Display
93.8	93.8
÷	93.8
2	46.9
÷	46.9
3.142	3.142
=	14.9 ...

Hence the radius is 14.9 cm.

Exercise 27.1

Taking $\pi = \frac{22}{7}$, find the circumference of each of the following circles:

1. Radius 21 cm.
2. Radius 350 cm.
3. Radius 14 ft.
4. Diameter 28 cm.
5. Diameter 560 ft.

Taking $\pi = 3.142$, find the circumference of each of the following circles:

6. Radius 43 cm.
7. Radius 3.16 m.
8. Diameter 85 in.
9. Diameter 42.7 mm.
10. Diameter 86.15 ft.
11. A circular flower bed has a diameter of 64 m. What is its circumference?
12. A circular pond has a circumference of 12.62 m. Calculate its radius.
13. Find the diameter of a circle whose circumference is 110 cm.
14. A garden roller has a diameter of 80 cm. If it is 1.5 m wide, what area of lawn does it roll in 70 revolutions?
15. A wheel has a diameter of 220 in. What is the distance around its rim?

Area of a Circle

It can be shown that

$$\text{Area of a circle} = \pi \times \text{Radius}^2$$

or

$$A = \pi r^2$$

Example 3

(a) Find the area of a circle whose radius is 3 cm.

$$A = (\pi \times 3^2)\,\text{cm}^2$$

$$= (3.142 \times 9)\,\text{cm}^2$$

$$= 28.3\,\text{cm}^2$$

(b) Calculate the area of a circle whose diameter is 28 cm.

$$A = (\pi \times 14^2)\,\text{cm}$$

$$= \frac{22}{7_1} \times \frac{14^2}{1} \times \frac{14}{1}\,\text{cm}^2$$

$$= (22 \times 2 \times 14)\,\text{cm}^2$$

$$= 616\,\text{cm}^2$$

(c) The area of a circle is 88 in^2. Find its radius.

$$A = \pi r^2$$

$$88 = \frac{22}{7}r^2$$

$$r^2 = \frac{88^4}{1} \times \frac{7}{22_1}$$

$$= 4 \times 7$$

$$= 28$$

$$r = \sqrt{28}$$

$$= 5.29$$

The circle has a radius of 5.3 in.

Area of an Annulus

An annulus consists of two concentric circles, as shown in Fig. 27.2.

Area of annulus = Area of outer circle
                  − Area of inner circle

$$= \pi R^2 - \pi r^2$$

$$= \pi(R^2 - r^2)$$

Fig. 27.2

Example 4

Find the area of an annulus which has an outer diameter of 20 cm and an inner diameter of 12 cm.

We are given that $R = \frac{20}{2} = 10$ and $r = \frac{12}{2} = 6$.

$$\begin{aligned}
A &= \pi(R^2 - r^2) \\
&= 3.142 \times (10^2 - 6^2) \\
&= 3.142 \times (100 - 36) \\
&= 3.142 \times 64 \\
&= 201
\end{aligned}$$

Hence the area of the annulus is 201 cm².

Exercise 27.2

Taking $\pi = \frac{22}{7}$, find the area of each of the following circles:

1. Radius 7 cm.   2. Radius 140 mm.
3. Radius 28 in.   4. Diameter 42 m.
5. Diameter 35 ft.

Taking $\pi = 3.142$, find the area of each of the following circles:

6. Radius 4 cm.   7. Radius 70 in.
8. Diameter 14 m.   9. Diameter 50 ft.
10. Diameter 130 mm.

11. An annulus has an outer diameter of 12 in and an inner diameter of 18 in. Calculate its area.

12. A copper pipe has a bore of 32 mm and an outside diameter of 42 mm. Calculate the cross-sectional area of the pipe.

13. A pond with a diameter of 36 m has a path 1 m wide around its circumference. Calculate the area of the path.

14. A hollow shaft has an outside diameter of 3.25 in and an inside diameter of 2.50 in. Calculate the cross-sectional area of the shaft.

15. A hollow shaft has a cross-sectional area of 8.68 cm² and a bore of 0.75 cm diameter. Find the outside diameter of the shaft.

Sector of a Circle

The area and the length of arc of a sector of a circle depend upon the angle that the sector subtends at the centre of the circle (Fig. 27.3).

Fig. 27.3

Referring to Fig. 27.3, where θ is measured in degrees:

Length of minor
arc of the sector $= 2 \times \pi \times \text{Radius} \times \dfrac{\theta}{360}$

$$= \pi \times \text{Diameter} \times \dfrac{\theta}{360}$$

or, as a formula

$$l = 2\pi r \times \dfrac{\theta}{360}$$

or

$$l = \pi d \times \dfrac{\theta}{360}$$

Area of sector $= \pi \times \text{Radius}^2 \times \dfrac{\theta}{360}$

or

$$A = \pi r^2 \times \dfrac{\theta}{360}$$

Example 5

Find the length of the minor arc and the area of the sector of a circle which subtends an angle of $120°$ at the centre if its radius is 8 cm.

The sector is shown in Fig. 27.4.

Fig. 27.4

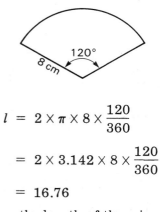

$$l = 2 \times \pi \times 8 \times \dfrac{120}{360}$$

$$= 2 \times 3.142 \times 8 \times \dfrac{120}{360}$$

$$= 16.76$$

Hence the length of the minor arc is 16.76 cm.

$$A = 3.142 \times 8^2 \times \dfrac{120}{360}$$

$$= 67.0$$

Therefore the area is 67.0 cm^2.

Find the length of arc and the area for each of the following sectors of a circle:

1. Radius 4 cm; sector angle $45°$.

2. Radius 10 cm; sector angle $90°$.

3. Radius 3 in; sector angle $120°$.

4. If an arc 7 cm long subtends an angle of $45°$ at the centre of the circle, what is the radius of the circle?

5. In marking out the plan of part of a building, a line 8 m long is pegged down at one end. Then with the line held horizontally and taut, the free end is swung through an angle of $57°$. Calculate the distance moved by the free end of the line and find the area swept out.

6. A bay window is semicircular in plan. It is to be covered with lead. If the radius of the bay is 2.4 yd, calculate the area of lead required.

7. Calculate the area of the cross-section shown in Fig. 27.5.

Fig. 27.5

Areas of Composite Shapes

Many shapes are composed of straight lines and arcs of circles. The areas of such shapes are found by splitting up the shape into figures such as rectangles, triangles and sectors of circles.

Example 6

Find the cross-sectional area of the steel section shown in Fig. 27.6.

Fig. 27.6

The shape consists of a semicircle and a rectangle.

$$\text{Area of semicircle} = \tfrac{1}{2} \times \pi \times \text{Radius}^2$$
$$= \tfrac{1}{2} \times 3.142 \times 2^2$$
$$= 6.28$$

$$\text{Area of rectangle} = \text{Length} \times \text{Width}$$
$$= 4 \times 3$$
$$= 12$$

$$\text{Total area of cross-section} = 6.28\,\text{cm}^2$$
$$+ 12\,\text{cm}^2$$
$$= 18.28\,\text{cm}^2$$

Example 7

Find the shaded area in Fig. 27.7.

Fig. 27.7

The shaded area is calculated by subtracting the area of the quarter-circle from the area of the square.

$$\text{Area of square} = 7^2$$
$$= 49$$

$$\text{Area of quarter-circle} = \tfrac{1}{4} \times \pi \times \text{Radius}^2$$
$$= \tfrac{1}{4} \times 3.142 \times 7^2$$
$$= 38.5$$

$$\text{Shaded area} = 49\,\text{cm}^2 - 38.5\,\text{cm}^2$$
$$= 10.5\,\text{cm}^2$$

Exercise 27.4

Find the areas of the shapes drawn in Fig. 27.8. They are all composed of squares or rectangles and parts of circles.

Fig. 27.8

1.

2.

3.

Find the areas of the shaded portions of the shapes shown in Fig. 27.9.

Fig. 27.9

4.

5.

6.

7.

8.

Section A

1. A wheel has a radius of 35 cm. What is the distance round its rim? (Take $\pi = \frac{22}{7}$)

2. A circular flower bed measures 7 yd across its centre. How far is it round its edge? (Take $\pi = \frac{22}{7}$)

3. A pipe has an outside diameter of 14 cm and an internal diameter of 9 cm. What is the area of its cross-section? (Take $\pi = 3.14$)

4. Find the area of a circle whose radius is 21 in. (Take $\pi = \frac{22}{7}$)

5. A steel bar has a diameter of 6.5 cm. Find the area of its cross-section. Take $\pi = 3.14$ and round off the answer to the nearest tenth.

Section B

1. Calculate the area of a circle with a diameter of 6 cm. (Take $\pi = 3.14$)

2. The circumference of a circle is 88 in. Taking $\pi = \frac{22}{7}$, find the radius of the circle.

3. (a) Calculate the area of the sector shown in Fig. 27.10, stating your answer correct to 2 significant figures. (Take $\pi = 3.14$)

   (b) What is the length of arc, correct to 2 significant figures?

Fig. 27.10

4. Calculate the area of a semicircle which has a radius of 14 cm. Take $\pi = \frac{22}{7}$.

5. A circle has a circumference of 176 in. Taking $\pi = \frac{22}{7}$, find its radius.

6. Calculate the shaded area of the shape shown in Fig. 27.11.

Fig. 27.11

7. The circumference of a circle is 44 cm. Find the length of an arc of this circle formed by an angle of 45° at the centre.

8. Calculate the radius of a circle which has an area of 616 cm².

9. A pipe has an internal diameter of 8 in and an external diameter of 12 in. Calculate the cross-sectional area of the pipe.

10. Fig. 27.12 shows a metal framework made of thin wire. ABCD is a square and A, B, C and D are the centres of four circles which have equal radii. Taking $\pi = 3.142$ and given that the radius of each circle is 3 cm, calculate

    (a) the circumference of one of the circles

    (b) the area of one of the circles

    (c) the total length of the straight pieces of metal in the framework

    (d) the total length of metal used

Fig. 27.12

Multi-Choice Questions 27

1. Calculate the length of the perimeter of the quadrant of the circle shown in Fig. 27.13.

   A $5\frac{1}{2}$ cm   B 9 cm

   C $12\frac{1}{2}$ cm   D 22 cm

Fig. 27.13

2. Fig 27.14 represents a circular running track whose circumference is 450 metres. If O is the centre of the track and a runner travels 90 metres along the track from A to B, find the size of the angle AOB.

   A 36°   B 45°   C 72°   D 90°

Fig. 27.14

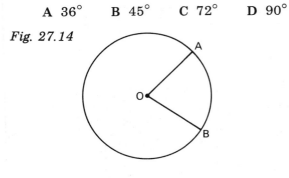

3. Taking $\pi = \frac{22}{7}$, the area of a circle whose radius is $3\frac{1}{2}$ cm is

   A 38.5 cm²   B 22 cm²

   C 19.25 cm²   D 11 cm²

4. The length of arc which subtends an angle of 120° at the centre of a circle of radius 6 cm is

   A 2π cm   B 4π cm

   C 20π cm   D 24π cm

5. Fig. 27.15 represents part of a play-ground. Find its perimeter, taking $\pi = \frac{22}{7}$.

   A 38 m   **B** 41 m   **C** 45 m   **D** 49 m

Fig. 27.15

6. In Fig. 27.16, the area of the circle is 27 cm². What is the area of the shaded sector?

   A 9 cm²   **B** 11.25 cm²

   C 12.5 cm²   **D** 15.5 cm²

Fig. 27.16

7. The fraction of the circle which has been shaded in Fig. 27.17 is

   A $\frac{5}{24}$ **B** $\frac{1}{4}$ **C** $\frac{1}{2}$ **D** $\frac{17}{24}$

Fig. 27.17

8. Taking $\pi = \frac{22}{7}$, the length of the minor arc of the sector shown in Fig. 27.18 is

   A $7\frac{1}{3}$ cm   **B** $25\frac{2}{3}$ cm

   C 44 cm   **D** 51.3 cm

Fig. 27.18

Mental Test 27

1. Find the circumference of a circle whose radius is 7 cm.

2. Find the circumference of a circle whose diameter is 28 in.

3. Find the area of a circle whose radius is 7 cm.

4. A circle has a circumference of 24 in. What is the length of an arc of this circle which subtends an angle of 60° at the centre of this circle?

5. A circle has an area of 48 cm². Find the area of a sector of this circle which subtends an angle of 120° at the centre.

6. Find the area of an annulus whose outside radius is 8 cm and whose inside radius is 4 cm. Give the answer as a multiple of π.

7. A circle has a radius of 8 cm. What is the length of an arc of this circle which subtends an angle of 45° at the centre of the circle? Leave your answer as a multiple of π.

8. A circle has a radius of 6 cm. Find the area of a sector of this circle which subtends an angle of 120° at the centre. Leave your answer as a multiple of π.

9. Taking $\pi = \frac{25}{8}$, find the area of a circle whose radius is 4 cm.

10. Taking $\pi = \frac{25}{8}$, calculate the circumference of a circle whose diameter is 32 cm.

28 Solid Figures

Introduction

The plane figures discussed in Chapter 26, such as rectangles and parallelograms, have two dimensions, namely length and breadth. They have no height (or thickness).

Solid figures have three dimensions, namely length, breadth and height (or thickness).

Types of Solid Figures

(1) A **sphere** is a circular solid. Examples are a football and a ball bearing (Fig. 28.1).

Fig. 28.1

(2) **Prisms** are solid figures which have a uniform cross-section. There are various types of prisms:

(a) **The cuboid** is a rectangular solid (Fig. 28.2). That is, it has a cross-section which is a rectangle. An example of a cuboid is a box.

(b) **The cube** (Fig. 28.3) has all its edges of equal length, and each of its faces is a square. A sugar lump is an example of a cube.

Fig. 28.2   *Fig. 28.3*

(c) **The cylinder** has a constant cross-section which is a circle. It could be called a circular prism. Most tins are cylindrical in shape (Fig. 28.4).

Fig. 28.4

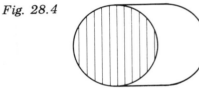

(d) A **triangular prism** has a uniform cross-section which is a triangle (Fig. 28.5). A ridge tent is an example of a triangular prism.

Fig. 28.5

(e) **Hexagonal prisms** have a uniform cross-section which is a hexagon (Fig. 28.6).

Fig. 28.6

(f) Irregular prisms have a cross-section with no well-known shape (Fig. 28.7). Some of the steel sections used in the construction industry fall into this category (Fig. 28.8).

Fig. 28.7

Front face

Fig. 28.8

(3) A **pyramid** is a solid figure which stands on a flat base which may be triangular, square, rectangular or polygonal. It tapers to a point which means that each of its sides is a triangle. Fig. 28.9 shows a triangular pyramid and a rectangular pyramid.

Fig. 28.9

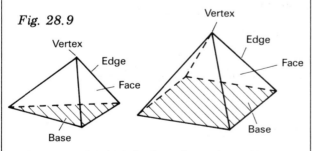

(4) A **regular tetrahedron** is a triangular pyramid with all its faces equilateral triangles (Fig. 28.10). A tetrahedron has therefore all its edges of equal length.

Fig. 28.10

(5) A **cone** is a solid figure with a circular base which tapers to a point (Fig. 28.11).

Fig. 28.11

Exercise 28.1

Write down the number of faces possessed by the following solid figures:

1. Cuboid.

2. Hexagonal prism.

3. Tetrahedron.

4. Cube.

5. Triangular prism.

6. Square pyramid.

Write down the number of edges on each of the following solid figures:

7. Cube.

8. Rectangular pyramid.

9. Triangular prism.

10. Tetrahedron.

11. Draw three different types of prism and label them carefully.

12. Draw three different types of pyramid and label them.

Nets

Suppose that we want to make a cube out of cardboard. We need a pattern giving us the shape of the cardboard to make the cube. As shown in Fig. 28.12, the shape of the pattern is six squares. This shape can be folded to make the cube.

Fig. 28.12

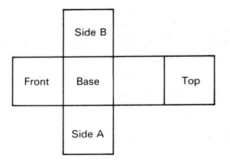

It is possible for there to be more than one net for a solid object. For instance the cube in Fig. 28.12 can also be made from the net shown in Fig. 28.13.

Fig. 28.13

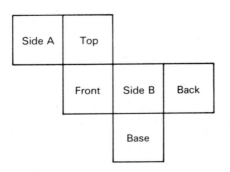

Example 1

Sketch the net of the triangular prism shown in Fig. 28.14.

Fig. 28.14

The net is sketched in Fig. 28.15. It consists of three rectangles representing the base and two sides and two triangles representing the two ends.

Fig. 28.15

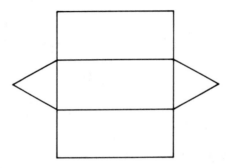

Nets of Curved Surfaces

The net for a cylinder without a top and a bottom is shown in Fig. 28.16. The length of the net is equal to the circumference of the cylinder.

Fig. 28.16

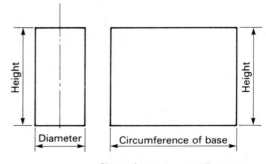

Circumference = π × diameter

Example 2

A cone has a base which is a circle 10 cm in diameter and a height of 15 cm. Draw a net for this cone.

The net is shown in Fig. 28.17 and it is seen to be the sector of a circle. Note that the arms of the sector are equal to the slant height of the cone and that the length of the arc is equal to the circumference of the base.

Fig. 28.17

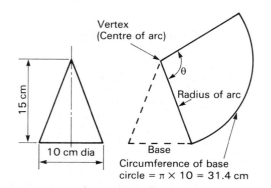

If the slant height is l, then by Pythagoras

$$l^2 = 5^2 + 15^2$$
$$= 25 + 225$$
$$= 250$$
$$l = \sqrt{250}$$
$$= 15.8$$

If the length of arc is C then

$$C = 2 \times \pi \times r$$
$$= 2 \times 3.142 \times 5$$
$$= 31.4$$

If the sector angle is θ then

$$C = 2 \times \pi \times l \times \frac{\theta}{360}$$

$$31.4 = 2 \times 3.142 \times 15.8 \times \frac{\theta}{360}$$

$$31.4 = 99.3 \times \frac{\theta}{360}$$

$$\theta = \frac{31.4 \times 360}{99.3}$$

$$= 114°$$

Exercise 28.2

Sketch the nets of the following solid figures:

1. A cuboid 8 cm long, 3 cm wide and 4 cm high.

2. A triangular prism whose ends are right-angled triangles of base 3 cm and height 4 cm and whose length is 6 cm.

3. A pyramid with a square base of side 5 cm and a vertical height of 8 cm.

4. A cube with an edge of 4 cm.

5. A cylinder with a height of 5 cm and a diameter of 3 cm.

6. A cone with a vertical height of 8 cm and a base diameter of 7 cm. Calculate the slant height of this cone and mark it on your diagram.

7. The diagrams in Fig. 28.18 show the nets of various solids. Name the solids.

Fig. 28.18

(a)

(b)

(c)

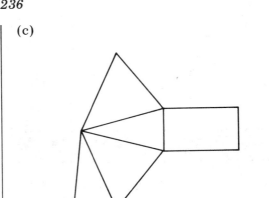

Surface Area

We often need to find the surface area of solid figures such as cylinders and rectangular blocks. The total surface area of such figures is composed of the lateral surface area plus the area of the ends. For a cylinder the lateral surface is the curved surface and for a rectangular block it is the top, bottom and the two sides.

The surface area can be found by drawing a net of the object and working out the areas of the various shapes making up the net.

Example 3

A cuboid is 8 cm long, 5 cm wide and 4 cm high. Draw its net and hence find its total surface area.

> The net is shown in Fig. 28.19 and the total surface area is found by adding together the areas of the six rectangles making up the net.

Fig. 28.19

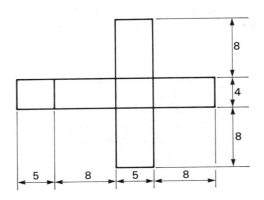

All dimensions in centimetres

$$\text{Total surface area} = 2 \times 5 \times 4$$
$$+ 2 \times 4 \times 8$$
$$+ 2 \times 5 \times 8$$
$$= 40 + 64 + 80$$
$$= 184$$

Hence the total surface area is $184\,\text{cm}^2$.

We could rewrite the terms comprising the total surface area as follows (see Fig. 28.20):

$$\text{Total surface area} = (2 \times 4 + 2 \times 5) \times 8$$
$$+ 2 \times (5 \times 4)$$
$$= \text{Perimeter of end}$$
$$\times \text{Length}$$
$$+ \text{Area of ends}$$
$$= \text{Lateral surface area}$$
$$+ \text{Area of ends}$$

Fig. 28.20

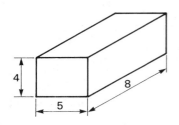

The lateral surface area of any solid figure with a constant cross-section (i.e. cuboid, prism and cylinder) can be found by multiplying the perimeter of the cross-section by the length of the solid:

Lateral
surface area = Perimeter of cross-section
$\qquad$ × Length of solid

Example 4

(a) Find the total surface area of a cylinder whose diameter is 28 cm and whose height is 20 cm.

Perimeter of cross-section = Circumference

$\qquad = \pi \times \text{Diameter}$

$\qquad = \frac{22}{7} \times 28\,\text{cm}$

$\qquad = 88\,\text{cm}$

Lateral surface area $= (88 \times 20)\,\text{cm}^2$

$\qquad = 1760\,\text{cm}^2$

Area of one end $= \pi \times \text{Radius}^2$

$\qquad = \frac{22}{7} \times 14^2$

$\qquad = 616\,\text{cm}^2$

Total surface area $= 1760\,\text{cm}^2$

$\qquad + (2 \times 616)\,\text{cm}^2$

$\qquad = 1760\,\text{cm}^2$

$\qquad + 1232\,\text{cm}^2$

$\qquad = 2992\,\text{cm}^2$

(b) Fig. 28.21 shows a triangular prism. Find its lateral surface area.

Fig. 28.21

Perimeter of cross-section $= (6 + 5 + 5)\,\text{cm}$

$\qquad = 16\,\text{cm}$

Lateral
surface area $=$ Perimeter of cross-section
$\qquad$ × Length

$\qquad = (16 \times 12)\,\text{cm}^2$

$\qquad = 192\,\text{cm}^2$

Exercise 28.3

1. A room is 5 m long, 4 m wide and 3 m high. Calculate the total surface area of the walls.

2. A rectangular block of wood is 5 cm long, 6 cm wide and 4 cm high. Calculate its total surface area.

3. A cylinder has a diameter of 28 cm and a height of 30 cm. Calculate its total surface area.

4. A triangular prism is shown in Fig. 28.22. Calculate

    (a) its lateral surface area

    (b) the total surface area

Fig. 28.22

5. Find the curved surface area of a cylinder whose diameter is 7 cm and whose height is 8 cm.

6. Calculate the total surface area of a cube whose edge is 6 cm long.

7. A water tank has a cross-section in the form of a rectangle with semicircular ends as shown in Fig. 28.23. The tank is 7 m long. Find its total surface area.

Fig. 28.23

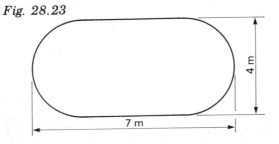

8. A tin is in the form of a cuboid and it is made from sheet metal. It is 9 cm long, 6 cm wide and 5 cm high. If it has no top, calculate the area of sheet metal needed to make it.

Units of Volume

Volume is measured by seeing how many cubic units the solid contains. The cubic centimetre (abbreviation cm^3) is the volume contained inside a cube of side 1 cm.

Similarly a cubic metre (abbreviation m^3) is the volume contained inside a cube of side 1 m.

In the imperial system, volumes are measured in cubic inches (in^3) and cubic feet (ft^3).

Volume of a Cuboid

The cuboid shown in Fig. 28.24 has been divided into three layers of small cubes, each having a volume of $1 cm^3$.
There are 5×4 cubes in each layer and therefore the total number of small cubes is $5 \times 4 \times 3 = 60$.
Therefore the cuboid has a volume of $60 cm^3$.

Fig. 28.24

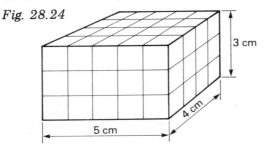

What we have done to find the volume of the cuboid is to multiply the length by the breadth by the height. This rule applies to any cuboid, and hence

Volume of cuboid = Length × Breadth
× Height

Since the area of the end of a cuboid is breadth by height we can also write

Volume of cuboid = Area of the end
× Length

This statement is true for any solid which has the same cross-section throughout its length. As a formula we write

$$V = Al$$

where V is the volume of the solid, A the area of the cross-section and l the length of the solid.

Example 5

(a) Find the volume of a cuboid which is 15 in long, 8 in wide and 4 in high.

Volume of a cuboid = $(15 \times 8 \times 4) in^3$
= $480 in^3$

(b) A tank is in the form of a cuboid. It is 120 cm long, 80 cm wide and 50 cm high. Calculate the volume of the tank in cubic metres.

In problems of this kind it is best to convert all the dimensions to metres before attempting to find the volume. Thus

Length = 120 cm
= 1.20 m
Breadth = 80 cm
= 0.8 m
Height = 50 cm
= 0.5 m
Volume of tank = $(1.2 \times 0.8 \times 0.5) m^3$
= $0.48 m^3$

(c) A block of wood has the cross-section shown in Fig. 28.25. If it is 9 m long find its volume, stating the answer in cubic metres.

Fig. 28.25

Because the length is given in metres we first convert all the other dimensions into metres.

$$\text{Area of cross-section} = (\tfrac{1}{2} \times \pi \times 0.15^2)\,\text{m}^2$$
$$+ (0.2 \times 0.3)\,\text{m}^2$$
$$= (0.035 + 0.06)\,\text{m}^2$$
$$= 0.095\,\text{m}^2$$

$$\text{Volume} = \text{Area of cross-section} \times \text{Length}$$
$$= (0.095 \times 9)\,\text{m}^3$$
$$= 0.855\,\text{m}^3$$

Volume of a Cylinder

Since a cylinder has a constant cross-section which is a circle (Fig. 28.26)

$$\text{Volume of cylinder} = \text{Area of cross-section} \times \text{Height}$$
$$= \pi \times \text{Radius}^2 \times \text{Height}$$

As a formula

$$V = \pi r^2 h$$

Fig. 28.26

Example 6

(a) Find the volume of a cylinder which has a radius of 14 in and a height of 8 in.

$$V = \pi r^2 h$$
$$= (\tfrac{22}{7} \times 14^2 \times 8)\,\text{cm}^3$$
$$= 4928\,\text{in}^3$$

(b) A pipe has the dimensions shown in Fig. 28.27. Calculate the volume of metal in the pipe.

Fig. 28.27

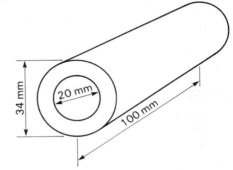

Since the cross-section is an annulus

$$\text{Area of cross-section} = \pi(R^2 - r^2)$$
$$= 3.142 \times (17^2 - 10^2)\,\text{mm}^2$$
$$= 3.142 \times (289 - 100)\,\text{mm}^2$$
$$= (3.142 \times 189)\,\text{mm}^2$$
$$= 594\,\text{mm}^2$$

$$\text{Volume} = \text{Area of cross-section} \times \text{Length}$$
$$= (594 \times 100)\,\text{mm}^3$$
$$= 59\,400\,\text{mm}^3$$

Exercise 28.4

1. Find the volume of a rectangular block of wood which is 8 in long, 5 in wide and 4 in high.

2. Fig. 28.28 shows the cross-section of a steel beam which is 8 m long. Find its volume in cubic metres.

Fig. 28.28

3. Calculate the volume of a cylinder which has a diameter of 7 m and a height of 8 m.

4. A hole 40 mm diameter is drilled in a plate which is 50 mm thick. Find the volume of metal removed in drilling the hole.

5. A rectangular tank is 500 cm long, 300 cm wide and 400 cm high. Calculate its volume in cubic metres.

6. A block of wood has the cross-section shown in Fig. 28.29. If it is 8 cm long, find its volume in cubic centimetres.

Fig. 28.29

7. Calculate the volume of a metal tube which has an inside diameter of 14 in and an outside diameter of 28 in, and a length of 30 in.

8. Fig. 28.30 shows a triangular prism. Calculate its volume.

Fig. 28.30

Volume of a Cone

The volume of a cone (Fig. 28.31) is one-third of the volume of an equivalent cylinder (i.e. a cylinder with the same radius as the base radius of the cone and having the same height). That is

Volume of cone $= \frac{1}{3} \times \pi \times \text{Radius}^2 \times \text{Height}$

or $\qquad V = \frac{1}{3}\pi r^2 h$

Fig. 28.31

Example 7

Find the volume of a cone which has a base radius of 6 cm and a height of 7 cm.

$$V = \frac{1}{3}\pi r^2 h$$
$$= \frac{1}{3} \times \frac{22}{7} \times 6^2 \times 7 \text{ cm}^3$$
$$= 264 \text{ cm}^3$$

Volume of a Pyramid

The volume of a pyramid (Fig. 28.32) is one-third of the volume of an equivalent prism. Hence

$$\text{Volume of a pyramid} = \tfrac{1}{3} \times \text{Area of base} \times \text{Height}$$

or
$$V = \tfrac{1}{3}Ah$$

Fig. 28.32

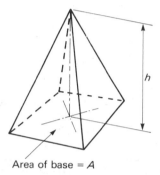

Area of base = *A*

Example 8

Find the volume of a pyramid whose base is a square of 8 in side and whose vertical height is 9 in.

$$\text{Area of base} = 8 \times 8 \text{ in}^2$$
$$= 64 \text{ in}^2$$
$$\text{Volume} = \tfrac{1}{3} \times 64 \times 9 \text{ in}^3$$
$$= 192 \text{ in}^3$$

The Sphere

It can be shown that

$$\text{Volume of a sphere} = \tfrac{4}{3} \times \pi \times \text{Radius}^3$$
or
$$V = \tfrac{4}{3}\pi r^3$$
$$\text{Surface area of a sphere} = 4 \times \pi \times \text{Radius}^2$$
or
$$A = 4\pi r^2$$

Example 9

A ball has a diameter of 14 cm. Calculate the volume and the surface area of the ball.

$$V = \tfrac{4}{3}\pi r^3$$
$$= \tfrac{4}{3} \times \tfrac{22}{7} \times 7^3 \text{ cm}^3$$
$$= 1437 \text{ cm}^3$$
$$A = 4\pi r^2$$
$$= 4 \times \tfrac{22}{7} \times 7^2 \text{ cm}^2$$
$$= 616 \text{ cm}^2$$

Exercise 28.5

1. Find the volume of a cone whose base radius is 28 cm and whose height is 10 cm.

2. A cone has a base diameter of 10 in and height of 8 in. Calculate its volume.

3. A pyramid has a rectangular base 4 cm by 3 cm and a vertical height of 6 cm. Calculate its volume.

4. A pyramid has a square base of side 8 ft and an altitude of 5 ft. What is its volume?

5. A sphere has a radius of $3\tfrac{1}{2}$ cm. Calculate its volume and its surface area.

6. A concrete dome is in the shape of a hemisphere. Its internal and external diameters are 4.8 m and 5 m, respectively. Calculate the volume of concrete used in its construction.

7. A small roof is in the form of a pyramid. The base is a regular pentagon whose area is 15 m², and its vertical height is 4 metres. Calculate the volume enclosed by the roof.

8. Fig. 28.33 shows a flask which may be considered as a sphere with a cylindrical neck. Calculate the volume of the flask.

Fig. 28.33 ⟶

3 cm

35 cm

20 cm

Capacity

The capacity of a container is the volume of liquid that it will hold. Capacity is sometimes measured in the same units as volume (i.e. cubic centimetres or cubic metres) but more often it is measured in litres (abbreviation ℓ) such that

$$1 \text{ litre} = 1000 \text{ cm}^3$$

Small capacities are often measured in centilitres (cℓ) or millilitres (mℓ).

$$100 \text{ cℓ} = 1000 \text{ mℓ}$$
$$= 1 \text{ ℓ}$$
$$1 \text{ mℓ} = 1 \text{ cm}^3$$

In the imperial system, capacity is measured in fluid ounces, pints and gallons.

$$20 \text{ fluid ounces (fl oz)} = 1 \text{ pint (pt)}$$
$$8 \text{ pints (pt)} = 1 \text{ gallon (gal)}$$

Example 10

(a) How many 5-millilitre doses of medicine can be obtained from a cylindrical medicine bottle which has a diameter of 5 cm and a height of 12 cm?

Volume of bottle $= \pi \times \text{Radius}^2 \times \text{Height}$
$$= 3.142 \times 2.5^2 \times 12 \text{ cm}^3$$
$$= 236 \text{ cm}^3$$
$$= 236 \text{ mℓ}$$

Number of doses $= 236 \div 5$
$$= 47.2$$

Number of full doses $= 47$

(b) A rectangular water tank is 4 m long, 2 m wide and 3 m high. When it is full how many litres of water does it hold?

Volume of tank $= (400 \times 200 \times 300) \text{ cm}^3$
$$= 24\,000\,000 \text{ cm}^3$$

Capacity of the tank $= (24\,000\,000 \div 1000) \text{ ℓ}$
$$= 24\,000 \text{ ℓ}$$

Exercise 28.6

1. A small can in the form of a cylinder has a radius of 7 cm and a height of 8 cm. How many litres will the can hold?

2. A rectangular tank is 3 m long, 1 m wide and 0.5 m tall. How many litres of liquid will it hold when full?

3. A basin may be considered to be a hemisphere of diameter 28 cm. How many litres of water will it hold when full?

4. A cylindrical garden pool has a radius of 3 m. How many litres of water will it take to fill the pool to a depth of $1\frac{1}{2}$ m?

5. A rectangular medicine bottle is 8 cm wide, 4 cm long and 12 cm high. How many centilitres will it hold when full?

6. A conical wine glass is 4.5 cm in diameter and 6 cm high. What is its capacity in millilitres? How many such glasses would be filled from a bottle of wine containing 70 centilitres?

7. An ice cream carton is cylindrical in shape. It is 6 cm in diameter and 8 cm tall. How many litres of ice cream are needed to fill 50 of these cartons?

8. A petrol storage tank is rectangular in shape and measures 3 m by 2 m by 1.5 m. Calculate the capacity of the tank in litres.

Similar Solids

For solids to be similar they must be wide, long and tall in the same proportions.

For instance, the diagram (Fig. 28.34) shows two child's building bricks. Looking at the diagram you might say that the larger brick is three times bigger than the smaller brick, and indeed all the lengths are three times bigger. However, 9 small bricks would sit on top of the larger brick and it would take 27 small bricks to make a cube as big as the larger brick.

Fig. 28.34

Notice that the length is 3 times bigger.

The surface area is $3 \times 3 = 9$ times bigger.

The volume and weight is $3 \times 3 \times 3 = 27$ times bigger.

From this example we see that:

(1) The volume of similar solids vary as the cube of the corresponding dimensions.

(2) The surface areas of similar solids vary as the square of the corresponding dimensions.

Fig. 28.35 shows two cylinders. If the ratio of corresponding dimensions is the same then the cylinders are similar. In the diagram

$$\frac{\text{Height of cylinder A}}{\text{Height of cylinder B}} = \frac{\text{Diameter of cylinder A}}{\text{Diameter of cylinder B}}$$

and hence cylinders A and B are similar.

Fig. 28.35

It follows that

$$\frac{\text{Volume of cylinder A}}{\text{Volume of cylinder B}}$$

$$= \left(\frac{\text{Height of cylinder A}}{\text{Height of cylinder B}}\right)^3$$

$$= \left(\frac{\text{Diameter of cylinder A}}{\text{Diameter of cylinder B}}\right)^3$$

$$\frac{\text{Surface area of cylinder A}}{\text{Surface area of cylinder B}}$$

$$= \left(\frac{\text{Height of cylinder A}}{\text{Height of cylinder B}}\right)^2$$

$$= \left(\frac{\text{Diameter of cylinder A}}{\text{Diameter of cylinder B}}\right)^2$$

Example 11

(a) The volume of a cone whose height is 135 mm is 1090 mm³ and its surface area is 680 mm². Find the volume and surface area of a similar cone which has a height of 72 mm.

$$\frac{\text{Volume of smaller cone}}{\text{Volume of larger cone}}$$

$$= \left(\frac{72}{135}\right)^3$$

$$= 0.1517$$

Volume of smaller cone

$$= 0.1517 \times \text{Volume of larger cone}$$

$$= 0.1517 \times 1090 \text{ cm}^3$$

$$= 165.4 \text{ cm}^3$$

$$\frac{\text{Surface area of smaller cone}}{\text{Surface area of larger cone}}$$

$$= \left(\frac{72}{135}\right)^2$$

$$= 0.2844$$

Surface area of smaller cone

$$= 0.2844 \times \text{Surface area of larger cone}$$

$$= 0.2844 \times 680 \text{ mm}^2$$

$$= 193 \text{ mm}^2$$

(b) The model of a lorry is made on a scale of 1 in 10.

   (i) The windscreen on the model has an area of 100 cm². Calculate the area, in square centimetres, of the windscreen of the lorry.

  (ii) The fuel tank of the lorry, when full, holds 100 litres. Calculate the capacity of the fuel tank on the model.

The lorry and its model are similar solids. Each dimension on the lorry is 10 times larger than the corresponding dimension on the model.

(i) $\dfrac{\text{Area of windscreen on the lorry}}{\text{Area of windscreen on the model}}$

$$= 10^2$$

$$= 100$$

Area of windscreen on the lorry

$$= 100 \times \text{Area of windscreen on the model}$$

$$= 100 \times 100 \text{ cm}^2$$

$$= 10\,000 \text{ cm}^2$$

(ii) $\dfrac{\text{Capacity of fuel tank on model}}{\text{Capacity of fuel tank on lorry}}$

$$= \left(\frac{1}{10}\right)^3$$

$$= \frac{1}{1000}$$

Capacity of fuel tank on model

$$= \frac{1}{1000} \times \text{Capacity of fuel tank on lorry}$$

$$= \frac{1}{1000} \times 100 \text{ litres}$$

$$= 0.1 \text{ litre}$$

Exercise 28.7

1.   Two tins of beans are similar in shape but the larger tin is twice as tall as the smaller tin.

    (a) If the surface area of the larger tin is 200 cm², what is the surface area of the smaller tin?

    (b) The volume of the smaller tin is 35 cm³. What is the volume of the larger tin?

2.   Two cubes are made of the same type of wood. The larger cube is 8 times heavier than the smaller cube. If the smaller cube has an edge 3 in long, what is the length of the edge on the larger cube?

3. A hemisphere has a radius of 2 cm and a volume of 6 cm³. Work out the volume of a hemisphere with a radius of 3 cm.

4. The volume of a cone of height 14.2 in is 210 in³. Find the volume of a similar cone whose height is 9.4 in.

5. Fig. 28.36 shows two bottles which are similar in shape. Bottle A is $2\frac{1}{2}$ times as tall as bottle B and holds 27 pints of liquid. How many pints does bottle B hold?

Fig. 28.36

6. A model aircraft, similar to the full-size aircraft, is made to a scale of 1 in 20. The model has a tail 0.15 m high, a wing area of 0.28 m² and a cabin volume of 0.016 m³. Find the corresponding figures for the full size aircraft.

7. A sphere has a radius of 3 ft. Calculate its surface area and its volume. Hence find the surface area and the volume of a sphere with a radius of 5 ft.

8. The curved surface of a cone has an area of 20.5 cm². Calculate the curved surface area of a similar cone whose height is $1\frac{1}{2}$ times greater than that of the first cone.

Miscellaneous Exercise 28

Section B

1. A piece of alloy is melted down to form 48 toy soldiers. how many similar soldiers of twice the height could be made from the same piece of metal?

2. Fig. 28.37 shows a large greenhouse. Calculate **(a)** the area of one end, **(b)** the volume of air contained in the greenhouse, **(c)** the perimeter of one end, **(d)** the total area of glass comprising both ends, both sides and the roof.

Fig. 28.37

3. Fig. 28.38 shows two solid wooden blocks taken from a child's building outfit. Each block is a right prism.

Fig. 28.38

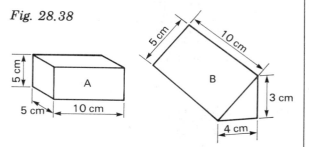

Prism A is a rectangular block 10 cm long, its uniform cross-section being a square of side 5 cm. Prism B is a triangular prism 10 m long, its uniform cross-section being a right-angled triangle of sides 3, 4 and 5 cm. Calculate

(a) the volume of prism A

(b) the total surface area of prism A

(c) the total surface area of prism B

4. Which of the diagrams in Fig. 28.39 are nets of

   (a) a cube   (b) a tetrahedron

Fig. 28.39

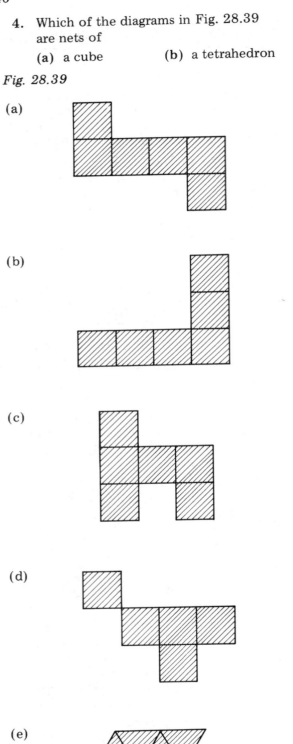

(a)

(b)

(c)

(d)

(e)

(f)

(g)

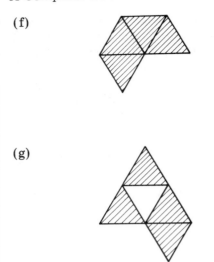

5. Fig. 28.40 shows three similar rectangular boxes.

   (a) The dimensions of box A are 1.6 in by 3.5 in by 5.3 in. Calculate its volume.

   (b) In box C the dimensions are double the dimensions of box A. Calculate the volume of box C.

   (c) If the ratio of the surface area of box B to the surface area of box A is 9 to 4, find the dimensions of box B.

Fig. 28.40

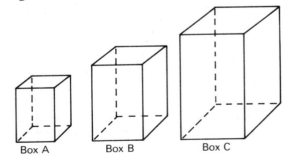

Box A   Box B   Box C

6. (a) A rectangular tank with a base 72 cm long and 45 cm wide contains, when full, 81 litres of water. Calculate the volume of the tank in cubic centimetres.

   (b) Find the height of the tank.

7. A pipe 20 m long and with 70 cm outside diameter is to be laid under a road. A trench 1.2 m wide and 1.5 m deep is dug across the road, and the pipe is to be laid in concrete as shown in Fig. 28.41. Calculate, to the nearest cubic metre, the volume of concrete that will be needed for the job.

Fig. 28.41

1.50 m

70 cm

Concrete

1.20 m

8. Plastic rain-water guttering is made in rectangular sections (type A) and semi-circular sections (type B) in 2 m lengths and with the other dimensions shown in Fig. 28.42.

Fig. 28.42

TYPE A

2 m

7 cm 14 cm

TYPE B

2 m

14 cm

(a) Calculate the area in square centimetres of the flat piece of plastic needed to make a length of type A guttering.

(b) Find the area in square centimetres of the flat piece of plastic needed to make a length of type B guttering.

(c) If the ends of each section were blocked up and the gutters were filled with water, find the volume of water, in litres, in each type of guttering.

9. Which is the better buy: a packet of soap powder which is 28 cm high costing 48p, or a geometrically similar packet of the same powder which is 35 cm high and costs 90p?

Multi-Choice Questions 28

Fig. 28.43 represents a solid rectangular block. Use this diagram to answer questions 1, 2 and 3.

1. What is the total length, in centimetres, of the edges of the block?

   A 13   B 16   C 36   D 52

2. What is the volume, in cubic centimetres, of the block?

   A 16   B 20   C 52   D 80

3. What is the total surface area of the block in square centimetres?

   A 16   B 80   C 96   D 112

Fig. 28.43

5 cm

4 cm

4 cm

4. Soup is sold in a closed cylindrical can whose diameter is 7 cm and whose height is 10 cm. How much soup does the can hold? (Take $\pi = \frac{22}{7}$.)

   A less than 0.4 litre

   B $\frac{1}{2}$ litre

   C 1 litre

   D more than 1 litre

5. What is the surface area of a cube having an edge of 6 cm?

   A 6 cm^2 B 36 cm^2

   C 144 cm^2 D 216 cm^2

6. A rectangular block is 13 cm long, 5 cm wide and 2 cm high. Its volume, in cubic centimetres, is

   A 20   B 65   C 80   D 130

7. A cylindrical jug with a base area of 30 cm^2 contains 450 cm^3 of water. The depth of water in the jug is

   A 5 cm   B 15 cm

   C 30 cm   D 450 cm

8. The total surface area of a cube is 150 cm^2. What is the length of an edge of the cube?

   A 5 cm   B 12$\frac{1}{2}$ cm

   C 25 cm   D 150 cm

9. The circumference of a cylindrical lawn roller is 90 cm and it is 40 cm wide. What is the area of lawn rolled when the roller makes 100 revolutions?

   A 9 m^2 B 36 m^2

   C 56 m^2 D 113 m^2

10. A cone has a base diameter of 6 cm and a vertical height of 8 cm. Calculate the volume of the cone, in cubic centimetres, leaving π in your answer.

    A 8π B 24π C 72π D 96π

11. A cylinder has a base radius of 7 cm and a height of 10 cm. Taking π as $\frac{22}{7}$, calculate the curved surface area of this cylinder.

    A 280 cm^2 B 440 cm^2

    C 880 cm^2 D 1540 cm^2

12. The number of bottles holding 8 cm^3 that can be filled from 5 litres of liquid is

    A 6.25   B 62.5   C 625   D 6250

13. The ratio of the volumes of two similar cylinders is 8 : 27. What is the ratio of their surface areas?

    A 8 : 27   B 2 : 3

    C 4 : 9   D none of these

14. The radius of a cylinder whose height is 60 cm and whose volume is 120π cubic centimetres is

    A 2 cm   B $\sqrt{2}$ cm

    C 4 cm   D 2$\sqrt{2}$ cm

15. The number of little sweets in a packet 10 cm high is 200. The approximate number in a similar box 8 cm high will be

    A 160   B 128   C 120   D 100

Mental Test 28

1. Find the volume of a cuboid whose length is 5 in, whose width is 4 in and whose height is 2 in.

2. A triangular prism has a constant cross-section whose area is 5 cm^2. If the prism is 20 cm long, find its volume.

3. A match box is 7 cm wide, 3 cm high and 10 cm long. What is its volume?

4. A block of wood has a cross-section which is a square of 5 in side. If the block is 10 in long, calculate its volume.

5. Find the lateral surface area for the block of wood in question 4.

6. A cylinder has a cross-sectional area of 12 cm^2. If it is 8 cm high, determine its volume.

7. A cylinder has a diameter of 7 cm and a height of 5 cm. Find its curved surface area.

8. A cylinder has a base radius of 5 cm and a height 4 cm. Find its volume, stating the answer as a multiple of π.

9. A cone has a base area of 21 cm² and a height of 4 cm. What is its volume?

10. A pyramid has a base 36 cm² in area and a height of 10 cm. What is the volume of the pyramid?

11. A sphere has a radius of 3 cm. Find its surface area as a multiple of π.

12. The ratio of the volumes of two similar cylinders is $1:8$. What is the ratio of their base diameters?

Trigonometry

Notation for a Right-angled Triangle

Look at the right-angled triangle in Fig. 29.1 and note the special names given to its sides.

Fig. 29.1

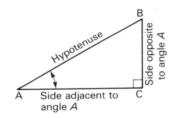

For the angle A:

Hypotenuse: the side AB opposite to the right angle.

Opposite: the side BC opposite to the angle A.

Adjacent: the side AC adjacent to the angle A as it forms the angle A with the hypotenuse.

For the angle B (Fig. 29.2):

Hypotenuse: the side AB opposite to the right angle.

Opposite: the side AC opposite the angle B.

Adjacent: the side BC adjacent to the angle B as it forms the angle B with the hypotenuse.

Fig. 29.2

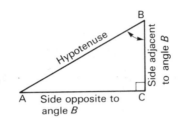

The Trigonometrical Ratios

In any right-angled triangle the following three ratios between the sides of the triangle can be formed:

$$\frac{\text{Opposite}}{\text{Hypotenuse}}$$ called the sine of the angle

$$\frac{\text{Adjacent}}{\text{Hypotenuse}}$$ called the cosine of the angle

$$\frac{\text{Opposite}}{\text{Adjacent}}$$ called the tangent of the angle

These three ratios are called the **trigonometrical ratios**. For any angle in a triangle which is not a right angle, each of these ratios has a definite value. These values are given in the tables of the trigonometrical ratios.

The abbreviations **sin**, **cos** and **tan** are usually used for sine, cosine and tangent respectively.

The Sine of an Angle

In Figs. 29.1 and 29.2

$$\sin A = \frac{\text{Side opposite } A}{\text{Hypotenuse}}$$

$$= \frac{BC}{AB}$$

$$\sin B = \frac{\text{Side opposite } B}{\text{Hypotenuse}}$$

$$= \frac{AC}{AB}$$

Example 1

Find, by drawing, the value of sin 30°.

Draw the lines AX and AY which intersect at A so that $\angle YAX = 30°$, as shown in Fig. 29.3. Along AX measure off AC equal to 1 unit (say 10 cm) and

from C draw CB perpendicular to AY. Measure BC.

BC will be found to be 0.5 unit (5 cm in this case).

$$\sin 30° = \frac{BC}{AC} = \frac{0.5}{1} = 0.5$$

Fig. 29.3

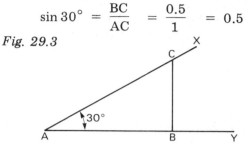

It is possible to find the sine of any angle by the drawing method shown in Example 1, but this is inconvenient and not very accurate. Tables of sines of angles have been calculated (Table 29.1) and these allow us to find the sine of any angle.

TABLE 29.1   PART OF THE TABLE OF NATURAL SINES   Mean differences

°	0′ 0.0°	6′ 0.1°	12′ 0.2°	18′ 0.3°	24′ 0.4°	30′ 0.5°	36′ 0.6°	42′ 0.7°	48′ 0.8°	54′ 0.9°	1′	2′	3′	4′	5′
0	0.0000	0.0017	0.0035	0.0052	0.0070	0.0087	0.0105	0.0122	0.0140	0.0157	3	6	9	12	15
1	0.0175	0.0192	0.0209	0.0227	0.0244	0.0262	0.0279	0.0297	0.0314	0.0332	3	6	9	12	15
2	0.0349	0.0366	0.0384	0.0401	0.0419	0.0436	0.0454	0.0471	0.0488	0.0506	3	6	9	12	15
3	0.0523	0.0541	0.0558	0.0576	0.0593	0.0610	0.0628	0.0645	0.0663	0.0680	3	6	9	12	15
4	0.0698	0.0715	0.0732	0.0750	0.0767	0.0785	0.0802	0.0819	0.0837	0.0854	3	6	9	12	14
5	0.0872	0.0889	0.0906	0.0924	0.0941	0.0958	0.0976	0.0993	0.1011	0.1028	3	6	9	12	14
6	0.1045	0.1063	0.1080	0.1097	0.1115	0.1132	0.1149	0.1167	0.1184	0.1201	3	6	9	12	14
7	0.1219	0.1236	0.1253	0.1271	0.1288	0.1305	0.1323	0.1340	0.1357	0.1374	3	6	9	12	14
8	0.1392	0.1409	0.1426	0.1444	0.1461	0.1478	0.1495	0.1513	0.1530	0.1547	3	6	9	11	14
9	0.1564	0.1582	0.1599	0.1616	0.1633	0.1650	0.1668	0.1685	0.1702	0.1719	3	6	9	11	14
10	0.1736	0.1754	0.1771	0.1788	0.1805	0.1822	0.1840	0.1857	0.1874	0.1891	3	6	9	11	14
11	0.1908	0.1925	0.1942	0.1959	0.1977	0.1994	0.2011	0.2028	0.2045	0.2062	3	6	9	11	14
12	0.2079	0.2096	0.2113	0.2130	0.2147	0.2164	0.2181	0.2198	0.2215	0.2232	3	6	9	11	14
13	0.2250	0.2267	0.2284	0.2300	0.2317	0.2334	0.2351	0.2368	0.2385	0.2402	3	6	8	11	14
14	0.2419	0.2436	0.2453	0.2470	0.2487	0.2504	0.2521	0.2538	0.2554	0.2571	3	6	8	11	14
15	0.2588	0.2605	0.2622	0.2639	0.2656	0.2672	0.2689	0.2706	0.2723	0.2740	3	6	8	11	14
16	0.2756	0.2773	0.2790	0.2807	0.2823	0.2840	0.2857	0.2874	0.2890	0.2907	3	6	8	11	14
17	0.2924	0.2940	0.2957	0.2974	0.2990	0.3007	0.3024	0.3040	0.3057	0.3074	3	6	8	11	14
18	0.3090	0.3107	0.3123	0.3140	0.3156	0.3173	0.3190	0.3206	0.3223	0.3239	3	6	8	11	14
19	0.3256	0.3272	0.3289	0.3305	0.3322	0.3338	0.3355	0.3371	0.3387	0.3404	3	5	8	11	14
20	0.3420	0.3437	0.3453	0.3469	0.3486	0.3502	0.3518	0.3535	0.3551	0.3567	3	5	8	11	14
21	0.3584	0.3600	0.3616	0.3633	0.3649	0.3665	0.3681	0.3697	0.3714	0.3730	3	5	8	11	14
22	0.3746	0.3762	0.3778	0.3795	0.3811	0.3827	0.3843	0.3859	0.3875	0.3891	3	5	8	11	13
23	0.3907	0.3923	0.3939	0.3955	0.3971	0.3987	0.4003	0.4019	0.4035	0.4051	3	5	8	11	13
24	0.4067	0.4083	0.4099	0.4115	0.4131	0.4147	0.4163	0.4179	0.4195	0.4210	3	5	8	11	13
25	0.4226	0.4242	0.4258	0.4274	0.4289	0.4305	0.4321	0.4337	0.4352	0.4368	3	5	8	11	13
26	0.4384	0.4399	0.4415	0.4431	0.4446	0.4462	0.4478	0.4493	0.4509	0.4524	3	5	8	10	13
27	0.4540	0.4555	0.4571	0.4586	0.4602	0.4617	0.4633	0.4648	0.4664	0.4679	3	5	8	10	13
28	0.4695	0.4710	0.4726	0.4741	0.4756	0.4772	0.4787	0.4802	0.4818	0.4833	3	5	8	10	13
29	0.4848	0.4863	0.4879	0.4894	0.4909	0.4924	0.4939	0.4955	0.4970	0.4985	3	5	8	10	13

Decimals of a Degree

The sine table (29.1) has been worked in degrees and minutes and in degrees and decimals of a degree. To convert an angle given in degrees and minutes to degrees and decimals of a degree we divide the minutes by 60, because 60 minutes = 1 degree.

Example 2

Convert $28°42'$ into degrees and decimals of a degree.

$$28°42' = 28\tfrac{42}{60}°$$
$$= 28.7°$$

To convert an angle given in degrees and decimals of a degree multiply the decimals by 60 to obtain an angle in degrees and minutes.

Example 3

Convert $59.8°$ into degrees and minutes.

$$59.8° = 59°(0.8 \times 60)'$$
$$= 59°48'$$

Reading the Table of Sines of Angles

(1) To find $\sin 17°$:
The sine of an angle with an exact number of degrees is given in the column headed 0. Thus $\sin 17° = 0.2924$.

(2) To find $\sin 17.8°$:
The value is found under the column headed 0.8°. Thus $\sin 17.8° = 0.3057$.

(3) To find $17°18'$:
The value is found under the column headed 18'. Thus $\sin 17°18' = 0.2974$.

(4) To find $\sin 17°27'$:
If the number of minutes is not an exact multiple of 6 we make use of the figures in the last five columns.
Now $\sin 17°24' = 0.2990$, and $27'$ is $3'$ more than $24'$.

Looking in the column headed $3'$ we find the number 8. This is **added** on to the value of $\sin 17°24'$ to give $\sin 17°27' = 0.2998$.

(5) To find the angle whose sine is 0.1711:
Look in the table of sines to find the nearest number *lower* than 0.1711. This is found to be 0.1702, which corresponds to an angle of $9°48'$. Now 0.1702 is 0.009 less than 0.1711 so we look in the last five columns in the row marked $9°$ and find 9 in the column headed $3'$.

The angle whose sine is 0.1711 is $9°48' + 3' = 9°51'$, i.e. $\sin 9°51' = 0.1711$.

1. Find, by drawing, the sines of
   (a) $45°$ (b) $60°$

2. Convert to degrees and decimals of a degree
   (a) $18°24'$ (b) $38°42'$
   (c) $79°54'$

3. Convert to degrees and minutes
   (a) $43.2°$ (b) $54.7°$
   (c) $43.1°$

4. Find the sines of the angles
   (a) $14°$ (b) $73°$
   (c) $25.2°$ (d) $59.8°$
   (e) $54°12'$ (f) $79°42'$
   (g) $19°53'$ (h) $76°14'$

5. Use the sine tables to write down the angles whose sines are
   (a) 0.9135 (b) 0.9816
   (c) 0.7349 (d) 0.9943
   (e) 0.8112 (f) 0.1452

Calculations on the Sine of an Angle

Example 4

(a) Find the length of the side BC in Fig. 29.4.

Fig. 29.4

Since BC is the side opposite to the angle A and AC is the hypotenuse

$$\frac{BC}{AC} = \sin 42°$$

$$BC = AC \times \sin 42°$$

$$= 12 \times 0.6691 \text{ cm}$$

$$= 8.03 \text{ cm}$$

(b) Find the length of AC in Fig. 29.5.

Fig. 29.5

AB is the side opposite $\angle C$ and AC is the hypotenuse.

$$\frac{AB}{AC} = \sin C$$

$$AC = \frac{AB}{\sin C}$$

$$= \frac{5}{\sin 17.8°}$$

$$= \frac{5}{0.3057}$$

$$= 16.4 \text{ cm}$$

(c) Find the angles A and B in Fig. 29.6

Fig. 29.6

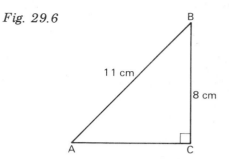

Since BC is the side opposite angle A and AB is the hypotenuse

$$\sin A = \frac{BC}{AB}$$

$$= \frac{8}{11}$$

$$= 0.7273$$

From the sine tables

$$\angle A = 46.7°$$

$$\angle B = 90° - 46.7°$$

$$= 43.3°$$

Exercise 29.2

1. Find the length of the sides marked x in Fig. 29.7.

Fig. 29.7

2. Find the size of the angles marked y in Fig. 29.8

Fig. 29.8

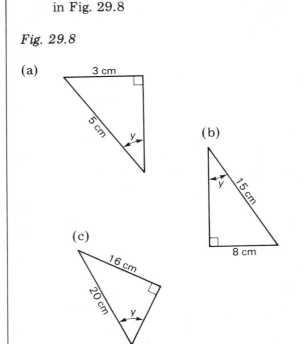

(a)

(b)

(c)

3. In the triangle ABC, $\angle B = 90°$, $\angle C = 26°21'$ and $b = 13.4$ cm. Calculate the length of the side c of the triangle.

4. In triangle ABC, $\angle C = 90°$, $\angle A = 69.3°$ and $a = 3.4$ cm. Calculate the length of the side c of the triangle.

5. An equilateral triangle has a vertical height of 20 cm. Calculate the length of the equal sides.

6. Calculate the length of the equal sides of an isosceles triangle whose altitude is 15 cm and whose equal angles are $48°36'$.

7. The equal sides of an isosceles triangle are each 12 cm long. If the altitude of the triangle is 8 cm, find the size of the three angles of the triangle.

8. In the triangle ABC, $\angle A = 90°$, $b = 10.8$ cm and $a = 12.3$ cm. Find the size of the angles B and C.

The Cosine of an Angle

In Figs. 29.1 and 29.2

$$\cos A = \frac{\text{Side adjacent to } \angle A}{\text{Hypotenuse}}$$

$$= \frac{AC}{AB}$$

$$\cos B = \frac{\text{Side adjacent to } \angle B}{\text{Hypotenuse}}$$

$$= \frac{BC}{AB}$$

The cosine of an angle may be found by drawing, the construction being similar to Fig. 29.3. However, tables of cosines are available which are used in a similar way to the sine tables except that the mean differences are **subtracted**.

Example 5

(a) Find the length of the side BC in Fig. 29.9

Fig. 29.9

Since BC is the side adjacent to $\angle C$ and AC is the hypotenuse

$$\frac{BC}{AC} = \cos C$$

$$BC = AC \cos C$$

$$= 16 \times \cos 18.9°$$

$$= 16 \times 0.9461$$

$$= 15.14$$

Therefore BC = 15.14 cm.

(b) In triangle ABC, $\angle A = 90°$, $\angle C = 25°$ and $b = 21$ cm. Find the length of the side a.

> The triangle is drawn in Fig. 29.10 where it will be seen that b is the side adjacent to the angle C whilst a is the hypotenuse.

Fig. 29.10

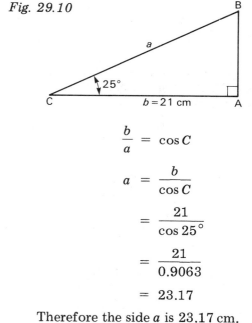

$$\frac{b}{a} = \cos C$$

$$a = \frac{b}{\cos C}$$

$$= \frac{21}{\cos 25°}$$

$$= \frac{21}{0.9063}$$

$$= 23.17$$

Therefore the side a is 23.17 cm.

(c) In Fig. 29.11, find the size of the angle B.

Fig. 29.11

> Since AB is the side adjacent to the angle B and BC is the hypotenuse

$$\cos B = \frac{AB}{BC}$$

$$= \frac{9.1}{16.2}$$

$$= 0.5617$$

$$\angle B = 55°50'$$

Exercise 29.3

1. Use the cosine tables to find the values of
    (a) $\cos 18°$ **(b)** $\cos 76°$
    (c) $\cos 37.3°$ **(d)** $\cos 81.6°$
    (e) $\cos 54°25'$ **(f)** $8°52'$

2. Find the length of the sides marked p in Fig. 29.12.

Fig. 29.12

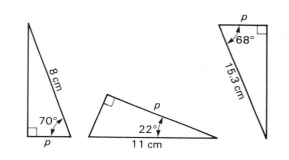

3. Find the angles marked q in Fig. 29.13.

Fig. 29.13

4. Calculate the vertical height of an isosceles triangle with an apex angle of $42°$ and equal sides 8 cm long.

5. The base of an isosceles triangle is 12 cm long and the equal sides are 15 cm long. Find the three angles of the triangle.

6. Use the cosine tables to write down the angles whose cosines are

   (a) 0.9063   (b) 0.9816

   (c) 0.5105   (d) 0.2045

   (e) 0.7946   (f) 0.3716

7. In triangle ABC, $\angle C = 90°$, $\angle B = 35°48'$ and $a = 9\,\text{cm}$. Find the length of the side c.

8. In Fig. 29.14, find $\angle BAC$ and the length of BD.

Fig. 29.14

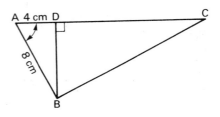

The Tangent of an Angle

In Figs. 29.1 and 29.2:

$$\tan A = \frac{\text{Side opposite angle } A}{\text{Side adjacent to angle } A}$$

$$= \frac{BC}{AC}$$

$$\tan B = \frac{\text{Side opposite angle } B}{\text{Side adjacent to angle } B}$$

$$= \frac{AC}{BC}$$

The tangent of an angle may be found by drawing, using a method similar to that shown in Example 1. However, tables of tangents are available and these are used in a similar way to the sine tables.

Example 6

(a) In Fig. 29.15, calculate the length of BC.

Fig. 29.15

BC is the side adjacent to the angle C and AB is the side opposite the angle C. Therefore

$$\frac{AB}{BC} = \tan C$$

$$BC = \frac{AB}{\tan C}$$

$$= \frac{25}{\tan 68°}$$

$$= \frac{25}{2.475}$$

$$= 10.10$$

Therefore BC is 10.10 cm.

(b) In triangle ABC, $\angle B = 90°$, $a = 8.7\,\text{cm}$ and $c = 11.2\,\text{cm}$. Find the angle A.

The triangle is sketched in Fig. 29.16, where it can be seen that a is the side opposite $\angle A$ and c is the side adjacent to $\angle A$. Hence

$$\tan A = \frac{a}{c}$$

$$= \frac{8.7}{11.2}$$

$$= 0.7768$$

$$\angle A = 37°50'$$

Fig. 29.16

Exercise 29.4

1. Use the tangent tables to write down the values of

    (a) $\tan 27°$ (b) $\tan 78°$

    (c) $\tan 87.3°$ (d) $\tan 15.4°$

    (e) $\tan 8°17'$ (f) $\tan 56°28'$

2. Use the tangent tables to write down the angles whose tangents are

    (a) 0.3057   (b) 5.1446

    (c) 0.2180   (d) 0.9590

    (e) 0.5154   (f) 1.706

3. Find the length of the sides marked x in Fig. 29.17.

Fig. 29.17

(a)   (b)   (c)

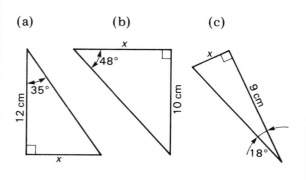

4. Find the size of the angles marked p in Fig. 29.18.

Fig. 29.18

(a)   (b)   (c)

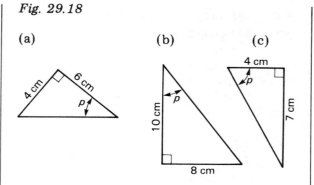

5. An isosceles triangle has a base 12 cm long, and the equal angles are each 54°. Calculate the altitude of the triangle.

6. In the triangle ABC, $\angle B = 90°$, $\angle C = 57°$ and $c = 6.1$ cm. Calculate the length a.

7. Calculate the distance marked a in Fig. 29.19.

Fig. 29.19

8. In Fig. 29.20, calculate the distance marked q.

Fig. 29.20

Angles of Elevation and Depression

If you look upwards at an object, the angle formed between the horizontal and your line of sight is called the **angle of elevation** (Fig. 29.21).

Fig. 29.21

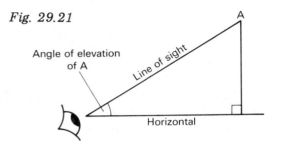

Example 7

A boy 1.5 m tall is 25 m away from a tower as shown in Fig. 29.22. The angle of elevation of the tower is 18°. How high is the tower?

Fig. 29.22

In the diagram, ∠TSR is the angle of elevation to the top of the tower. TU is the tower and SR = 25 m.

$$\frac{TR}{SR} = \tan\angle TSR$$

$$TR = SR \times \tan\angle TSR$$

$$= 25 \times \tan 18°$$

$$= 25 \times 0.3249$$

$$= 8.12$$

$$UT = 8.12\,m + 1.5\,m$$

$$= 9.62\,m$$

If you look downwards at an object, the angle formed between the horizontal and your line of sight is called the **angle of depression** (Fig. 29.23).

Fig. 29.23

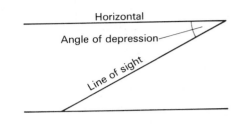

Example 8

A man standing on top of a cliff 50 m high is in line with two buoys whose angles of depression are 18° and 20°. Calculate the distance between the buoys.

The problem is illustrated in Fig. 29.24 where the buoys are C and D and the observer is A.

Fig. 29.24

In triangle ABC,
∠BAC = 90° − 20° = 70°

$$\frac{BC}{AB} = \tan\angle BAC$$

$$BC = AB \tan\angle BAC$$

$$= 50 \times \tan 70°$$

$$= 137.4\,m$$

In triangle ABD,

$$\angle BAD = 90° - 18° = 72°$$

$$\frac{BD}{AB} = \tan \angle BAD$$

$$BD = AB \tan \angle BAD$$

$$= 50 \times \tan 72°$$

$$= 153.9\,m$$

$$CD = BD - BC$$

$$= 153.9\,m - 137.4\,m$$

$$= 16.5\,m$$

Therefore the distance between the buoys is 16.5 m.

Altitude of the Sun

The **altitude of the sun** is the angle of elevation of the sun (Fig. 29.25).

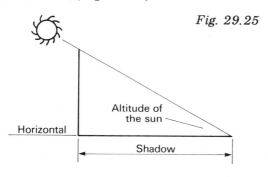

Fig. 29.25

Example 9

A flagpole is 11 m tall and it stands vertically. Calculate the length of shadow that it will cast when the altitude of the sun is 43°.

Fig. 29.26

In Fig. 29.26

$$\frac{BC}{AB} = \tan \angle CAB$$

$$AB = \frac{BC}{\tan \angle CAB}$$

$$= \frac{11}{\tan 43°}$$

$$= 11.8$$

Therefore the length of shadow cast by the flagpole is 11.8 m.

Exercise 29.5

1. From a point on level ground the angle of elevation of a post is 15°. If the post is 18 m from the point of observation, calculate the height of the post.

2. A man 1.6 m tall observes the angle of elevation of the top of a spire to be 43°. If he is standing 28 m from the foot of the spire, calculate the height of the spire.

3. A man 1.8 m tall stands 20 m from the foot of a tower which is 19.8 m high. Find the angle of elevation of the top of the tower from his eye.

4. A man lying down on top of a cliff 35 m high observes the angle of depression of a buoy to be 20°. If he is in line with the buoy, calculate the distance between the buoy and the foot of the cliff, which may be assumed to be vertical.

5. A flagpole is 17 m high. Calculate the length of shadow it casts when the altitude of the sun is 64°.

6. A pole casts a shadow 20 m long when the altitude of the sun is 49°. Calculate the height of the pole.

7. A man standing on top of a mountain observes the angle of depression of a steeple to be 41°. If the mountain is 500 m high, how far is the steeple from the mountain?

8. In Fig. 29.27, a vertical cliff is 50 m high. The cliff is observed from a boat which is 60 m from the foot of the cliff. Calculate the angle of elevation of the top of the cliff from the boat.

Fig. 29.27

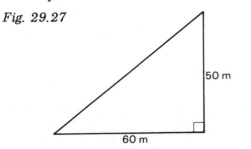

9. A person standing 30 m away from the foot of a tower observes the angles of elevation to the top and bottom of a flagpole standing on the tower to be 64° and 66° respectively. Calculate the height of the flagpole.

10. A surveyor stands 120 m from the foot of a tower on which an aerial stands. He measures the angles of elevation to the top and bottom of the aerial to be 59° and 57° respectively. What is the height of the aerial?

Area of a Triangle

Three formulae are commonly used for finding the area of a triangle.

(1) Given the base and the altitude (vertical height, Fig. 29.28):

$$\text{Area} = \tfrac{1}{2}bh$$

See also Chapter 26.

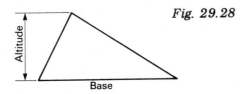

Fig. 29.28

(2) Given any two sides and the included angle (Fig. 29.29):

$$\text{Area} = \tfrac{1}{2}bc \sin A$$

$$\text{Area} = \tfrac{1}{2}ac \sin B$$

$$\text{Area} = \tfrac{1}{2}ab \sin C$$

Fig. 29.29

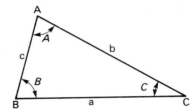

(3) Given all three sides (see also Chapter 26):

$$\text{Area} = \sqrt{s(s-a)(s-b)(s-c)}$$

where $s = \tfrac{1}{2}$ the perimeter

$$= \tfrac{1}{2}(a+b+c)$$

Example 10

(a) Find the areas of the triangles shown in Fig. 29.30.

Fig. 29.30

In each of the triangles shown in the diagram, the base is taken as being the side whose length is given, and the vertical height (altitude) is measured perpendicular to this side.

Area of
each triangle $= \tfrac{1}{2} \times \text{Base} \times \text{Altitude}$

$$= \tfrac{1}{2} \times 30 \times 20 \text{ m}^2$$

$$= 300 \text{ m}^2$$

(b) Find the area of the triangle shown in Fig. 29.31.

Fig. 29.31

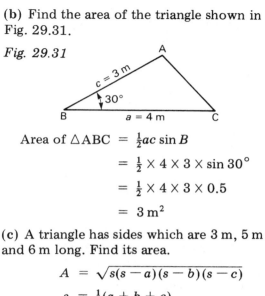

Area of $\triangle ABC = \frac{1}{2}ac \sin B$

$$= \frac{1}{2} \times 4 \times 3 \times \sin 30°$$

$$= \frac{1}{2} \times 4 \times 3 \times 0.5$$

$$= 3\,\text{m}^2$$

(c) A triangle has sides which are 3 m, 5 m and 6 m long. Find its area.

$$A = \sqrt{s(s-a)(s-b)(s-c)}$$

$$s = \frac{1}{2}(a+b+c)$$

$$= \frac{1}{2} \times (3+5+6)$$

$$= 7\,\text{cm}$$

$$A = \sqrt{7 \times (7-3) \times (7-5) \times (7-6)}$$

$$= \sqrt{7 \times 4 \times 2 \times 1}$$

$$= \sqrt{56}$$

$$= 7.48\,\text{m}^2$$

Area of a Polygon

The area of a regular polygon may be found by dividing the polygon into a number of equal triangles (the number of triangles equals the number of sides). Then by finding the area of one of these triangles the area of the polygon is determined.

Example 11

The base of a steeple is a regular octagon (eight-sided figure) of side 4 m. Find its area.

In Fig. 29.32, the octagon is divided into eight equal triangles. AOB is one of these triangles. In $\triangle AOB$

$$\angle AOB = \frac{360°}{8} = 45°$$

Since $\triangle AOB$ is isosceles

$$\angle AON = \frac{45°}{2}$$

$$= 22°30'$$

$$AN = \frac{AB}{2}$$

$$= \frac{4}{2}$$

$$= 2\,\text{m}$$

In $\triangle AON$

$$\frac{AN}{ON} = \tan \angle AON$$

$$ON = \frac{AN}{\tan \angle AON}$$

$$= \frac{2}{\tan 22°30'}$$

$$= 4.828\,\text{m}$$

Area of $\triangle AOB = \frac{1}{2} \times 4 \times 4.828$

$$= 9.656\,\text{m}^2$$

Area of octagon $= 8 \times 9.656\,\text{m}^2$

$$= 77.25\,\text{m}^2$$

Fig. 29.32

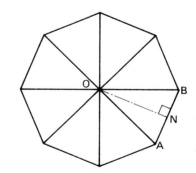

Exercise 29.6

1. Find the area of a triangle whose base is 7.5 cm and whose vertical height is 5.9 cm.

2. Find the area of an isosceles triangle whose equal sides are 8.2 cm long and whose base is 9.6 cm.

3. An equilateral triangle has an area of 15.6 cm². Calculate the lengths of its sides.

4. A triangle has sides of length 39.3 cm and 41.5 cm. If the angle between these sides is 41°30', find the area of the triangle.

5. Fig. 29.33 shows a playground. Find its area.

Fig. 29.33

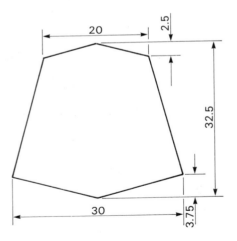

All dimensions are in metres

6. Calculate the area of triangle ABC if $a = 3$ m, $c = 6$ m and $\angle B = 63°44'$.

7. A triangle has sides 4 m, 7 m and 8 m long. Find its area.

8. Find the area of the plot of land with dimensions shown in Fig. 29.34.

Fig. 29.34

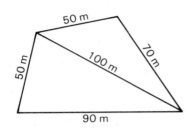

9. Find the area of a regular hexagon (six-sided figure) whose perpendicular distance between opposite sides is 4 cm.

10. A regular octagon (eight-sided figure) has sides 6 m long. Calculate its area.

Bearings

The four cardinal directions are north, south, east and west (Fig. 29.35). Directions between the cardinal points are called **bearings**.

Fig. 29.35

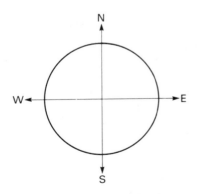

The usual way of stating a bearing is to measure the angle from north in a clockwise direction. Three figures are always stated, north being 000°. The figure 005° is written instead of 5°, 027° instead of 27°. East is 090°, south is 180° and west is 270°.

Some typical bearings are shown in
Fig. 29.36.

Fig. 29.36

Example 12

P is a point due east of a harbour H and Q
is a point on the coast 10 km due south of
H. If the distance PQ is 14 km, find the
bearing of Q from P.

Method 1 (by drawing)

First choose a suitable scale to represent
the distances HP and HQ. In Fig. 29.37,
a scale of 1 cm to 2 km has been used.
Remembering that east has a bearing of
090° and south a bearing of 180°, the
scale drawing will look like the diagram.
The reflex angle at P gives the bearing
of Q from P. By using a protractor
this angle is found to be 224°, and this
is the bearing of Q from P.

Fig. 29.37

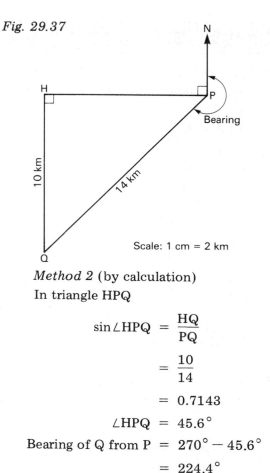

Scale: 1 cm = 2 km

Method 2 (by calculation)
In triangle HPQ

$$\sin \angle HPQ = \frac{HQ}{PQ}$$

$$= \frac{10}{14}$$

$$= 0.7143$$

$$\angle HPQ = 45.6°$$

$$\text{Bearing of Q from P} = 270° - 45.6°$$

$$= 224.4°$$

Example 13

The bearing of a point B from a point A is
065°. What is the bearing of A from B?

The situation is shown in Fig. 29.38, where
it can be seen that the bearing of A from B
is 245°.

Fig. 29.38

Exercise 29.7

1. Fig. 29.39 shows part of a map. Using three-digit numbers, state the bearings of
   (a) Beeson from Newtown
   (b) Dayton from Beeson

Fig. 29.39

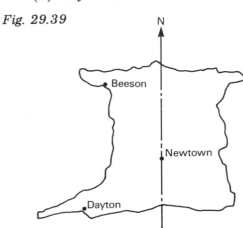

2. Fig. 29.40 shows a map of a small island. Write down the bearings of the following places from Ganley airport:
   (a) Speightstown
   (b) Bridgetown
   (c) Panton

Fig. 29.40

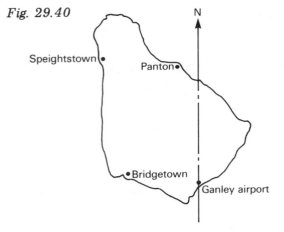

3. The bearing of Manley from Trestor is 155°. What is the bearing of Trestor from Manley?

4. A ship is on a bearing of 068° from a lighthouse. What is the bearing of the lighthouse from the ship?

5. B and C (Fig. 29.41) are both 100 km from A. C is on a bearing of 225° from B. Calculate, or find by scale drawing
   (a) the bearing of A from B
   (b) the size of the angle ABC

Fig. 29.41

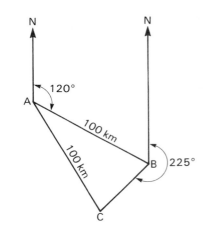

6. From Fig. 29.42, find the bearing of X from Y.

Fig. 29.42

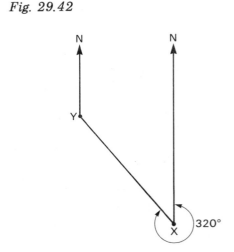

7. The distances from town A to towns B and C are 160 km and 340 km respectively. The bearing of B from A is 346° and the bearing of C from A is 035°. Make a scale drawing of this information and from it find

   (a) the distance between B and C

   (b) the bearing of C from B

8. A boat leaves a harbour H on a course of 120° and sails 80 km in this direction until it reaches a point B.

   (a) Using a scale of 1 cm to 10 km, make a scale drawing of the movement. Hence find
   (i) how far B is east of H,
   (ii) what distance B is south of H.

   (b) Find, by using trigonometry,
   (i) how far B is east of H,
   (ii) how far B is south of H.

Solid Geometry

Problems with solid figures are solved by choosing suitable right-angled triangles in different planes. It is essential to make a clear three-dimensional drawing in order to find these triangles.

Example 14

Fig. 29.43 shows a cuboid. Calculate the length of the diagonal AG.

Fig. 29.43

Fig. 29.44 shows that in order to find AG we must use the right-angled triangle AGE. GE is the diagonal of the base rectangle EFGH.

Fig. 29.44

In triangle EFG, EF = 8 cm, GF = 6 cm and ∠EFG = 90°. By Pythagoras' theorem

$$EG^2 = EF^2 + GF^2$$
$$= 8^2 + 6^2$$
$$= 64 + 36$$
$$= 100$$
$$EG = \sqrt{100}$$
$$= 10$$

In triangle AGE, AE = 4 cm, EG = 10 cm and ∠AEG = 90°. By Pythagoras' theorem

$$AG^2 = AE^2 + EG^2$$
$$= 4^2 + 10^2$$
$$= 16 + 100$$
$$= 116$$
$$AG = \sqrt{116}$$
$$= 10.77 \text{ cm}$$

Example 15

Fig. 29.45 shows a pyramid with a square base. The base has sides 6 cm long and the edges of the pyramid VA, VB, VC and VD are each 10 cm long. Find the vertical height of the pyramid.

Fig. 29.45

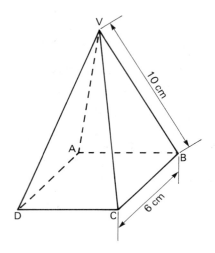

The right-angled triangle VBE (Fig. 29.46) allows the vertical height of the pyramid to be found. However, we must first find BE from the right-angled triangle BEF.

Fig. 29.46

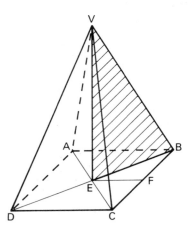

In triangle BEF, BF = EF = 3 cm and ∠BFE = 90°. By Pythagoras' theorem

$$BE^2 = BF^2 + EF^2$$
$$= 3^2 + 3^2$$
$$= 9 + 9$$
$$= 18$$
$$BE = \sqrt{18}$$
$$= 4.243 \text{ cm}$$

In triangle VBE, BE = 4.243 cm, VB = 10 cm and ∠VEB = 90°.

By Pythagoras' theorem

$$VE^2 = VB^2 - BE^2$$
$$= 10^2 - 4.243^2$$
$$= 100 - 18$$
$$= 82$$
$$VE = \sqrt{82}$$
$$= 9.055 \text{ cm}$$

Exercise 29.8

1. Fig. 29.47 shows a cuboid.

(a) Sketch the triangle EHF and calculate the length of FH.

(b) Sketch the triangle HBF adding known dimensions. Calculate the length of the diagonal BH.

Fig. 29.47

2. Fig. 29.48 shows a pyramid on a square base of side 8 cm. The altitude of the pyramid is 12 cm.

   (a) Calculate EF.

   (b) Draw the triangle VEF and add known dimensions.

   (c) Find the angle VEF.

   (d) Calculate the slant height VE.

   (e) Calculate the area of triangle VAD.

   (f) Calculate the complete surface area of the pyramid.

Fig. 29.48

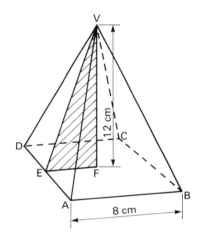

3. Fig. 29.49 shows a pyramid on a rectangular base. Calculate the length of VA.

Fig. 29.49

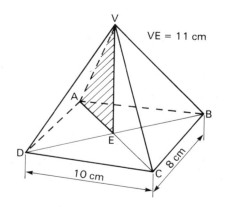

4. Fig. 29.50 shows a triangular prism with the face YDC inclined as shown in the diagram. Find the angle that the sloping face YDC makes with the base.

Fig. 29.50

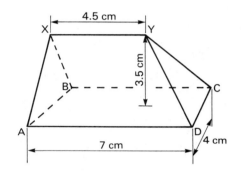

5. Fig. 29.51 shows a pyramid on a square base with VA = VB = VC = VD = 5 cm. Calculate the vertical height of the pyramid (i.e. VE).

Fig. 29.51

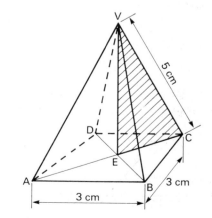

6. Fig. 29.52 shows a wooden wedge. Find the length of EA and the size of the angle marked q. The end faces ADF and EBC are equilateral triangles.

Fig. 29.52

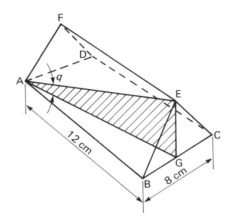

7. Fig. 29.53 shows a shed with a slanting roof ABCD. The rectangular base ABEF rests on level ground and the shed has three vertical sides.

(a) Determine the angle between the slanting roof and the ground.

(b) The roof is to be covered with roofing felt. What area of roofing felt is required?

(c) Calculate the volume of the shed in cubic metres.

Fig. 29.53

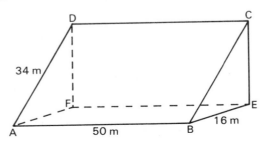

Miscellaneous Exercise 29

1. Fig. 29.54 represents a ramp. The distance AC is 5 m and the distance AB is 1 m. Use the trigonometrical tables to find (a) $\angle ACB$, (b) the distance BC, giving your answer correct to 3 significant figures.

Fig. 29.54

2. In the triangle ABC (Fig. 29.55) AB = AC = 10 cm, $\angle ABC = 50°$ and AN is perpendicular to BC. Calculate (a) the size of $\angle BAC$, (b) the length of BC, (c) the length of AN.

Fig. 29.55

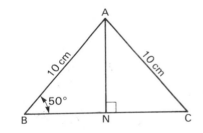

3. Given Fig. 29.56, calculate (a) the length of CD, (b) $\tan \angle ABD$, (c) the size of $\angle ABD$.

Fig. 29.56

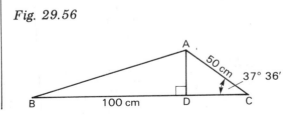

4. Fig. 29.57 represents a vertical tower TB standing on horizontal ground AB with AB = 100 m. The angle of elevation of the top of the tower T from A is 31.3°.

   (a) Calculate the height of the tower TB.

   (b) A man standing at A walks 20 m directly towards B. Calculate the angle of elevation of T from his new position.

Fig. 29.57

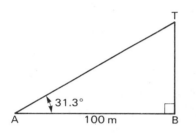

5. In Fig. 29.58, the area of the rectangle ABCD is 62 cm², AD = 8 cm, ED = 12.9 cm and ∠ADE = 90°.

   (a) Find the length of AB.

   (b) Calculate the area of triangle ADE.

   (c) Find the size of ∠AED.

Fig. 29.58

6. (a) Find x if $\sin x = 0.3891$.

   (b) For what angle is 0.4000 the cosine?

7. Fig. 29.59 represents a coastguard at C looking out to sea from the top of a cliff which is 100 m high. He sights a rowing boat at A and observes that it makes an angle of depression of 15°. If BC represents the vertical cliff, find the distance of the rowing boat from the foot of the cliff.

Fig. 29.59

8. Fig. 29.60 shows a vertical mast PQ which stands on horizontal ground. A straight wire 20 m long runs from P at the top of the mast to a point R on the ground which is 10 m from the foot of the mast.

   (a) Calculate the angle of inclination of the wire to the ground (i.e. find ∠PRQ).

   (b) Calculate the height of the mast.

Fig. 29.60

9. The bearing of point A from point B is 207°. What is the bearing of B from point A?

10. Point C is 50 km from point A on a bearing of 320°. Point B is 60 km from A on a bearing of 230°. Make a scale drawing showing the relative positions of A, B and C using a scale of 1 cm to 10 km. Hence find the distance between the points B and C and state the bearing of B from C.

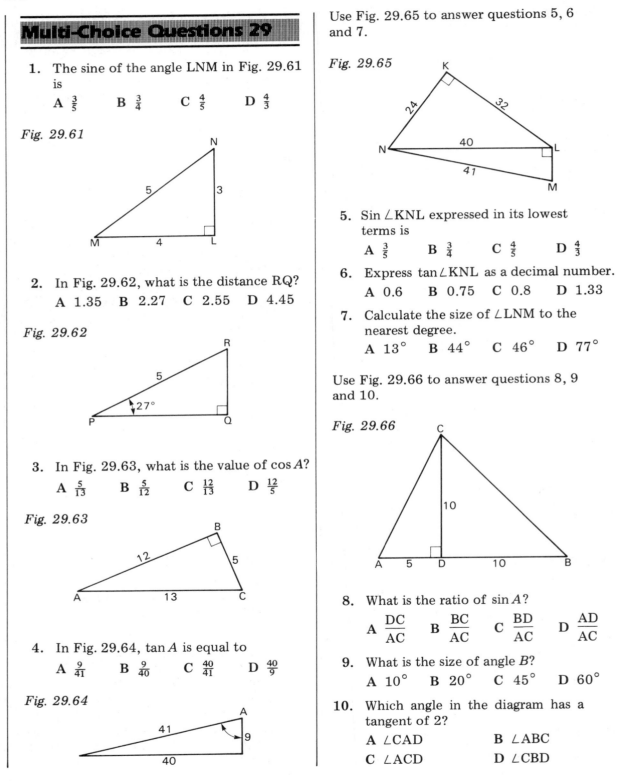

Multi-Choice Questions 29

1. The sine of the angle LNM in Fig. 29.61 is

   A $\frac{3}{5}$ B $\frac{3}{4}$ C $\frac{4}{5}$ D $\frac{4}{3}$

Fig. 29.61

2. In Fig. 29.62, what is the distance RQ?

   A 1.35   B 2.27   C 2.55   D 4.45

Fig. 29.62

3. In Fig. 29.63, what is the value of cos A?

   A $\frac{5}{13}$ B $\frac{5}{12}$ C $\frac{12}{13}$ D $\frac{12}{5}$

Fig. 29.63

4. In Fig. 29.64, tan A is equal to

   A $\frac{9}{41}$ B $\frac{9}{40}$ C $\frac{40}{41}$ D $\frac{40}{9}$

Fig. 29.64

Use Fig. 29.65 to answer questions 5, 6 and 7.

Fig. 29.65

5. Sin ∠KNL expressed in its lowest terms is

   A $\frac{3}{5}$ B $\frac{3}{4}$ C $\frac{4}{5}$ D $\frac{4}{3}$

6. Express tan ∠KNL as a decimal number.

   A 0.6   B 0.75   C 0.8   D 1.33

7. Calculate the size of ∠LNM to the nearest degree.

   A 13°   B 44°   C 46°   D 77°

Use Fig. 29.66 to answer questions 8, 9 and 10.

Fig. 29.66

8. What is the ratio of sin A?

   A $\dfrac{DC}{AC}$ B $\dfrac{BC}{AC}$ C $\dfrac{BD}{AC}$ D $\dfrac{AD}{AC}$

9. What is the size of angle B?

   A 10°   B 20°   C 45°   D 60°

10. Which angle in the diagram has a tangent of 2?

   A ∠CAD   B ∠ABC

   C ∠ACD   D ∠CBD

Use Fig. 29.67 to answer questions 11, 12 and 13.

Fig. 29.67

11. If ∠ABC is a right angle, what is the value of x?

   A 135°   B 105°   C 75°   D 45°

12. The length of DE in centimetres is

   A 4.24   B 3.00   C 2.12   D 1.5

13. The length of CD in centimetres is

   A 3 tan 30°   B 3 tan 60°
   C 3 sin 30°   D 3 cos 30°

14. Fig. 29.68 shows a rhombus with sides 20 cm long, and ∠PSR = 60°. The height h in centimetres, to the nearest centimetre, is

   A 10   B 14   C 17   D 20

Fig. 29.68

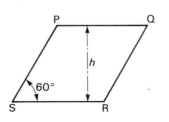

15. In Fig. 29.69, what is the bearing of P from Q?

   A 025°   B 065°   C 245°   D 295°

Fig. 29.69

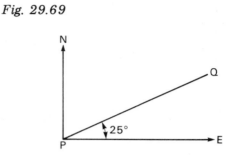

Use Fig. 29.70 to answer questions 16 and 17.

16. The bearing of B from A is

   A 030°   B 060°   C 120°   D 150°

17. The bearing of C from A is

   A 045°   B 135°   C 225°   D 315°

Fig. 29.70

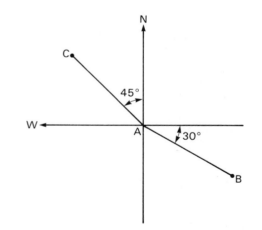

18. In Fig. 29.71, PQRS is a horizontal rectangular parade ground 80 m by 60 m. PT is a vertical pole 47 m high. To the nearest degree, the angle of elevation from T to R is

   A 25°   B 28°   C 30°   D 38°

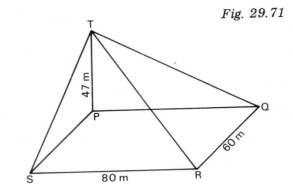

Fig. 29.71

19. From a place 400 m north of X, a man walks eastwards to a place Y which is 800 m from X. What is the bearing of X from Y?

A 270° B 240° C 210° D 180°

20. In the triangle LMN, LM = 4 cm, MN = 7 cm and ∠LMN = 30°. What is the area of triangle LMN?

A 14 cm²   B 9 cm²

C 7 cm²   D 4 cm²

Sets

Collections

A set is a collection of objects, numbers, ideas, etc. Some sets have special names, such as a pack of cards, a pride of lions, a flock of sheep, a fleet of ships, and so on. However, many sets do not have special names because the objects in the set are seldom considered collectively; for example, the objects in a girl's handbag.

Elements

The different objects, numbers, etc., which form a set are called the **members** or **elements** of the set, just as the players who belong to a cricket team are called members of the team.

Describing Sets

A set may be described by one of the following methods:

  (i) By listing all of the elements; for instance, $A = \{1, 3, 5, 7, 9\}$. The order of the elements does not matter and each element is listed once only.

 (ii) By listing only enough of the elements to indicate the pattern and showing that the pattern continues by using dots. Thus $B = \{2, 3, 4, 5, \ldots\}$.

(iii) By description, such as {all even numbers}.

(iv) By using an algebraic expression, such as $C = \{x : 0 \leqslant x \leqslant 8,\ x \text{ an integer}\}$ which denotes a set of elements x such that x is an integer whose value lies between 0 and 8, i.e. $C = \{0, 1, 2, 3, 4, 5, 6, 7, 8\}$. This notation is called **set builder notation**.

Types of Sets

A **finite set** is one in which all the elements are listed, such as $\{2, 3, 4, 5, 6\}$.

An **infinite set** is one in which it is impossible to list all the members. Thus if $B = \{$all even numbers$\}$, it is not possible to list all of the elements and so we write $B = \{2, 4, 6, 8, \ldots\}$ where the dots mean 'and so on'.

The **null set** contains no elements. It is represented by $\emptyset$ or by $\{\ \}$. The set of whole numbers between 0.3 and 0.7 is null and so is the set of prime numbers between 8 and 10.

Membership of a Set

The symbol $\in$ means 'is a member of'. Thus if $A = \{2, 4, 6, 8, 10\}$. The fact that 6 is a member of A is written $6 \in A$.

The symbol $\notin$ means 'is not a member of'. Thus, since 3 is not a member of A, we write $3 \notin A$.

Exercise 30.1

Write down the members of the following sets:

1. { prime numbers less than 20 }

2. { multiples of 5 up to 30 }

3. { odd numbers between 8 and 16 }

4. { even numbers between 1 and 13 }

State which of the following sets are finite, infinite or null:

5. $A = \{$ even numbers $\}$

6. $B = \{1, 3, 5, 7, 9\}$

7. $C = \{$ letters of the alphabet $\}$

8. $D = \{$ even numbers divisible by 3 $\}$

9. $E = \{$ people who have swum the Atlantic Ocean $\}$

10. $F = \{$ points on a straight line $\}$

State which of the following statements are true:

11. $3 \in \{$ prime factors of 30 $\}$

12. marlin $\notin \{$ fish $\}$

13. rectangle $\in \{$ quadrilaterals $\}$

14. octagon $\in \{$ triangles $\}$

15. polygon $\notin \{$ solid figures $\}$

16. $5 \notin \{$ odd numbers $\}$

Write the following statements with the correct symbol $\in$ or $\notin$ in place of $\circ$:

17. horse $\circ \{$ animals with horns $\}$

18. cricket $\circ \{$ outdoor games $\}$

19. $\frac{5}{2} \circ \{$ improper fractions $\}$

20. $25 \circ \{$ multiples of 3 $\}$

Subsets

Consider the set $P = \{2, 5, 7, 8, 9, 11, 12\}$. A diagram of this set can be made by enclosing the members in a ring, as shown in Fig. 30.1.

Fig. 30.1

2   5   7   8
9   11   12

Now consider the set $E = \{2, 8, 12\}$. We see that all of the members of E are also members of P. We say that E is a subset of P and we write $E \subset P$. We can show this in a diagram by drawing a ring within a ring (Fig. 30.2).

Fig. 30.2

It is quite common for a set to contain several subsets. For instance, if $Q = \{$ quadrilaterals $\}$, $R = \{$ rectangles $\}$, $S = \{$ squares $\}$ and $P = \{$ parallelograms $\}$, then R, S and P are all subsets of Q, as illustrated in Fig. 30.3.

Fig. 30.3

Diagrams such as Figs. 30.1, 30.2 and 30.3 are called **Venn diagrams**.

The null set and the set itself are considered to be subsets of a given set.

Example 1

List all the subsets of $\{a, b, c\}$.

The subsets are $\varnothing$, $\{a\}$, $\{b\}$, $\{c\}$, $\{a, b\}$, $\{a, c\}$, $\{b, c\}$ and $\{a, b, c\}$.

Every possible subset of a given set, except the set itself, is called a **proper subset**.

The Number of Subsets

If there are n elements in a set, then the number of subsets N is given by

$$N = 2^n$$

Example 2

A set has 5 elements. How many subsets can be formed from it?

The number of subsets is

$$N = 2^5 = 32$$

Equality of Sets

As stated previously, the order in which the elements of a set are written does not matter. Thus the set $\{1, 3, 5, 7\}$ is the same as the set $\{3, 5, 7, 1\}$.

Two sets are said to be **equal** if their elements are identical. Thus if $A = \{3, 5, 7, 9\}$ and $B = \{7, 5, 9, 3\}$, then $A = B$.

The Universal Set

The **universal set** for any particular problem is the set that contains all the available elements for that problem.

Suppose that in a class in a school we are asked to form sets of those who (i) like playing cricket, (ii) wear spectacles and (iii) cycle to school. Each of these sets is a subset of $\{\text{class}\}$. The class in this case is the universal set.

The symbol $\mathscr{E}$ is generally used to denote the universal set but sometimes the letter $\mathscr{U}$ is used.

Example 3

If $\mathscr{E} = \{1, 2, 3, 4, 5, 6, 7\}$, write down the subset $E = \{\text{even numbers}\}$ and represent the relationship in a Venn diagram.

$$E = \{2, 4, 6\}$$

The diagram is shown in Fig. 30.4 where it will be seen that the universal set is represented by a rectangle and the subset is shown by a closed curve.

Fig. 30.4

Complement

The **complement** of a set A is the set of elements of $\mathscr{E}$ which do not belong to A. Thus is $\mathscr{E} = \{1, 3, 5, 7, 9\}$ and $A = \{1, 9\}$ then the complement of A, A', is $\{3, 5, 7\}$. In Fig. 30.5 the shaded portion represents A'.

Fig. 30.5

Exercise 30.2

1. If
   $A = \{3, 4, 5, 7, 8, 9, 10, 12, 13, 15, 16\}$,
   list the subsets whose members are:

   (a) all the odd numbers of A

   (b) all the even numbers of A

   (c) all the prime numbers of A

   (d) all the numbers in A that are divisible by 4.

2. $B = \{m, n, p, q\}$. Write down all the subsets of B. How many subsets are there?

3. If $P = \{2, 4, 8, 16, 32\}$, $Q = \{2, 4, 7, 8\}$ and $R = \{2, 16\}$, state which of the following statements are correct:

   (a) $P \subset Q$ (b) $Q \subset P$

   (c) $R \subset P$ (d) $R \subset Q \subset P$

4. Below are eight sets. Connect the appropriate sets by the symbol ⊂.

    {whole numbers between 3 and 25}

    {all cutlery}

    {all outdoor games}

    {knife, fork}

    {all domestic animals}

    {9, 15, 20, 21}

    {football, cricket, tennis}

    {sheep, goats, cows, dogs, cats}.

5. A set has 6 elements. How many subsets can be formed from it?

6. Draw a Venn diagram to represent the set {2, 3, 5, 8, 9, 12}.

7. Draw a Venn diagram to represent the relationship between the two sets {all triangles} and {acute-angled triangles}.

8. If $\mathscr{E} = \{2, 5, 8, 9, 11\}$ and $M = \{2, 5, 8\}$, write down the elements of M', the complement of M, and represent M' in a Venn diagram.

9. Figure 30.6 shows a Venn diagram representing the universal set $\mathscr{E}$ and two subsets A and B. Write down the elements of A' and B'.

Fig. 30.6

10. $\mathscr{E} = \{$letters of the alphabet$\}$. Give the subsets of

    (a) vowels

    (b) letters after t

    (c) consonants up to n

Relations between Number Systems

As shown in Chapter 1,

counting numbers or **natural numbers** are the numbers 1, 2, 3, 4, 5, 6, . . .

whole numbers include natural numbers and also zero. Hence whole numbers are the numbers 0, 1, 2, 3, 4, 5, . . .

integers are whole numbers but they also include negative numbers. Some examples of integers are $-27, -5, 0, 8$ and 32.

Other types of numbers include:

rational numbers which are numbers that can be expressed as fractions with integer numerators (top part) and non-zero integer denominators.

Types of rational numbers are as follows:

(i) proper fractions, such as $\dfrac{1}{4}, \dfrac{3}{7}$ and $-\dfrac{19}{32}$

(ii) improper fractions, such as $\dfrac{5}{3}, \dfrac{15}{11}$ and $-\dfrac{84}{79}$

(iii) whole numbers, such as 8, because 8 can be written as $\dfrac{8}{1}$

(iv) decimal numbers, such as 0.73, because 0.73 can be written as $\dfrac{73}{100}$

(v) recurring decimals, such as $0.\dot{1}$ and $0.\dot{1}\dot{8}$, since they represent the fractions $\dfrac{1}{9}$ and $\dfrac{2}{11}$, respectively

(vi) zero because it can be the numerator (top part) of a fraction whose denominator is not zero: thus $\dfrac{0}{6}$ and $\dfrac{0}{15}$ are rational numbers whose value is 0.

irrational **numbers,** which cannot be expressed as the ratio of two integers. Thus $\sqrt{6}$ and $\sqrt{13}$ are two examples of irrational numbers. Note that not all square roots are irrational. For instance

$$\sqrt{16} = 4 = \frac{4}{1} \text{ and}$$

$$\sqrt{0.25} = 0.5 = \frac{1}{2}$$

real numbers, which include all the rational and irrational numbers.

If $\quad N = \{\text{natural numbers}\}$

$\quad\quad W = \{\text{whole numbers}\}$

$\quad\quad Z = \{\text{integers}\}$

$\quad\quad Q = \{\text{rational numbers}\}$

and $\quad R = \{\text{real numbers}\}$

then $N \subset W \subset Z \subset Q \subset R$.

That is, all natural numbers are included in the set of whole numbers which are, in turn, included in the set of integers which are, in turn, included in the set of rational numbers which are, in turn, included in the set of real numbers.

Set Builder Notation

An algebraic expression may be used to denote the elements of a set. For instance, $A = \{x : 3 \leqslant x \leqslant 8, \ x \in Z\}$ is read as 'the set of integers x such that x lies between 3 and 8 inclusive'. Thus $A = \{3, 4, 5, 6, 7, 8\}$. When a set is described using an algebraic expression, we are using **set builder notation**

Example 4

Show that $\{x : -3 \leqslant x \leqslant 2, \ x \in Z\} \subset \{x : -5 \leqslant x \leqslant 4, \ x \in Z\}$

$\{x : -3 \leqslant x \leqslant 2\} = \{-3, -2, -1, 0, 1, 2\}$

$\{x : -5 \leqslant x \leqslant 4\} = \{-5, -4, -3, -2, -1,$
$\quad\quad\quad\quad\quad\quad\quad\quad 0, 1, 2, 3, 4\}$

$\{x : -3 \leqslant x \leqslant 2\} \subset \{x : -5 \leqslant x \leqslant 4\}$

1. Write the following numbers in rational form:

    (a) $2\frac{1}{2}$ $\quad$ (b) $-4\frac{1}{3}$ $\quad$ (c) 5

    (d) 0.89 $\quad$ (e) $\sqrt{36}$

2. Draw a number line. On it show the positions of the numbers $-2, -1\frac{1}{4},$ 1.3 and 2.8.

3. Write down all the elements of A where $A = \{x : 3 \leqslant x \leqslant 9, \ x \in W\}$.

4. Show that $\{x : -2 \leqslant x \leqslant 4, \ x \in Z\} \subset \{x : -4 \leqslant x \leqslant 7, \ x \in Z\}$.

5. Using a single Venn diagram, represent the following:

    $\mathcal{E} = \{x : 3 \leqslant x \leqslant 10, \ x \in W\}$

    $A = \{x : 5 \leqslant x \leqslant 8, \ x \in W\}$

    $B = \{6, 7, 9\}$

6. Draw a Venn diagram to represent the following three sets:

    $\mathcal{E} = \{x : 1 \leqslant x \leqslant 10, \ x \in N\}$

    $P = \{x : 2 \leqslant x \leqslant 7, \ x \in N\}$

    $Q = \{x : 1 \leqslant x \leqslant 5, \ x \in N\}$

Intersection

The intersection of two sets, A and B, is the set of elements which are members of *both* A and B. $A \cap B$ means the intersection of the sets A and B.

If $A = \{3, 5, 8\}$ and $B = \{2, 3, 4, 5\}$, then the intersection of A and B is $A \cap B = \{3, 5\}$. The shaded portion of Fig. 30.7 represents the intersection of A and B.

Fig. 30.7

$A \cap B = \{3, 5\}$

Union

The union of two sets, A and B, is the set of *all* the elements contained in A *and* all those in B. $A \cup B$ means the union of the sets A and B.

If $A = \{1, 3, 5\}$ and $B = \{2, 3, 5, 6, 8\}$, then the union of A and B is $A \cup B = \{1, 2, 3, 5, 6, 8\}$.
The shaded portion of Fig. 30.8 represents the union of A and B.

Fig. 30.8

$$A \cup B = \{1, 2, 3, 5, 6, 8\}$$

Example 5

If $\mathscr{E} = \{1, 2, 3, 4, 5, 6, 7, 8, 9, 10, 11, 12\}$, $A = \{2, 3, 4, 5\}$ and $B = \{2, 4, 6, 8, 10\}$, draw a Venn diagram to represent this information. Hence write down the elements of

(a) A', (b) $A \cap B$, (c) $A \cup B$, (d) B'.

The Venn diagram is shown in Fig. 30.9.

Fig. 30.9

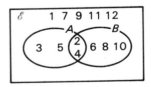

(a) $A' = \{1, 6, 7, 8, 9, 10, 11, 12\}$

(b) $A \cap B = \{2, 4\}$

(c) $A \cup B = \{2, 3, 4, 5, 6, 8, 10\}$

(d) $B' = \{1, 3, 5, 7, 9, 11, 12\}$

1. Figure 30.10 is a Venn diagram which shows the universal set and two subsets A and B. Write down all the elements of

   (a) $\mathscr{E}$ (b) A (c) B

   (d) A' (e) B' (f) $A \cap B$

   (g) $A \cup B$ (h) $(A \cup B)'$

Fig. 30.10

2. Figure 30.11 shows a Venn diagram representing the universal set $\mathscr{E}$ and two subsets X and Y. Draw a similar diagram and insert on it the elements to represent $\mathscr{E} = \{1, 2, 3, 4, 5, 6, 7, 8, 9, 10\}$, $X = \{2, 4, 6, 7, 8\}$ and $Y = \{1, 3, 4, 5, 7, 10\}$.

Fig. 30.11

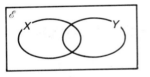

3. Use set notation to describe the shaded portions of the Venn diagrams shown in Fig. 30.12.

Fig. 30.12

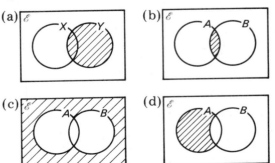

4. The Venn diagram in Fig. 30.13 shows two sets, A and B, which are subsets of the universal set $\mathscr{E}$. Write down all the elements of

(a) $\mathscr{E}$ (b) A (c) B

(d) A' (e) B' (f) $A \cup B$

(g) $A \cap B$

Fig. 30.13

5. Figure 30.14 shows three Venn diagrams. Draw diagrams similar to these to represent each of the following:

(a) $\mathscr{E} = \{5, 6, 7, 8, 9, 10, 11, 12\}$, $X = \{6, 7, 9, 10, 11\}$, $Y = \{6, 7\}$

(b) $\mathscr{E} = \{2, 3, 4, 5, 6, 7, 8, 9\}$, $X = \{5, 6\}$, $Y = \{7, 8, 9\}$

(c) $\mathscr{E} = \{2, 3, 4, 5, 6, 7, 8, 10\}$, $X = \{2, 3, 4, 6\}$, $Y = \{3, 4, 7, 8\}$

Fig. 30.14

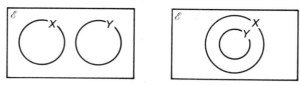

The Number of Elements in a Set

The number of elements in a set A is denoted by $n(A)$. Thus if $A = \{2, 4, 6, 8\}$ $n(A) = 4$ because the set A contains four elements.

Example 6

Figure 30.15 shows two intersecting sets A and B. The figures give the number of elements in each region. Determine

(a) $n(A)$ (b) $n(B)$

(c) $n(A \cap B)$ (d) $n(A \cup B)$

Fig. 30.15

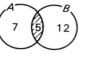

(a) $n(A) = 7 + 5 = 12$

(b) $n(B) = 5 + 12 = 17$

(c) Since the shaded area in the diagram represents $A \cap B$, $n(A \cap B) = 5$

(d) $n(A \cup B) = 7 + 5 + 12 = 24$

Problems on the Intersection of Sets

Example 7

Draw a Venn diagram to represent the two intersecting sets $A = \{2, 3, 4, 5, 6\}$ and $B = \{1, 3, 4, 7\}$. Hence find $n(A)$, $n(B)$, $n(A \cap B)$ and $n(A \cup B)$.

The Venn diagram is shown in Fig. 30.16. We see that

$n(A) = 5$

$n(B) = 4$

$A \cap B = \{3, 4\}$;   hence

$n(A \cap B) = 2$

$A \cup B = \{1, 2, 3, 4, 5, 6, 7\}$

hence $n(A \cup B) = 7$

Fig. 30.16

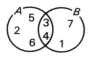

Example 8

In a school of 150 pupils, 85 take physics and 115 take mathematics. Each pupil takes at least one of these subjects. How many take both?

Refer to Fig. 30.17.

Fig. 30.17

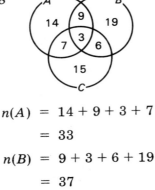

M = {pupils who take mathematics}
$n(M) = 115$

P = {pupils who take physics}
$n(P) = 85$

$M \cap P$ = {pupils who take both subjects}
$n(M \cap P) = x$

$M \cup P$ = {number of pupils in the school}
$n(M \cup P) = 150$

$$115 - x + x + 85 - x = 150$$
$$200 - x = 150$$
$$x = 50$$

Hence the number of pupils who take both subjects is 50.

Example 9

The entries in Fig. 30.18 show the number of elements in the various regions. Find $n(A)$, $n(B)$, $n(C)$, $n(A \cap B)$, $n(A \cap C)$, $n(B \cap C)$, $n(A \cap B \cap C)$ and $n(A \cup B \cup C)$.

Fig. 30.18

$$n(A) = 14 + 9 + 3 + 7$$
$$= 33$$
$$n(B) = 9 + 3 + 6 + 19$$
$$= 37$$

$$n(C) = 7 + 3 + 6 + 15$$
$$= 31$$
$$n(A \cap B) = 9 + 3$$
$$= 12$$
$$n(A \cap C) = 7 + 3$$
$$= 10$$
$$n(B \cap C) = 3 + 6$$
$$= 9$$
$$n(A \cap B \cap C) = 3$$
$$n(A \cup B \cup C) = 14 + 9 + 3 + 7 + 19$$
$$+ 6 + 15$$
$$= 73$$

Example 10

The following information concerns a sports club which has 200 members:

   50 play cricket only
   11 play football only
   40 play cricket and football only
   21 play hockey and cricket only
    5 play football and hockey only
   14 play all three games

If every member of the club plays at least one of these games, find the number of members who play hockey only.

The Venn diagram is shown in Fig. 30.19 where H = members who play hockey
C = members who play cricket
F = members who play football.

Fig. 30.19

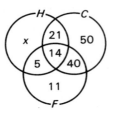

If x is the number who play hockey only, then

$$x + 21 + 14 + 5 + 40 + 50 + 11 = 200$$

$$x + 141 = 200$$

$$x = 59$$

Hence 59 members play hockey only.

Correspondence and Equivalence

When two sets have the same number of elements, **one to one correspondence** is said to exist. When two sets possess one to one correspondence they are said to be **equivalent**. One to one correspondence is often indicated by a two-way arrow.

Thus if the set A is equivalent to the set B, we write $A \leftrightharpoons B$.

If $X = \{1, 2, 3, 4, 5\}$ and $Y = \{a, b, c, d, e\}$ then $X \leftrightharpoons Y$ because $n(X) = n(Y)$.

Exercise 30.5

1. Figure 30.20 shows two intersecting sets, X and Y. The entries give the number of elements in each region. Find

    (a) $n(X)$ (b) $n(Y)$

    (c) $n(X \cap Y)$ (d) $n(X \cup Y)$

Fig. 30.20

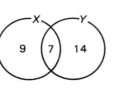

2. The number of elements in sets A and B are shown in Fig. 30.21. If $n(A \cup B) = 30$, find $n(A)$, $n(B)$ and $n(A \cap B)$.

Fig. 30.21

3. 120 candidates in an examination took history and geography, every student taking at least one of these subjects. Figure 30.22 shows the number of students who took history only and geography only. How many students took both subjects and how many, in total, took history?

Fig. 30.22

4. Two sets P and Q intersect. If $n(P) = 20$ and $n(Q) = 30$, find $n(P \cap Q)$ if $n(P \cup Q) = 45$.

5. In a class of 25 students, 18 take mathematics and 12 take physics. How many take

    (a) mathematics only

    (b) physics only

    (c) both subjects

6. The entries in Fig. 30.23 show the number of elements in each region. Find

    (a) $n(X)$ (b) $n(Y)$ (c) $n(Z)$

    (d) $n(X \cap Y)$ (e) $n(X \cap Z)$

    (f) $n(Y \cap Z)$ (g) $n(X \cap Y \cap Z)$

    (h) $n(X \cup Y \cup Z)$

Fig. 30.23

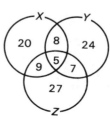

7. 100 cars were checked for defective brakes, lights and tyres. The results of the check are shown in Fig. 30.24. Find:

    (a) the total number of cars with defective brakes

    (b) the number of cars with only defective tyres

    (c) the number of cars with defective lights and defective tyres

    (d) the number of cars with all three defects

    (e) the number of cars with at least one defect

Fig. 30.24

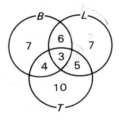

8. A newsboy delivers copies of three weekly papers on his round. They are *Churchman, Villager* and *Shopper*. Figure 30.25 shows the number of deliveries. Find:

    (a) the number of houses taking all three papers

    (b) the number of houses taking *Churchman* and *Shopper* but not *Villager*

    (c) the number of houses taking *Shopper*

Fig. 30.25

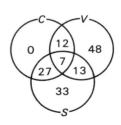

9. The Venn diagram in Fig. 30.26 shows the number of people who read the papers *Sun, Moon* and *Globe*.

    (a) How many people read only one of these papers?

    (b) How many people read all three papers?

Fig. 30.26

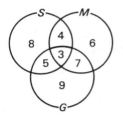

10. State which of the following are true statements and which are false:

    (a) $\{a, b, c\} \Leftrightarrow \{2, 3, 4\}$

    (b) $\{1, 4, 7\} = \{4, 7, 1\}$

    (c) $\{wood, electricity, gas, oil\} \Leftrightarrow \{girl, boy, man, woman\}$

    (d) $\{mongoose, rat, mouse, lion, bear, tiger\} = \{chair, table, wardrobe, bed, settee\}$

11. Below are two lots of four sets. Using a two-way arrow, write down the sets in lot 1 which are equivalent to those in lot 2.

    Lot 1: $\{chisel, saw, hammer\}$
    $\{m, n, o, p, q, r\}$
    $\{rectangle, trapezium, square, parallelogram, kite\}$
    $\{rule, compass, protractor, set square\}$

    Lot 2: $\{7, 8, 9, 10, 11\}$
    $\{cylinder, sphere, cuboid, cone\}$
    $\{marlin, king fish, tuna fish, flying fish, eel\}$
    $\{spade, trowel, fork\}$

Miscellaneous Exercise 30

Section A

1. Given that

   $\mathscr{E} = \{1, 2, 3, 4, 5, 6, 7, 8, 9, 10\}$

   $T = \{1, 3, 4, 5, 8\}$

   $S = \{3, 5, 7, 8\}$

   (a) Draw a Venn diagram similar to Fig. 30.27 and shade $T \cap S$.

   (b) List the members of the set $(T \cap S)'$

Fig. 30.27

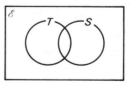

2. $\mathscr{E} = \{\text{letters of the alphabet}\}$. Write down the subsets of

   (a) vowels

   (b) letters after t

3. The universal set is

   $\mathscr{E} = \{1, 2, 3, 4, 5, 6, 7, 8\}$

   $A = \{3, 6, 7, 8\}$

   $B = \{2, 3, 5, 7\}$

   (a) List the set $A \cap B$.

   (b) List the set A'.

   (c) List the set $(A \cap B)'$.

4. Using the Venn diagram (Fig. 30.28) write down all the elements of

   (a) the universal set

   (b) the set A

   (c) the set B'

   (d) the set $A \cap B$

   (e) the set $A \cup B$

Fig. 30.28

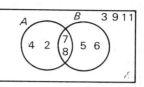

5. Draw a Venn diagram to represent the following information:

   $\mathscr{E} = \{A, B, C, D, E, F, G, H\}$

   $X = \{A, B, C, D\}$

   $Y = \{C, D, E, F, G\}$

   Use the diagram to list $(X \cap Y)'$.

Section B

1. $\mathscr{E} = \{\text{prime numbers}\}$

   $T = \{\text{prime factors of 30}\}$

   $S = \{\text{prime factors of 70}\}$

   (a) List the members of the sets T and S.

   (b) Draw a diagram similar to Fig. 30.29 and shade $S \cap T'$. Write down the only member of this set.

Fig. 30.29

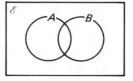

2. Copy the Venn diagram shown in Fig. 30.30 and shade A'.

Fig. 30.30

3. State which of the diagrams of Fig. 30.31 show

   (a) $B \subset A$

   (b) $A \cap B = \emptyset$

   (c) $A \cap B' = A$

   (d) $A \cap B = A$

Fig. 30.31

(i)   (ii)

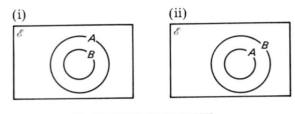

(iii)

4. $\mathscr{E} = \{1, 2, 3, 4, 5, 6, 7, 8, 9, 10\}$

   $A = \{2, 4, 6, 8, 10\}$

   $B = \{1, 2, 3, 5, 7\}$

   (a) Write down the elements of $A \cap B$.

   (b) Write down the elements of A'.

   (c) Find the number of elements in $A' \cup B'$, i.e. find $n(A' \cup B')$.

5. Copy the Venn diagram of Fig. 30.32 and shade the region $A \cap B'$.

Fig. 30.32

6. If $A = \{G, E, O, F\}$ and $Z = \{B, O, Y, C, T\}$, list the elements of the set $A \cup Z$.

7. If $P = \{a, b, c, d, e, f, g\}$ and $Q = \{b, x, y, d\}$, write down the set $P \cap Q$.

8. If $A = \{$positive multiples of 3$\}$ and $B = \{$all the factors of 36$\}$, list the members of (a) B,   (b) $A \cap B$.

9. T is the set of factors of 12 and F is the set of multiples of 5 less than 28. List the elements of (a) T,   (b) F, (c) $T \cap F$.

10. The Venn diagram in Fig. 30.33 shows the number of elements in each of two sets X and Y which are subsets of the universal set $\mathscr{E}$. Use this diagram to answer the following:

    (a) Write down $n(X)$.

    (b) Write down $n(X \cap Y)$.

    (c) Write down $n(X \cap Y')$.

    (d) Write down $n(X \cup Y)'$.

Fig. 30.33

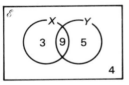

11. The Venn diagram in Fig. 30.34 represents the subsets A and B of the universal set $\mathscr{E}$. List the members of

    (a) $A \cap B$ (b) $(A \cup B)'$

Fig. 30.34

12. If $\mathscr{E} = \{1, 2, 3, 4, 5, 6, 7, 8, 9, 10\}$, $A = \{2, 4, 6, 8, 10\}$ and $B = \{1, 2, 5, 10\}$, write down:

    (a) the value of $n(B)$

    (b) the elements of $A \cap B$

    (c) the elements of A'

    (d) the elements of $(A \cup B)'$

Multi-Choice Questions 30

The following information is to be used in Questions 1 to 4:

$$\mathscr{E} = \{1, 2, 3, 4, 5, 6, 7\}$$
$$X = \{1, 2, 6, 7\}$$
$$Y = \{2, 3, 4, 5, 6\}$$
$$Z = \{4, 5, 6, 7\}$$

1. $X \cap Y$ is
   A {2, 6}   B Ø
   C {3, 4, 5}   D {1, 2, 4, 5, 6, 7}

2. $X \cup Z$ is
   A {2, 6}   B Ø
   C {3, 4, 5}   D {1, 2, 4, 5, 6, 7}

3. X' is
   A {2, 6}   B Ø
   C {3, 4, 5}   D {1, 2, 4, 5, 6, 7}

4. $Y \cap Y'$ is
   A {2, 6}   B Ø
   C {3, 4, 5}   D {1, 2, 4, 5, 6, 7}

5. Which of the diagrams in Fig. 30.35 shows the relation $P \subset Q$?

Fig. 30.35

A

B

C
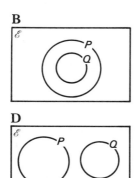

D

6. The shaded region in the Venn diagram of Fig. 30.36 may be described in set notation by
   A $X \cup Y$ B $X \cap Y$
   C $X' \cup Y'$ D $(X \cup Y)'$

Fig. 30.36

7. The shaded part of Fig. 30.37 represents
   A $P \cup Q$ B $P \cap Q$
   C P' D Q'

Fig. 30.37

8. The shaded part of Fig. 30.38 represents
   A $P \cup Q$ B $P \cap Q$
   C Q D Q'

Fig. 30.38

9. The shaded part of Fig. 30.39 represents
   A $R \cup S$ B $R \cap S$
   C R' D S'

Fig. 30.39

10. The shaded region of Fig. 30.40
represents

    A $R \cap S$ B $(R \cap S)'$

    C $R \cup S$ D $(R \cup S)'$

Fig. 30.40

11. If $X = \{4, 6, 8, 10\}$, $Y = \{3, 6, 9\}$ and
$Z = \{1, 2, 3, 6\}$ which of the following
are correct statements?

    I $(X \cap Y) \subset Z$

    II $(Y \cap Z) \subset X$

    III $(X \cap Z) \subset Y$

A I and II only   B I and III only

C II and III only   D I, II and III

12. The numbers in Fig. 30.41 give the
number of elements in each region.
$n(P \cap Q \cap R) =$

    A 2   B 4   C 10   D 11

Fig. 30.41

Maps and Scale Drawings

Scales of Maps and Drawings

When drawings of large objects such as houses and ships are to be made, they are usually drawn to a scale, for example 1 cm = 10 m. This means that a distance of 40 m on the ground would be represented on the drawing by a distance of 4 cm.

On maps and drawings, scales are often expressed as a ratio, for instance 1 : 1000. This means that a distance of 2000 m on the ground would be represented by a distance of 2 m on the map or drawing.

Example 1

A map is drawn to a scale of 1 cm = 50 km. Express this scale as a ratio.

$$1 \, cm = 50 \, km$$
$$= 50\,000 \, m$$
$$= 5\,000\,000 \, cm$$

Hence the scale is 1 : 5 000 000.

Example 2

A map is drawn to a scale of 1 : 25 000. Measured on the map a road has a length of 20 cm. What is the actual length of the road in metres.

1 cm on the map represents 25 000 cm = 250 m on the ground.

Therefore 20 cm on the map represents
$$20 \times 250 \, m = 5000 \, m \text{ on the ground}$$

Example 3

The scale of a map is 1 cm = 2 km. The distance between two towns is 8 km. What does this distance measure on the map?

2 km is represented by 1 cm

8 km is represented by $\frac{8}{2} \times 1 \, cm = 4 \, cm$

Exercise 31.1

1. The scale of a map is 1 : 20 000. What distance in metres does 8 cm represent?

2. The scale of a map is 1 : 100 000. The actual distance between two towns shown on the map is 25 km. What is the distance in centimetres on the map between the two towns?

3. The drawing of a house is made to a scale of 1 cm = 2 m.

   (a) What is this scale as a ratio?

   (b) On the drawing, the kitchen measures 2 cm by 3 cm. What is the actual size of the kitchen?

4. A map is drawn to a scale of 2 cm = 1 km.

   (a) Express this scale as a ratio.

   (b) The distance between two points on the map is 10.8 cm. How far apart in kilometres are the two points on the ground.

5. The distance between two towns A and B is measured as 8.1 cm on a map drawn to a scale of $1 : 5\,000\,000$. What is the actual distance in kilometres between the two towns?

6. Fig. 31.1 shows a map of a tropical island called Palm Island.

   (a) Measure the distance between Markton and Johnstown and hence estimate the actual distance between the two places.

   (b) The map shows a road connecting Bridges and St. Thomas via Johnstown. What is the length of the road, in centimetres, measured on the map? How far is it by road between Bridges and St. Thomas?

Fig. 31.1

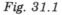

Plans of Houses and Buildings

Example 4

Fig. 31.2 shows the plan of part of a school.

(a) How many windows are shown?

(b) How many doors are shown?

(c) Calculate the area of each of the two rooms.

Fig. 31.2

Scale 1:500

(a) There are 4 windows in the corridor, 1 window in room A and 1 window in room B, giving a total of 6 windows.

(b) There is 1 door in room A and 1 door in room B giving a total of 2 doors.

(c) By measuring the drawing we find that room A has a length of 2.5 cm and a width of 2 cm. The actual measurements are:

$$\text{Length} = 2.5 \times 500 \text{ cm}$$
$$= 1250 \text{ cm}$$
$$= 12.5 \text{ m}$$
$$\text{Width} = 2 \times 500 \text{ cm}$$
$$= 1000 \text{ cm}$$
$$= 10 \text{ m}$$
$$\text{Area} = \text{Length} \times \text{Width}$$
$$= 12.5 \times 10 \text{ m}^2$$
$$= 125 \text{ m}^2$$

From the drawing, the length of room B is 2 cm and its width is also 2 cm.

$$\text{Length and Width} = 2 \times 500 \text{ cm}$$
$$= 1000 \text{ cm}$$
$$= 10 \text{ m}$$
$$\text{Area} = 10 \times 10 \text{ m}^2$$
$$= 100 \text{ m}^2$$

Hence the actual area of room A is 125 m^2 and the actual area of room B is 100 m^2.

Exercise 31.2

1. The plan of a school is drawn to a scale of 250:1. On the plan, a rectangular room has a length of 6 cm and a width of 5 cm. Calculate

   (a) the actual length and width of the room

   (b) the actual area of the room

2. The drawing of a house and garden is to a scale of 1 cm = 2 m. The length of the rectangular garden is shown as 8 cm and its width as 20 cm. Calculate the actual area of the garden in square metres.

3. Fig. 31.3 shows the plan of the ground floor of a house. Copy and complete the following table:

Name of room	Length (m)	Width (m)	Area (m²)
Lounge			
Dining room			
Kitchen			
Bathroom			

   The hall of the house is to be carpeted. What area of carpet is required?

Fig. 31.3

4. Fig. 31.4 represents the plan of a small apartment. Copy and complete the following table:

Name of room	Length (m)	Width (m)	Area (m²)
Lounge			
Hall			
Bedroom			
Kitchen			
Bathroom			

Fig. 31.4

5. Fig. 31.5 shows the plan of a garage.

Fig. 31.5

Scale 1:500

   (a) How many doors are there?

   (b) How many windows are there?

   (c) What is the actual internal length of the garage?

(d) What is the actual internal width of the garage?

(e) What is the actual floor area of the garage?

(f) The floor is to be made of concrete to a depth of 15 cm. What volume of concrete, in cubic metres, is required?

6. Fig. 31.6 shows the plan of a house and garden drawn to a scale of 1 : 2500. Find, by scaling, the dimensions of the plot and the plan area of the house. Hence calculate, in square metres, the area of the garden.

Fig. 31.6

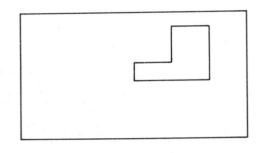

7. Fig. 31.7 represents the lounge and kitchen of a house.

   (a) The lounge floor is to be covered with a fitted carpet. Calculate, in square metres, the area of carpet required.

Fig. 31.7

(b) The kitchen floor is to be tiled with tiles 15 cm square. How many tiles are required?

Areas on Maps and Drawings

Consider a field which is shown on a map. The field itself is simply an **enlargement** of the representation of the field on the map. The scale factor of the enlargement is the scale, expressed as a ratio, to which the map has been drawn.

It will be remembered (from Chapter 26) that the ratio of the areas of similar shapes is equal to the square of the enlargement factor. Hence to find the actual area of the field we multiply the area on the map by the square of the scale.

Example 5

A drawing is made to a scale of 1 : 1000. The area of a plot of ground is shown as 40 cm^2 on the drawing. What is the actual area of the plot?

$$\text{Actual area of plot} = 40 \times 1000^2 \text{ cm}^2$$
$$= \frac{40 \times 1000^2}{100^2} \text{ m}^2$$
$$= 4000 \text{ m}^2$$

Example 6

The scale of a map is 1 cm to 5 km. What area does 8 cm^2 on the map represent on the ground?

$$1 \text{ cm} = 5 \text{ km}$$
$$= 5000 \text{ m}$$
$$= 500\,000 \text{ cm}$$
$$\text{Hence the map scale} = 1 : 500\,000$$
$$\text{Area on the ground} = 8 \times 500\,000^2 \text{ cm}^2$$
$$= \frac{8 \times 500\,000^2}{100^2 \times 1000^2} \text{ km}^2$$
$$= 200 \text{ km}^2$$

Exercise 31.3

1. The scale of a map is 1 : 10 000. What area, in square metres, does 4 cm² on the map represent?

2. The scale of a drawing is 2 cm to 1 m. The lounge of a house has an area of 48 cm² on the drawing. What is the actual area of the lounge?

3. The scale of a map is 1 cm to 2 km. How many square kilometres are represented by 5 cm² on the map?

4. The scale of a drawing is 5 cm to 1 km. A field is represented on the map by a trapezium (Fig. 31.8). What is the actual area of the field?

Fig. 31.8

5. A, B and C are three points on a map whose scale is 1 : 10 000.
   AB = 7 cm, AC = 5 cm and
   ∠ABC = ∠BAC = 50°. Calculate

   (a) the area of triangle ABC

   (b) the actual area represented by triangle ABC

6. A triangular plot of land when measured on a map has sides of 8 cm, 9 cm and 11 cm. If the scale of the map is 1 : 10 000, what is the actual area of the plot in hectares?
   (1 hectare = 10 000 m².)

Volumes and Scales

Engineers and architects often make models of houses and buildings before starting work on them. The object and its model are similar solids. Therefore to find the volume of the object we must multiply the volume of the model by the cube of the scale to which the model is made.

Example 7

A model of a school is made to a scale of 1 : 50. The volume of the model is 6 m³. What is the volume of the school?

$$\text{Volume of school} = 6 \times 50^3 \text{ m}^3$$
$$= 750\,000 \text{ m}^3$$

Exercise 31.4

1. A model of a library is made to a scale of 1 : 20. Its volume is 3 m³. Find the volume of the library.

2. The model of a train is made to a scale of 1 : 40. Its volume is 24 cm³. Find, in cubic metres, the actual volume of the train.

3. The fuselage of an aeroplane has a volume of 600 m³. A model of the aeroplane is made to a scale of 1 cm = 10 m. Find the volume of the fuselage on the model.

4. On the model of a church, the spire has a volume of 0.5 m³. If the model is made to a scale of 1 : 25, find the actual volume of the spire.

5. The model of a house consists of a cuboid 7 cm by 5 cm by 3 cm. If the scale of the model is 1 cm = 2 m, find the actual volume of the house.

Gradient

When used in connection with roads and railways, the gradient is used to denote the ratio of the vertical distance to the corresponding distance measured along the slope. Thus in Fig. 31.9,

$$\text{Gradient} = \frac{BC}{AC}$$

Fig. 31.9

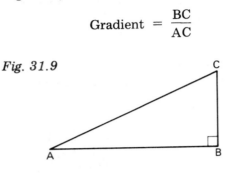

Example 8

(a) Two points P and Q are linked by a straight road with a uniform gradient. If Q is 50 m higher than P and the distance PQ measured along the road is 800 m, what is the gradient of the road?

Fig. 31.10

Referring to Fig. 31.10 we see that

$$\text{Gradient} = \frac{RQ}{PQ}$$

$$= \frac{50}{800}$$

$$= \frac{1}{16}$$

The gradient of the road is 1 in 16.

Nowadays gradients are often expressed as percentages, in which case

$$\text{Gradient} = \frac{1}{16} \times \frac{100}{1}$$

$$= 6\tfrac{1}{4}\%$$

(b) A man walks up a road whose gradient is 1 in 15. How much higher is he than when he started if he walks a distance of 3000 m?

Fig. 31.11

From Fig. 31.11,

$$\text{Gradient of road} = \frac{x}{3000}$$

$$\frac{x}{3000} = \frac{1}{15}$$

$$x = \frac{3000}{15}$$

$$= 200 \text{ m}$$

The man is 200 m higher than when he started.

Exercise 31.5

1. Two points A and B are connected by a straight road with a uniform gradient. If A is 40 m higher than B and the distance AB (measured along the road) is 1 km, what is the gradient of the road?

2. Two points P and Q whose heights differ by 50 m are connected by a straight road 2 km long. Find the gradient of the road.

3. A road has a gradient of 1 in 15. A man walks 1.5 km up the road. How much higher is he than when he first started?

4. A man walks up a road whose gradient is 1 in 30. He walks a distance of 900 m. How much higher is he than when he first started?

5. A road has a gradient of $12\frac{1}{2}\%$. What angle does the road make with the horizontal?

6. A mountain track is inclined at $12°$ to the horizontal. Express the gradient of the track as

   (a) $1 : n$ (b) a percentage

7. A woman walks 100 m up a slope whose gradient is 5% and then a further 80 m up a slope whose gradient is $12\frac{1}{2}\%$. How much higher is the woman than when she first started?

Contours

The contours on a map join places which have the same height. Fig. 31.12 shows a small hill. The bottom view, the plan, is the one which would be shown on a map.

Fig. 31.12

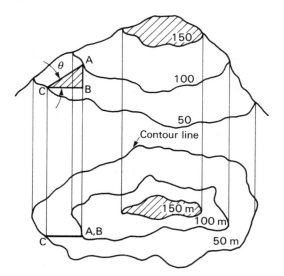

Suppose that A and C are two positions on a path up the hill, the path being taken as a straight line. It will be seen that C is on the 50 m contour and that A is on the 100 m contour. B is directly below A (within the hill) and on the same contour as C. A and B appear as the same point on the map but the height AB is 50 m. The distance AC measured on the map corresponds to the length BC on the elevation. The angle of elevation is θ and $\tan \theta = AB/BC$. The gradient is AB/AC which is $\sin \theta$.

Example 9

(a) Two points P and Q whose heights differ by 50 m are shown 0.60 cm apart on a map whose scale is $1 : 25\,000$. Calculate the angle of elevation of the line PQ.

$$0.60 \text{ cm} = 0.60 \times 25\,000 \text{ cm}$$

$$= \frac{0.6 \times 25\,000}{100} \text{ m}$$

$$= 150 \text{ m on the ground}$$

From Fig. 31.13

$$\tan \theta = \frac{RQ}{PR}$$

$$= \frac{50}{150}$$

$$= 0.3333$$

$$\theta = 18°26'$$

Therefore the angle of elevation of PQ is $18°26'$.

Fig. 31.13

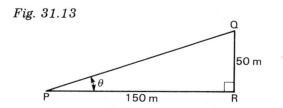

(b) Two points A and B are linked by a straight road with a uniform gradient. A is on the 100 m contour and B is on the 125 m contour. If the scale of the map is 10 cm to 1 km and the distance AB on the map is 4 cm, find the gradient of the road stating the answer **(i)** in the form $n:1$, **(ii)** as a percentage. The map is shown in Fig. 31.14.

Fig. 31.14

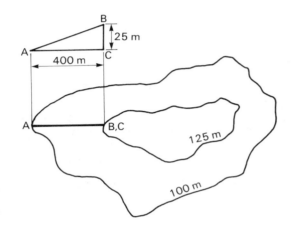

Since 10 cm = 1 km on the ground

$$1 \text{ cm} = 0.1 \text{ km}$$
$$= 100 \text{ m on the ground}$$

and 4 cm = 4×100 m

$$= 400 \text{ m on the ground}$$

In triangle ABC,

$$\tan \theta = \frac{BC}{AC}$$

$$= \frac{25}{400}$$

$$= 0.0625$$

$$\theta = 3°35'$$

$$\text{Gradient} = \sin \theta$$

$$= \sin 3°35'$$

$$= 0.0625$$

(i) As a ratio,

$$\text{Gradient} = 1:0.0625$$

$$= \frac{1}{0.0625}:1$$

$$= 16:1$$

(ii) As a percentage,

$$\text{Gradient} = 0.0625 \times 100\%$$

$$= 6.25\%$$

$$= 6\tfrac{1}{4}\%$$

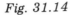

Exercise 31.6

1. Two points P and Q whose heights differ by 150 m are shown 0.8 cm apart on a on a map whose scale is 1:10 000. Calculate the angle of elevation of the line AB.

2. Two points P and Q whose heights differ by 150 m are shown 0.8 cm apart on a map whose scale is 1 cm = 500 m. Calculate the gradient of the line PQ, stating the answer

    (a) in the form $1:n$

    (b) as a percentage

3. A and B are two places on a map whose scale is 10 cm to 1 km. A is on the 500 m contour and B is on the 250 m contour. The distance AB measured on the map is 8 cm. Find the angle of elevation from B to A.

4. Fig. 31.15 is a scale diagram of a contour map. Use the map to find **(a)** the difference in height between A and B, **(b)** the horizontal distance, in metres, between A and B, **(c)** the average gradient between A and B, stating the answer as a percentage.

Fig. 31.15

Scale: 1 cm = 1 km

The contours are measured in metres.

5. Fig. 31.16 is a scale drawing (scale: 2 cm to 2 km) of part of a contour map of a given area.

(a) What is the difference in level between B and G?

(b) Find the difference in height between the highest and lowest points on the drawing.

(c) What is the horizontal distance between A and C and between D and J?

(d) Find the average angle of slope of the roadway ME.

Fig. 31.16

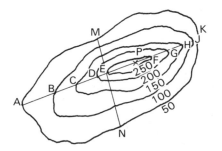

Miscellaneous Exercise 31

1. A map is drawn to a scale of $1:25\,000$. What length of road, in metres, is represented by 2 cm on the map?

2. A rectangular kitchen is $3\frac{1}{2}$ m wide, $4\frac{1}{2}$ m long and $2\frac{1}{2}$ m high. A model of the house is made to a scale of 1 cm to 2 m. Work out the dimensions of the kitchen on the model.

3. The building plans of a house are drawn to a scale of $1:50$.

   (a) Find the length of the house if it is represented by a length of 20 cm on the drawing.

   (b) The actual height of the house is to be 7 m. Find the distance, in centimetres, which represents this dimension on the plan.

   (c) On the plan, the area of the kitchen floor is 80 cm^2. Calculate the actual area of the kitchen floor, stating the answer in square metres.

4. A model of a public building is made to a scale of $1:20$.

   (a) The model has a length of 4 m. What is the actual length of the building?

   (b) The actual floor area of one of the rooms is 2000 m^2. What is the area on the model?

   (c) The volume of the model is 36 m^3. What is the actual volume, in cubic metres, of the building?

5. The plan of a house and garden is drawn to a scale of $1:60$.

   (a) Find the distance on the plan which represents an actual distance of 30 m.

   (b) On the plan, the garden measures 60 cm by 30 cm. What is the actual area, in square metres, of the garden?

   (c) On the plan, the volume of the house is shown to be 90 cm^3. What is the actual volume, in cubic metres, of the house?

6. Fig. 31.17 shows the plan of a bungalow. Copy and complete the following table:

Name of room	Length (m)	Breadth (m)	Area (m^2)
Lounge			
Dining room			
Bedroom 1			
Bedroom 2			
Hall			
Bathroom			
WC			
Kitchen			

How many windows and doors does the bungalow possess?

7. Two points P and Q are two places on a map whose scale is $1:10\,000$. P is on the 800 m contour and Q is on the 600 m contour. The distance PQ measured on the map is 10 cm.

(a) Find the angle of elevation from Q to P.

(b) Calculate the gradient of the line PQ in the form $1:n$.

(c) State the gradient as a percentage.

Fig. 31.17

Charts, Tables and Diagrams

Tables

Conversion tables of various types are used in business, science and engineering.

Ready reckoners provide a quick way of multiplying sums of money. The calculator has largely superseded the ready reckoner but many small businesses and shops still use it. Table 32.1 shows an extract from a ready reckoner giving the price of N articles at 26p each.

TABLE 32.1

N	£	N	£	N	£	N	£
22	5.72	64	16.64	106	27.56	525	136.50
23	5.98	65	16.90	107	27.82	550	143.00
24	6.24	66	17.16	108	28.08	600	156.00
25	6.50	67	17.42	109	28.34	625	162.50
26	6.76	68	17.68	110	28.60	650	169.00
27	7.02	69	17.94	111	28.86	700	182.00
28	7.28	70	18.20	112	29.12	750	195.00
29	7.54	71	18.46	113	29.38	800	208.00
30	7.80	72	18.72	114	29.64	900	234.00

Example 1

Use Table 32.1 to find the cost of
(a) 28 articles at 26p each,
(b) 579 articles at 26p each,
(c) $7\frac{1}{2}$ metres of cloth at 26p per metre,
(d) 72.9 kg of foodstuff at 26p per kilogram.

    (a) Directly from the table:

       Cost of 28 articles at 26p each $= £7.28$

(b) $579 = 550 + 29$

Cost of 550 articles at 26p each
$$= £143.00$$

Cost of 29 articles at 26p each
$$= £7.54 +$$

Cost of 579 articles at 26p each

$$= £150.54$$

(c) $7\frac{1}{2} = 7.5$

$$= \frac{750}{100}$$

From the table:

Cost of 750 metres of
cloth at 26p per metre $= £195.00$

Cost of $7\frac{1}{2}$ metres of
cloth at 26p per metre $= £195.00 \div 100$

$$= £1.95$$

(d) $72.9 = \frac{729}{10}$

$$= \frac{700 + 29}{10}$$

Cost of 700 kg of foodstuff
at 26p per kilogram $= £182.00$

Cost of 29 kg of foodstuff
at 26p per kilogram $= £7.54 +$

Cost of 729 kg of foodstuff
at 26p per kilogram $= £189.54$

Cost of 72.9 kg of foodstuff
at 26p per kilogram $= £189.54 \div 10$

$$= £18.95$$

Conversion tables are frequently used by
engineers and scientists. Table 32.2 shows
part of a temperature conversion table.

TABLE 32.2

TEMPERATURE	Degrees Fahrenheit to degrees Celsius									
°F	0	1	2	3	4	5	6	7	8	9
0	−17.8	−17.2	−16.7	−16.1	−15.6	−15.0	−14.4	−13.9	−13.3	−12.8
10	−12.2	−11.7	−11.1	−10.6	−10.0	−9.4	−8.9	−8.3	−7.8	−7.2
20	−6.7	−6.1	−5.6	−5.0	−4.4	−3.9	−3.3	−2.8	−2.2	−1.7
30	−1.1	−0.6	0	0.6	1.1	1.7	2.2	2.8	3.3	3.9
40	4.4	5.0	5.6	6.1	6.7	7.2	7.8	8.3	8.9	9.4
50	10.0	10.6	11.1	11.7	12.2	12.8	13.3	13.9	14.4	15.0
60	15.6	16.1	16.7	17.2	17.8	18.3	18.9	19.4	20.0	20.6
70	21.1	21.7	22.2	22.8	23.3	23.9	24.4	25.0	25.6	26.1
80	26.7	27.2	27.8	28.3	28.9	29.4	30.0	30.6	31.1	31.7
90	32.2	32.8	33.3	33.9	34.4	35.0	35.6	36.1	36.7	37.2
100	37.8	38.3	38.9	39.4	40.0	40.6	41.1	41.7	42.2	42.8
110	43.3	43.9	44.4	45.0	45.6	46.1	46.7	47.2	47.8	48.3
120	48.9	49.4	50.0	50.6	51.1	51.7	52.2	52.8	53.8	53.9
130	54.4	55.0	55.6	56.1	56.7	57.2	57.8	58.3	58.9	59.4
140	60.0	60.6	61.1	61.7	62.2	62.8	63.3	63.9	64.4	65.0
150	65.6	66.1	66.7	67.2	67.8	68.3	68.9	69.4	70.0	70.6
160	71.1	71.7	72.2	72.8	73.3	73.9	74.4	75.0	75.6	76.1
170	76.7	77.2	77.8	78.3	78.9	79.4	80.0	80.6	81.1	81.7
180	82.2	82.8	83.3	83.9	84.4	85.0	85.6	86.1	86.7	87.2
190	87.8	88.3	88.9	89.4	90.0	90.6	91.1	91.7	92.2	92.8
Interpolation:	deg F:	0.1	0.2	0.3	0.4	0.5	0.6	0.7	0.8	0.9
	deg C:	0.1	0.1	0.2	0.2	0.3	0.3	0.4	0.4	0.5

Example 2

Convert (a) 54°F to degrees Celsius,
(b) 122.7°F to degrees Celsius,
(c) 35°C to degrees Fahrenheit,
(d) 62.7°C to degrees Fahrenheit.

(a) We first find 50 in the first column
and move along this row until we find
the column headed 4. We find the
number 12.2 and hence 54°F = 12.2°C.

(b) To convert 122.7°F into degrees
Celsius we make use of the figures
given under the heading 'interpolation'.
From the table we find 122°F = 50°C.
Looking at the figures given at the foot
of the table we can find 0.7°F and this
is equivalent to 0.4°C which is shown
immediately below. Thus

$$122.7°F = 50°C + 0.4°C$$

$$= 50.4°C$$

(c) To convert degrees Celsius into
degrees Fahrenheit we use the table in
reverse. To convert 35°C into degrees
Fahrenheit we search in the body of
the table until we find the figure 35.
This occurs in the column headed 5 in
the row starting with 90. Hence 35°C
is equivalent to 95°F.

(d) To convert 62.7°C into degrees
Fahrenheit we look in the body of the
table for a number as close to 62.7 as
possible, but less than 62.7. This number is 62.2 corresponding to 144°F.
Now 62.2°C is 0.5°C less than 62.7°C.
Using the interpolation figures we see
that 0.5°C corresponds to 0.9°F.
Therefore

$$62.7°C = 62.2°C + 0.5°C$$

$$= 144°F + 0.9°F$$

$$= 144.9°F$$

Mileage charts are often set out like Table
32.3.

TABLE 32.3

Example 3

Use Table 32.3 to find the distance between
(a) Aberdeen and Carlisle,
(b) Barnstaple and Cambridge.

(a) To find the distance from Aberdeen
to Carlisle, find Aberdeen and Carlisle
in the sloping list. Move vertically down
the column for Aberdeen and horizontally along the row for Carlisle until the
two movements coincide. The figure
given in the table at this point is 212

and thus the distance between Aberdeen and Carlisle is 212 miles.

(b) Find Barnstaple in the sloping list and move down the third column. The row for Cambridge is the seventh row from the top. In the third column of the seventh row we find the figure 240 and hence the distance between Barnstaple and Cambridge is 240 miles.

Parallel Scale Conversion Charts

A system of parallel scales may be used when we wish to convert from one set of units to another related set. Fig. 32.1 is a chart relating degrees Fahrenheit and degrees Celsius.

Fig. 32.1

```
┌──────────────────────────────────────────────────┐
│ TEMPERATURE                                        │
│                                                    │
│  0°    32°       °Fahrenheit            212°       │
│     25°   50°  75° 100° 125° 150° 175° 200°        │
│                                                    │
│   −10°    10° 20° 30° 40° 50° 60° 70° 80° 90°      │
│ −17.8°  0°       °Centigrade             100°      │
└──────────────────────────────────────────────────┘
```

Example 4

Using the chart in Fig. 32.1 convert (a) 50°F to degrees Celsius, (b) 65°F to degrees Celsius, (c) 20°C to degrees Fahrenheit.

(a) From the chart we see that 50°F corresponds with 10°C and therefore 50°F is equivalent to 10°C.

(b) From the chart we see that 75°F corresponds roughly with 24°C. Note carefully that when using the chart we cannot expect accuracy greater than 1°.

(c) From the chart we see that 20°C is about 68°F.

Exercise 32.1

1. Use the ready reckoner on page 297 to find the cost of

   (a) 24 articles costing 26p each

   (b) 102 articles costing 26p each

   (c) 827 articles costing 26p each

   (d) 672 articles costing 26p each

   (e) $6\frac{1}{2}$ metres of materal costing 26p per metre

   (f) 9.3 lbs of fruit costing 26p per pound

   (g) 82.2 litres of fluid costing 26p per litre

2. Use the temperature conversion table on page 298 to convert

   (a) 83°F to degrees Celsius

   (b) 114.6°F to degrees Celsius

   (c) 55°C to degrees Fahrenheit

   (d) 68.7°C to degrees Fahrenheit

3. Use the mileage chart on page 298 to find the distances between

   (a) Aberdeen and Cardiff

   (b) Birmingham and Carmarthen

   (c) Bristol and Dorchester

4. Table 32.4 shows a comparison between gradients expressed as a ratio and gradients expressed as a percentage. Use the table to convert

   (a) a gradient of 1:7 to a percentage

   (b) a gradient of 7.3% to a ratio

TABLE 32.4

Gradients	
Ratio	%
1:3	33.3
1:4	25
1:5	20
1:6	16.7
1:7	14.3
1:8	12.5
1:9	11.1
1:10	10
1:11	9.1
1:12	8.3
1:13	7.7
1:14	7.1
1:15	6.7
1:16	6.3
1:17	5.9
1:18	5.6
1:19	5.3
1:20	5

5. Table 32.5 allows conversion from miles to kilometres (and also miles per hour to kilometres per hour) to be made. Use the table to convert

   (a) 15 miles to kilometres

   (b) 48.27 km/h to miles per hour

   (c) 58 miles to kilometres

   (d) 568 miles per hour to kilometres per hour

TABLE 32.5

Distance and Speed				
Miles to kilometres				
Miles per hour to kilometres per hour				
1	1.60	20	32.18	75	120.7
2	3.21	25	40.23	80	128.7
3	4.82	30	48.27	85	136.8
4	6.43	35	56.32	90	144.8
5	8.04	40	64.37	95	152.9
6	9.65	45	72.41	100	160.9
7	11.26	50	80.46	200	321.9
8	12.87	55	88.51	300	482.8
9	14.48	60	96.55	400	643.7
10	16.09	65	104.60	500	804.7
15	24.13	70	112.70	1000	1609.3

1 mile = 1.606344 km
1 kilometre = 0.621371 miles

6. Use the chart (Fig. 32.1) relating temperatures in degrees Fahrenheit to degrees Celsius to convert

   (a) 100°F to degrees Celsius

   (b) 80°C to degrees Fahrenheit

   (c) 64°F to degrees Celsius

   (d) 48°C to degrees Fahrenheit

7. Fig. 32.2 is a chart relating pounds avoirdupois to kilograms. Use the chart to convert

   (a) 3.4 lb to kilograms

   (b) 568 lb to kilograms

   (c) 1.8 kg to pounds

Fig. 32.2

8. Table 32.6 shows a comparison of tyre pressures. Convert (a) 22 lb/in² into kilograms per square centimetre, (b) 2.10 kg/cm² into pounds per square inch. (c) Estimate the tyre pressure in pounds per square inch corresponding to 3 kg/cm².

TABLE 32.6

Tyre Pressures					
Pounds per sq in to kg per sq cm					
16	1.12	26	1.83	40	2.80
18	1.26	28	1.96	50	3.50
20	1.40	30	2.10	55	3.85
22	1.54	32	2.24	60	4.20
24	1.68	36	2.52	65	4.55

Simple Bar Charts

The information in a simple bar chart is represented by a series of bars all of the same width. The bars may be drawn vertically or horizontally. The height (or length) of the bars represents the magnitude of the figures given. A simple bar chart shows clearly the size of each item of information but it is not easy to obtain the total of all the items from the diagram.

Example 5

A family spends its weekly income of £300 as follows:

Food	£100
Clothes	£50
Fuel	£40
Mortgage	£90
Other expenses	£20
Total	£300

Draw (a) a vertical bar chart, (b) a horizontal bar chart to represent this information.

Fig. 32.3 shows the information in the form of a vertical bar chart whilst Fig. 32.4 shows the information in the form of a horizontal bar chart.

Fig. 32.3

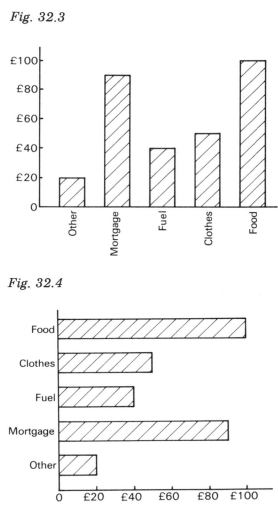

Fig. 32.4

Year	Population (millions)
1750	728
1800	906
1850	1171
1900	1608
1950	2504

When drawing a chronological bar chart, time is always marked off along the horizontal axis, as shown in Fig. 32.5. The chart clearly shows how the population of the world has increased over this period of 200 years.

Fig. 32.5

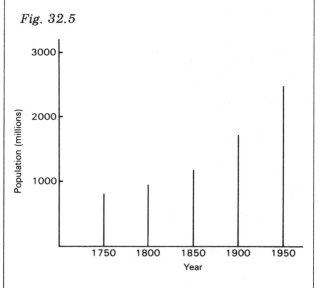

Chronological Bar Charts

A chronological bar chart compares quantities over periods of time. It consists of a number of vertical lines (see also page 310).

Example 6

The figures below give the world population in millions of people from 1750 to 1950. Draw a chronological bar chart to represent this information.

Proportionate Bar Charts

The proportionate bar chart relies on heights (if drawn vertically) or lengths (if drawn horizontally) to convey proportions of the whole. It should be the same width throughout its height or length.

The proportionate bar chart shows clearly the total of all the items, but it is rather difficult to obtain the proportion of each item accurately.

Example 7

The table below shows the number of people employed on various kinds of work in a factory. Represent this information in the form of a proportionate bar chart.

Type of personnel	Number employed
Unskilled workers	45
Craftsmen	27
Draughtsmen	5
Clerical workers	8
Total	85

Suppose that the total height of the proportionate bar chart is to be 6 cm. The heights of the component parts must first be calculated and then drawn accurately using a rule.

45 unskilled workers are represented by

$$\frac{45}{85} \times 6\,cm = 3.18\,cm$$

27 craftsmen are represented by

$$\frac{27}{85} \times 6\,cm = 1.91\,cm$$

5 draughtsmen are represented by

$$\frac{5}{85} \times 6\,cm = 0.35\,cm$$

8 clerical workers are represented by

$$\frac{8}{85} \times 6\,cm = 0.56\,cm$$

The proportionate bar chart is shown below.

Fig. 32.6

Pie Charts

A pie chart displays the proportion of a whole as a sector angle or sector area. The circle as a whole represents the total of the component parts.

Pie charts are very useful when component parts of a whole are to be represented, but it is not easy to discover the total quantity represented. Up to eight component parts can be represented, but above this number the chart loses its clarity and effectiveness.

Example 8

25 sixth form students were asked what sort of career they would like.

    8 chose work with animals.

    3 chose office work.

    2 chose teaching.

    5 chose outdoor work.

    7 said they did not know.

The first step is to calculate the angles at the centre of the pie chart. Since there are $360°$ in the circle we divide $360°$ by the total number of students quizzed. This gives the angle for one pupil. To find the angle for each section, we multiply the angle for one pupil by the number of pupils in the section and correct the product to the nearest degree. The work may be set out as follows:

$$360° \div 25 = 14.4°$$

Career	Number of pupils	Sector angle
Work with animals	8	$8 \times 14.4° = 115°$
Office work	3	$3 \times 14.4° = 43°$
Teaching	2	$2 \times 14.4° = 29°$
Outdoor work	5	$5 \times 14.4° = 72°$
Don't know	7	$7 \times 14.4° = 101°$

The pie chart is shown in Fig. 32.7.

Fig. 32.7

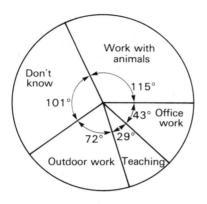

Line Graphs

Line graphs are sometimes used instead of chronological bar charts.

Example 9

Represent the information given in Example 2 in the form of a line graph.

> The line graph is shown in Fig. 32.8. As with chronological bar charts the year is always taken along the horizontal axis. The points are joined by straight lines because no information is given about the world population between the years given in the table.

Fig. 32.8

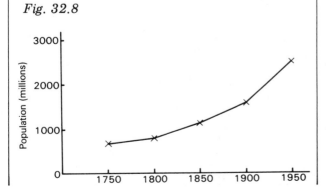

Pictograms

Pictograms are diagrams in the form of pictures which are used to present information to those who are unskilled in dealing with figures or to those who have only a limited interest in the topic depicted.

Example 10

The table below shows the output of bicycles for the years 1980 to 1984.

Year	1980	1981	1982	1983	1984
Output	2000	4000	7000	8500	9000

Represent this information in the form of a pictogram.

> The pictogram is shown in Fig. 32.9. It will be seen that each bicycle in the diagram represents an output of 2000 bicycles. Part of a symbol is shown in 1982, 1983 and 1984 to represent a fraction of 2000 but clearly this is not a precise way of representing the output.

Fig. 32.9

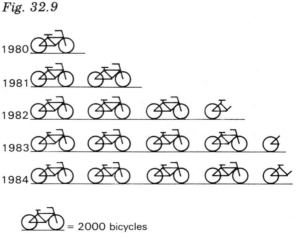

A method not recommended is shown in Fig. 32.10. Comparison is difficult because the reader is not sure whether to compare heights, areas or volumes.

Fig. 32.10

Sales of milk in 1960 and 1980 (millions of litres)

Exercise 32.2

1. Fig. 32.11 is a pictogram showing the method by which first year boys come to school. How many

   (a) come by bus

   (b) come by car

Fig. 32.11

WALKING 人 人 人 人 人 人 人

BUS 人 人 人 人 人 人 人 人

BICYCLE 人 人 ⫟

CAR 人 人 ○

人 Represents 5 boys

(head 1, arms and legs 1 each)

2. Draw a proportionate bar chart for the figures shown below which represent the way in which people travel to work in Gloucester.

Type of transport	Numbers using
Bus	780
Private motoring	420
Other (foot, bicycle, motor cycle, etc.)	160
Total	1360

3. Draw a proportionate bar chart 8 cm long to represent the following information, which relates to the way in which a family spends its income.

Item	Amount spent (£)
Food and drink	95
Housing	42
Transport	29
Clothing	33
Other	51
Total	250

4. The table below shows the results of a survey of the colours of doors on a housing estate. Draw a vertical bar chart to represent this information.

Colour of door	Number of houses
White	85
Red	17
Green	43
Brown	70
Blue	15

5. The figures below show the results of a survey to find the favourite sports of a group of boys. Draw a pie chart to represent this information.

Sport	Number of boys
Athletics	20
Cricket	58
Football	32
Hockey	10

6. The table below shows the output of bicycles from a certain factory for the years 1980–1984. Represent this information in the form of

(a) a chronological bar chart

(b) a line graph

Year	1980	1981	1982	1983	1984
Number of bicycles	2000	4000	7000	8500	9000

7. The table below shows the number of houses completed in a certain town for the years 1978–1983. Represent this information by means of

(a) a chronological bar chart

(b) a line graph

Year	1978	1979	1980	1981	1982	1983
Number of houses	81	69	73	84	80	120

8. The pie chart (Fig. 32.12) shows a local election result. There were three candidates, White, Green and Brown. The angle representing the votes cast for White is 144° and that for Green is 90°.

(a) Work out the angle representing the votes cast for Brown.

(b) Calculate the votes cast for each candidate if the total number voting was 20 000.

Fig. 32.12

9. A pie chart was drawn to show the number of tons of various crops grown by a group of farmers. The sector angles corresponding to each crop was as follows:

Crop	Sector angle (degrees)
Vegetables	137
Grain	93
Potatoes	80
Fruit	50

If the total amount grown was 500 tons, calculate the amount of each crop grown.

10. The information shown below gives the output of motor tyres made by the Treadwell Tyre Company for the first six months of 1985. Draw

(a) a chronological bar chart

(b) a line graph, to represent this information

Month	Jan	Feb	Mar	Apr	May	June
Output (thousands)	40	43	39	38	37	45

11. Fig. 32.13 shows how Mrs. Lever spends her housekeeping money. If she receives £50 per week, how much does she spend on each item shown in the diagram?

Fig. 32.13

Multi-Choice Questions 32

1. The bar chart in Fig. 32.14 shows the distribution of children in classes 1 to 6. Each child is given three exercise books for the term. If each book costs 20p, how much was spent altogether on exercise books?

   A £180   B £360   C £600   D £1800

Fig. 32.14

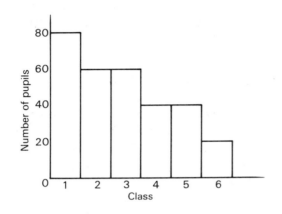

2. The pie chart (Fig. 32.15) illustrates the sports preferred by a group of male sixth-formers. What percentage preferred tennis?

   A 20%   B 25%   C 30%   D 50%

Fig. 32.15

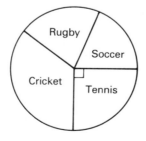

3. The bar chart (Fig. 32.16) shows the ... of the children in a school. How ... y children took part in the survey?

   ... 5   B 17   C 65   D 222

Fig. 32.16

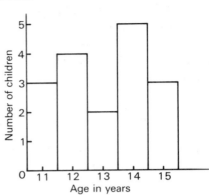

4. 2000 students were interviewed to find out which of four sports they preferred. The results were as follows:

Sport	Cricket	Tennis	Swimming	Athletics
Number of students	800	480	400	320

   A pie chart is to be drawn to represent this information. What will be the sector angle for swimming?

   A 95°   B 72°   C 65°   D 60°

5. The main leisure interests of 150 fifth year pupils are shown as percentages below:

Football	40%	Television	30%
Reading	10%	Scouts	8%
Other	12%		

   If a pie chart is drawn of this information, what size of angle will represent those pupils whose main interest is television?

   A 30°   B 72°   C 108°   D 120°

6. The table below shows the results of a survey to find the colour that the external woodwork was painted on the houses of a housing estate.

Colour	White	Red	Blue	Green
Number of houses	90	70	60	80

The information is to be illustrated in a proportionate bar chart which is to have a total height of 15 cm. What height must be used to represent the houses with green external paintwork?

A 4.5 cm   B 4.0 cm

C 3.5 cm   D 3.0 cm

7. There are 300 pupils in a school. The pie chart (Fig. 32.17) shows the proportion of children travelling to school by private bus, public transport, bicycle and on foot. How many travel to school by bicycle?

A 10   B 30   C 120   D 360

Fig. 32.17

8. Fig. 32.18 shows the frequency distribution of the marks obtained in a test. How many candidates took the test?

A 15   B 24   C 50   D 62

Fig. 32.18

9. Fig. 32.19 is a pie chart which shows the sales for a departmental store during one day. The total sales were £4000. What were the sales of clothing?

A £1200   B £1000

C £900   D £600

Fig. 32.19

Mental Test 32

1. Use the mileage chart shown in Table 32.7 to find the distance from

   (a) Exeter to Gloucester

   (b) Exeter to Lincoln

   (c) Hereford to Liverpool

   (d) Kendal to Manchester

TABLE 32.7

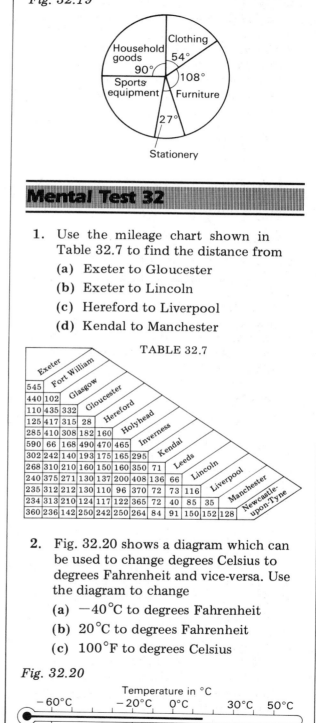

2. Fig. 32.20 shows a diagram which can be used to change degrees Celsius to degrees Fahrenheit and vice-versa. Use the diagram to change

   (a) $-40\,^\circ$C to degrees Fahrenheit

   (b) $20\,^\circ$C to degrees Fahrenheit

   (c) $100\,^\circ$F to degrees Celsius

Fig. 32.20

3. Fig. 32.21 is a simple bar chart which shows the way that commuters in the south-east region travel to work. Use the chart to find the number of commuters in the sample who travelled to work by **(a)** private motoring, **(b)** bus and underground, **(c)** British Rail.

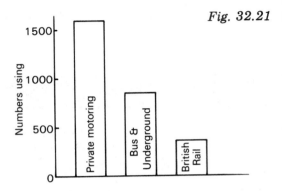

Fig. 32.21

4. The chronological bar chart in Fig. 32.22 shows the number of colour television sets sold in southern England during the period 1970–5. Use the diagram to estimate the numbers of television sets sold in the years 1972 and 1975.

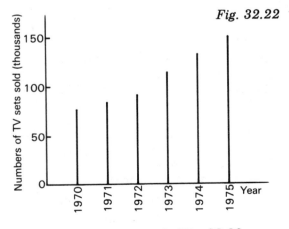

Fig. 32.22

5. The pie chart shown in Fig. 32.23 depicts the sales of the various departments of a large store. If the sales for the week depicted were £600 000, find value of

    the clothing sold

    the household goods sold

Fig. 32.23

6. The pictogram (Fig. 32.24) shows the sales of cars in the years 1975–1979.

   (a) In which year were the most cars sold?

   (b) How many cars were sold in 1978?

   (c) How many cars were sold in 1976?

   (d) Estimate the total number of cars sold in the years represented on the pictogram.

Fig. 32.24

7. Mrs. Wood has £100 per week for housekeeping. A pie chart is drawn which illustrates how she spends the money. If she spends £20 on food, work out the angle of the sector representing this amount.

Graphs

In newspapers, business reports and government publications great use is made of pictorial illustrations to help the reader understand the report. The most common form of illustration is the graph.

Axes of Reference

The first step in drawing a graph is to draw two lines at right angles to each other as shown in Fig. 33.1. These lines are called the **axes of reference**. The vertical axis is usually called the **y-axis** and the horizontal axis is called the **x-axis**.

Fig. 33.1

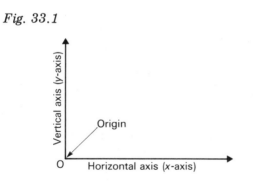

Scales

The number of units represented by a unit length along an axis is called the **scale**. For instance 1 cm could represent 2 units and we say that the scale is 1 cm = 2 units. The scales need not be the same along both axes.

The most useful scales are 1 cm to 1, 2, 5 and 10 units. Some multiples of these are also satisfactory, for instance 1 cm to 20, 50 and 100 units.

No matter what scale is chosen it is very important that the scale is easy to read. When graph paper is used the scale will depend upon the type of graph paper available and its size.

Cartesian Coordinates

Cartesian coordinates are used to mark the points on a graph. In Fig. 33.2, values of x are to be plotted against values of y. The point P has been plotted so that $x = 4$ and $y = 6$. The values of 4 and 6 are called the **rectangular coordinates** of P. For brevity the point P is said to be the point $(4, 6)$. Note carefully that the value of x is always given first.

Fig. 33.2

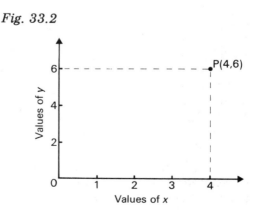

Drawing a Graph

Every graph shows a relationship between
two sets of numbers.

Example 1

Tom and Martha made a count of the number of motor cars which passed their school
between 10.00 a.m. and 11.00 a.m. and the
number of passengers, including the driver,
who travelled in each car. The results were
as follows:

Number of passengers	1	2	3	4
Number of cars	46	30	12	5

Draw a graph to represent this information.

One way of representing this information on a graph is to draw a bar chart
(Fig. 33.3). It consists of a number of
vertical lines whose heights represent
the number of cars carrying the number
of passengers given on the horizontal
axis.

Fig. 33.3

.tities like time, age, etc. which
ease at a steady rate are usually
en on the horizontal axis.

1. The children in a class at school are
given a test periodically. One child
obtained the following marks out of
50.

Number of test	1	2	3	4	5	6
Score	18	23	32	36	42	47

Draw a bar chart to represent this
information. Plot the scores on the
vertical axis.

2. The table below gives the temperature
at 12.00 noon on seven successive days.
Plot a bar chart to represent this
information. Take the days on the
horizontal axis.

Day	1	2	3	4	5	6	7
Temperature (°C)	15	21	26	24	17	25	28

3. The information given below shows the
number of television sets sold by a
wholesaler during the years 1980 to
1985.

Year	1980	1981	1982	1983	1984	1985
Number sold (thousands)	35	43	58	67	77	92

Draw a bar chart with the year taken
on the horizontal axis.

4. The table below shows the amount of
fruit grown by a farmer in seven
successive years.

Year	1	2	3	4	5	6	7
Fruit grown (tonnes)	23	45	34	63	32	54	70

Draw a bar chart with the year taken
on the horizontal axis.

5. In a primary school the ages of the children were as follows:

Age (years)	5	6	7	8	9	10
Number of children	21	19	36	42	28	32

Draw a bar chart taking age on the horizontal axis.

Example 2

The table below shows corresponding values of x and y.

x	0	2	4	6	8	10
y	4	10	16	22	28	34

Plot a graph of this information and use it to find the value of y when $x = 7$ and the value of x when $y = 19$.

It is usualy to plot x along the horizontal axis and y along the vertical axis. When the points are plotted it is seen that a straight line passes through all of them, as shown in Fig. 33.4.

Fig. 33.4

When a graph is a straight line or a smooth curve it can be used to find corresponding values of x and y not given in the table of values. Thus to find the value of y when $x = 7$, draw the vertical and horizontal lines shown in the diagram and it is found that when $x = 7$, $y = 25$. Also it is found that when $y = 19$, $x = 5$.

Using a graph in this way to find values not given in the original table is called **interpolation.**

Example 3

The table below shows corresponding values of P and Q.

P	5	6	7	8	9
Q	57	79	105	135	169

Plot P horizontally and draw the graph which shows the relationship between P and Q. From the graph find the value of Q when $P = 7.5$.

Since the values of P range from 5 to 9, we can make 5 our starting point on the horizontal axis. The values of Q start at 57 and hence it will be convenient to start at 40 on the vertical axis. By doing this we can have larger scales on both axes, thereby allowing us to plot the graph more accurately.

The graph is shown in Fig. 33.5. and it is a smooth curve. We can therefore use the graph to estimate values intermediate to those given in the original table.

From the graph we see that when $P = 7.5$, $Q = 120$.

Fig. 33.5

Fig. 33.6

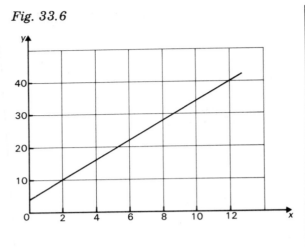

2. Fig. 33.7 shows a graph of distance (in metres) plotted against time (in seconds). From the graph find

   (a) the distances corresponding to times of 2 s, 3 s and 5 s

   (b) the times corresponding to distances of 4 m, 28 m and 54 m

Fig. 33.7

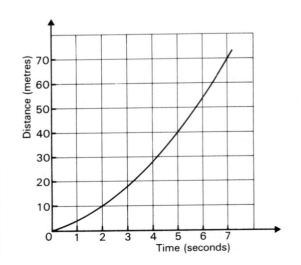

Exercise 33.2

1. Fig. 33.6 shows a graph with values of y plotted against values of x. Use the graph to find

   (a) the values of y corresponding to $x = 0$, $x = 4$, $x = 9$ and $x = 6.5$

   (b) the values of x corresponding to $y = 16$, $y = 22$, $y = 37$ and $y = 20.5$

3. The figures in the table below show corresponding values of the number of units of electricity used and the cost in pounds.

Units used	0	300	600	900	1200
Cost (£)	2	8	14	20	26

Draw a graph of this information taking the number of units used along the horizontal axis. Use a scale of 1 large square equals 200 units and 1 large square on the vertical axis to represent £5. Use your graph to find

(a) the cost of 800 units

(b) the number of units used when the cost is £12

4. The following table gives corresponding values of P and Q:

P	0	2	4	6	8	10
Q	2	12	22	32	42	52

Take P along the horizontal axis using a scale of 1 large square to represent 2 units. On the vertical axis use 1 large square to represent 10 units. Draw a graph to show the relationship between P and Q and use it to find the value of Q when $P = 5$ and to find the value of P when $Q = 37$.

5. The speed of a body (v metres per second) measured at various times (t seconds) was as follows:

t	2	4	6	8	10	12
v	6.4	7.7	9.0	10.3	11.7	13.0

Plot a graph of this information taking t on the horizontal axis. Use scales of 1 large square to represent 2 seconds on the horizontal axis and 1 large square to represent 2 metres per second on the vertical axis. Use your graph to estimate the speed when $t = 7$ seconds and the time at which $v = 9.5$ metres per second.

6. The table below shows corresponding values of x and y:

x	0	1	2	3	4	5
y	8	11	20	35	56	83

Taking x along the horizontal axis plot a graph showing the relationship between x and y. Suitable scales are: 1 large square to 1 unit horizontally and 1 large square to 10 units vertically. Use your graph to find the value of y when $x = 3.5$ and the value of x when $y = 27$.

7. A quantity of gas is contained in a cylinder and subjected to various pressures. The volume occupied by the gas under different pressures is as follows:

Pressure	1	2	3	4	5	6
Volume	2.4	1.2	0.8	0.6	0.48	0.4

Plot a grpah of volume against pressure with pressure on the horizontal axis. Use a scale of 1 large square to 1 unit on the horizontal axis and 1 large square to 0.5 units on the vertical axis. Use your graph to find

(a) the volume when the pressure is 3.5

(b) the pressure when the volume is 0.5

Graphs of Linear Equations

Consider the equation

$$y = 2x + 5$$

We can give x any value we please and so calculate a corresponding value for y.

When $x = 0$ $y = 2 \times 0 + 5 = 5$

When $x = 1$ $y = 2 \times 1 + 5 = 7$

When $x = 2$ $y = 2 \times 2 + 5 = 9$

and so on.

The value of y depends upon the value allocated to x. We therefore call y the **dependent variable** and x the **independent variable.**

When drawing a graph of an equation it is usual to take the values of the independent variable along the horizontal axis, and hence this is often called the x-axis. The values of the dependent variable are then taken along the vertical axis, usually called the y-axis.

In plotting graphs of equations we may have to include coordinates which are positive and negative. To represent these on the graph we make use of the number lines used for directed numbers as shown in Fig. 33.8.

Fig. 33.8

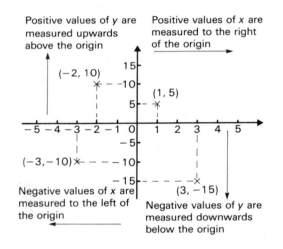

Positive values of y are measured upwards above the origin

Positive values of x are measured to the right of the origin

$(-2, 10)$

$(1, 5)$

Negative values of x are measured to the left of the origin

Negative values of y are measured downwards below the origin

$(-3, -10)$

$(3, -15)$

Example 4

Draw the graph of $y = 2x - 5$ for values of x between -3 and 4.

Having decided upon some values for x we calculate the corresponding values of y by substituting the chosen values of x in the given equation.
Thus when $x = -3$,

$$y = 2 \times (-3) - 5$$
$$= -6 - 5$$
$$= -11$$

For convenience the calculations are tabulated as shown below.

x	-3	-2	-1	0	1	2	3	4
$2x$	-6	-4	-2	0	2	4	6	8
-5	-5	-5	-5	-5	-5	-5	-5	-5
y	-11	-9	-7	-5	-3	-1	1	3

The graph is plotted in Fig. 33.9 and it is seen to be a straight line.

Fig. 33.9

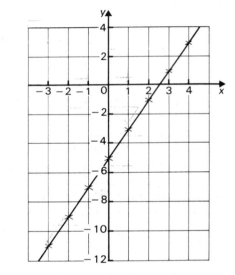

Equations of the type $y = 2x - 5$ where the powers of the variables x and y are 1, always give a graph which is a straight line. For this reason they are often called **linear equations.**

In order to draw the graph of a linear equation it is only necessary to plot two points. It is safer, however, to plot three points, the third point acting as a check on the other two.

Example 5

By means of a graph show the relationship between x and y in the equation $y = 5x + 3$. Plot the graph for values of x between -3 and 3.

Since $y = 5x + 3$ is a linear equation we need only take three points:

x	-3	0	3
y	-12	3	18

The graph is shown in Fig. 33.10.

Fig. 33.10

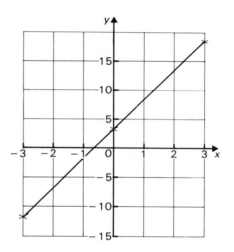

The Equation of a Straight Line

Every linear equation may be written in the standard form:

$$y = mx + c$$

Hence $y = 3x - 7$ is the standard form with $m = 3$ and $c = -7$.

The equation $y = 4 - 5x$ may be put in the standard form by rearranging it to give $y = -5x + 4$.

We then see that $m = -5$ and $c = 4$.

The Meaning of m and c in the Equation of a Straight Line

The point B is any point on the straight line shown in Fig. 33.11. It has the coordinates x and y. Point A is where the straight line cuts the x-axis and it has the coordinates $x = 0$ and $y = c$.

Fig. 33.11

BC/AC is called the gradient of the straight line.

Now

$$BC = \frac{BC}{AC} \times AC$$

$$= AC \times \text{the gradient of the straight line}$$

$$y = BC + CD$$

$$= BC + AO$$

$$= AC \times \text{the gradient of the straight line} + AO$$

$$= AC \times \text{the gradient of the straight line} + c$$

But

$$y = mx + c$$

Hence

$$m = \text{the gradient of the straight line}$$

$$c = \text{the intercept on the } y\text{-axis}$$

Fig. 33.12 shows the difference between a positive and a negative gradient.

Fig. 33.12

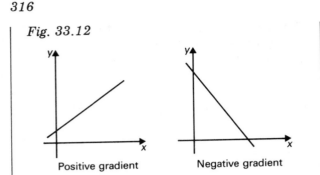

Positive gradient   Negative gradient

Example 6

The table below shows corresponding values of x and y. Plot a graph and hence find the equation connecting x and y.

x	3	5	8	12
y	13	17	23	31

The graph is shown in Fig. 33.13. It is seen to be a straight line which slopes upwards from left to right. Since the true origin is at the intersection of the two axes, c is the intercept on the y-axis, and from the graph we see that $c = 7$.

We now have to find the value of m. Since m is the gradient of the straight line we draw the right-angled triangle PQR, making the sides reasonably long because a small triangle will give less accurate results.

Using the scales of the x and y-axes, PQ = 16 units and RQ = 8 units. Hence

$$m = \frac{\text{Vertical height}}{\text{Base}}$$

$$= \frac{PQ}{RQ}$$

$$= \frac{16}{8}$$

$$= 2$$

Since the straight line slopes upwards from left to right m is positive. Therefore the equation connecting x and y is

$$y = 2x + 7$$

Fig. 33.13

Example 7

The table below shows corresponding values of x and y.

x	10	15	18	20
y	30	20	14	10

Plot the graph and hence find the equation connecting x and y.

The graph is shown in Fig. 33.14. It will be noticed that the starting points on the x and y-axes is 10.
Since the true origin is not shown on the graph we cannot find the value of c by direct measurement. We can, however, find the value of m by drawing the right-angled triangle PQR.

$$m = \frac{PQ}{QR}$$

$$= \frac{10}{5}$$

$$= 2$$

Since the straight line slopes downwards from left to right, the gradient is negative and $m = -2$. The equation of the line is

$$y = -2x + c$$

To find the value of c we choose a point, such as A, which lies on the straight line. In Fig. 33.14, the coordinates of A are $x = 16$ and $y = 18$. Substituting these values in the equation $y = -2x + c$ gives

$$18 = -2 \times 16 + c$$
$$18 = -32 + c$$
$$c = 50$$

Therefore the equation connecting x and y is

$$y = -2x + 50$$

Fig. 33.14

Example 8

(a) The straight line $y = 3x + c$ passes through the point $(2, 8)$.
Find the value of c.

When $x = 2$, $y = 8$, hence

$$8 = 3 \times 2 + c$$
$$8 = 6 + c$$
$$c = 2$$

(b) The straight line $y = mx - 3$ passes through the point $(4, -11)$.
Find the value of m.

When $x = 4$, $y = -11$, hence

$$-11 = 4m - 3$$
$$-8 = 4m$$
$$m = -2$$

Exercise 33.3

1. Draw the graph of $y = 3x + 5$ for values of x between 0 and 6.

2. Draw the graph of $y = 3 - 2x$ for values of x between 3 and 7.

3. Draw the graph of $y = 3x - 4$ for values of x between -4 and 3.

4. Draw the graph of $y = 3 - x$ for values of x between -2 and 6.

The following equations represent straight lines. State in each case the gradient of the line and the intercept on the y-axis.

5. $y = x + 3$ 6. $y = -3x + 4$

7. $y = 2 - 5x$ 8. $y = 4x - 3$

9. $y = -5x - 2$

10. Find the values of m and c if the straight line $y = mx + c$ passes through the point $(-2, 5)$ and has a gradient of 4.

11. Find the values of m and c if the straight line $y = mx + c$ passes through the point $(3, 4)$ and the intercept on the y-axis is -2.

12. The following table gives values of x and y which are connected by an equation of the type $y = mx + c$. Plot the graph and from it find the values of m and c.

x	4	6	10	12
y	16	22	34	40

Use scales of 1 large square to 2 units along the x-axis and 1 large square to 5 units on the y-axis.

13. Using the table below, plot a graph showing the relationship between x and y. Hence find the equation connecting x and y. Use scales of 1 large square to 2 units horizontally and 1 large square to 10 units vertically.

x	10	15	18	22
y	52	72	84	100

14. Using a scale of 1 large square to 1 unit on the horizontal axis and 1 large square to 2 units on the vertical axis draw a pair of axes on graph paper for values of x between -5 and 5 and for values of y between -10 and 10.

(a) Plot the points P(3, 10) and Q(-3, 4) and then join P and Q.

(b) Use your graph to find an equation of the line PQ in the form $y = ax + b$.

15. The table below shows corresponding values of P and Q. Plot a graph showing the relationship between P and Q. Suitable scales are 1 large square to 5 units on both axes. Take P along the horizontal axis and hence find the equation connecting P and Q.

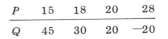

P	15	18	20	28
Q	45	30	20	-20

Experimental Data

One of the most important applications of the straight-line equation is in the determination of an equation connecting two quantities when values have been obtained in an experiment.

Example 9

In an experiment carried out with a lifting machine the effort E and the load W were found to have the values given in the table below.

W	15	25	40	50	60
E	2.75	3.80	5.75	7.00	8.20

Plot these figures and obtain the equation connecting E and W, which is thought to be of the type $E = aW + b$.

Note that if E and W are connected by an equation of the type $E = aW + b$, then the graph must be a straight line (here a and b have been used instead of m and c in the straight-line equation, which is not unusual). Note also that E is the dependent variable and W the independent variable, so that W is taken on the horizontal axis and E on the vertical axis.

On plotting the points (Fig. 33.15) it will be seen that they deviate slightly from a straight line; since the data are experimental we must expect errors in measurement and observation, and slight deviations from a straight line are to be expected. Although the straight line will not pass through some of the points, an attempt must be made to ensure an even spread of points above and below the line.

Fig. 33.15

To find the gradient of the line (i.e. the value of *a*) we draw the right-angled triangle PQR.

$$a = \frac{PQ}{RQ}$$

$$= \frac{4.4}{35}$$

$$= 0.13$$

To find the value of *b* we find the coordinates of the point P, which lies on the line. These are $W = 55$ and $E = 7.5$. Substituting these coordinates in $E = aW + b$ gives

$$7.5 = 55 \times 0.13 + b$$

$$7.5 = 7.15 + b$$

$$b = 0.35$$

Therefore the equation connecting *E* and *W* is

$$E = 0.13W + 0.35$$

Exercise 33.4

1. The following observed values of *P* and *Q* are supposed to be related by the linear equation $P = aQ + b$, but there are experimental errors. Find by plotting the graph the most probable values of *a* and *b*.

Q	2.5	3.5	4.4	5.8	7.5	9.6	12.0	15.1
P	13.6	17.6	22.2	28.0	35.5	47.4	56.1	74.6

2. In an experiment carried out with a machine the effort *E* and the load *W* were found to have the values given in the table below. The equation connecting *E* and *W* is thought to be of the type $E = aW + b$. By drawing a graph check if this is so and from the graph find the most probable values of *a* and *b*.

W	10	30	50	60	80	100
E	8.9	19.1	29	33	45	54

3. A test on a metal-filament lamp gave the following values of resistance *R* (ohms) at various voltages *V* (volts).

V	62	75	89	100	120
R	100	117	135	149	175

These results are expected to agree with an equation of the type $R = mV + c$. Test this assumption by drawing a graph, and from it find suitable values for *m* and *c*.

4. During an experiment to verify Ohm's law the following results were obtained:

E (volts)	0	1.0	2.0	2.5	3.7	4.1
I (amps)	0	0.24	0.50	0.63	0.92	1.05

Plot these values with *I* on the horizontal axis and find an equation connecting *E* and *I*.

5. The following figures show how the resistance of a conductor R (ohms) varies as the temperature t (°C) increases.

t	25	50	75	100	150
R	20.7	21.5	22.3	23.0	24.5

Plot these results and show that $R = at + b$.

Hence estimate values for a and b.

Miscellaneous Exercise 33

Section A

1. The graph shown in Fig. 33.16 shows the relation between the number of units used and the cost of a telephone bill. Use the graph to find

(a) the telephone bill when 360 units are used

(b) the number of units used when the bill is £14.75

(c) The bill is made up of a rental charge plus so much per unit used. What is the rental charge?

Fig. 33.16

Fig. 33.17

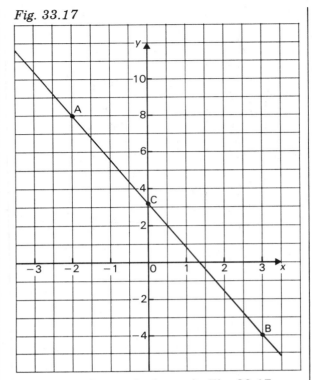

2. Copy the graph shown in Fig. 33.17.

(a) On your graph write down the coordinates of the points A, B and C.

(b) Plot the points D($-3, -2$) and E($3, 7$).

3. A car was tested for fuel consumption at various speeds.

Speed in mile/h	10	30	50	70	80
Fuel consumption in mile/gal	20	35	40	30	23

(a) Plot the points and draw a smooth curve through them. Suitable scales are one large square to represent 5 mile/gal along the horizontal axis and one large square to represent 20 mile/h on the vertical axis.

(b) Use your graph to answer the following questions:

(i) What is the fuel consumption when the speed is 40 miles/h?

(ii) At what speeds is the consumption of fuel 25 miles/gal?

(iii) What is the most economical speed?

4. This is part of a table to change kilograms into pounds.

kilograms	5	10	15	20	25	30
pounds	11	22	33	44	55	66

Using scales of one large square to represent 5 kg along the x-axis and one large square to represent 10 lb on the y-axis plot this information on a graph. Use your graph to convert

(a) 12 kg into pounds

(b) 35 lb into kilograms

5. Using a scale of one large square to 2 units on both the x and y-axes, construct a pair of axes on graph paper for values of x and y from -8 to $+8$. Then plot the points $A(-7, -4)$, $B(6, -4)$, $C(6, 5)$, $D(-4, 5)$. Join the points in alphabetical order with straight lines to make a quadrilateral ABCD. What is this quadrilateral called?

Section B

1. Fig. 33.18 shows the relation between the number of units of electricity used and the cost.

(a) Find the cost when 360 units are used.

(b) Find the number of units used when the cost is £17.

(c) What is the standing charge?

(d) What is the cost per unit used?

Fig. 33.18

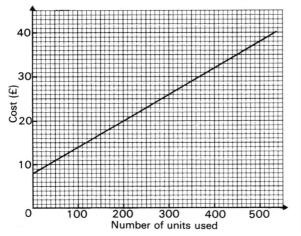

2. Using a scale of 1 large square to 1 unit on the x-axis and 1 large square to 2 units on the y-axis construct a pair of axes on graph paper for values of x between -4 and 5 and values of y between -10 and 10.

(a) On your graph plot the points $A(4, 8)$ and $B(-3, 1)$.

(b) Use your graph to find the coordinates of the point where the line AB cuts the y-axis.

(c) Find the gradient of the line AB.

(d) Write down the equation of the line AB in the form $y = ax + b$.

3. The equation of the straight line PQ is expressed in the form $y = mx + c$. Given that P is the point $(0, 3)$ and the gradient of PQ is 4:

(a) Write down the equation of the line PQ.

(b) The point R has its x-coordinate equal to -2. What is its y-coordinate?

4. In Fig. 33.19, RP is the line $y = 2x + 3$. If ON is 3 units, find the length of PN. What is the area of the trapezium AONP?

Fig. 33.19

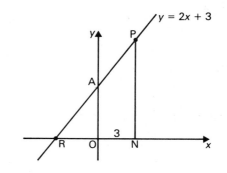

5. The line $y = 2x + c$ passes through the point $(3, 10)$. Find the value of c.

6. Using the graph of Fig. 33.20
   (a) find the coordinates of the point A
   (b) find the gradient of the line AB
   (c) write down the equation of the line AB

Fig. 33.20

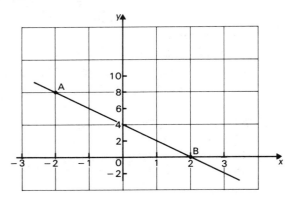

7. Fig. 33.21 shows a straight line passing through the points A and B. The equation of the line AB is $y = \frac{1}{2}x + 6$.

   (a) Write down the coordinates of the point A.

   (b) Find the coordinates of the point B.

   (c) Calculate the area of the triangle AOB.

Fig. 33.21

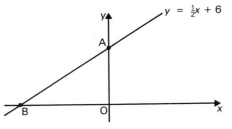

8. The equation of the line AB (Fig. 33.22) is $y = -\frac{1}{2}x + c$.
   (a) Read off the value of c.
   (b) Calculate the distance OB.

Fig. 33.22

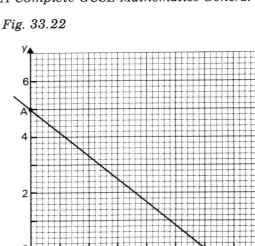

Multi-Choice Questions 33

1. Consider the four graphs shown in Fig. 33.23. Which graph represents the equation $y = 1 - x$?

Fig. 33.23

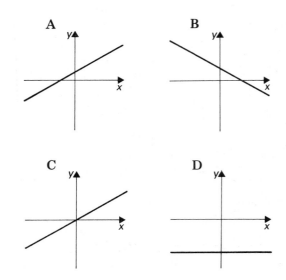

2. The gradient of the line drawn in Fig. 33.24 is

A 2   B $1\frac{1}{2}$ C 1   D $\frac{1}{2}$

Fig. 33.24

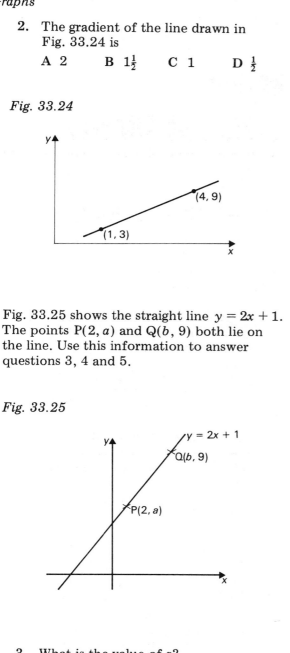

Fig. 33.25 shows the straight line $y = 2x + 1$. The points P(2, a) and Q(b, 9) both lie on the line. Use this information to answer questions 3, 4 and 5.

Fig. 33.25

3. What is the value of a?

A 1   B 2   C 3   D 5

4. What is the value of b?

A 4   B 5   C 6   D 16

5. What is the intercept on the y-axis?

A −1   B $\frac{1}{2}$ C 1   D 2

6. In Fig. 33.26, BM = 13. What is the length of OM?

A $3\frac{1}{2}$ B 5   C $6\frac{1}{2}$ D 8

Fig. 33.26

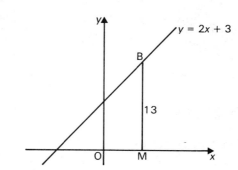

7. The equation of the straight line shown in Fig. 33.27 is

A $y = \frac{1}{2}x - 1$ B $y = \frac{1}{2}x - 2$
C $y = \frac{1}{2}x + 1$ D $y = 2x + 1$

Fig. 33.27

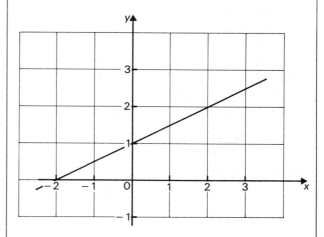

8. The straight line $y = 2x - 3$ passes through one of the following points. Which one?

A (0, −3)   B (0, 3)
C (−3, 0)   D (3, 0)

Further Graphical Work

Functions

Consider the equation

$$y = 3x^2 - 4x + 5$$

We can give x any value we like and so find the corresponding value of y. Thus

if $x = 2$,

$$y = 3 \times 2^2 - 4 \times 2 + 5$$

$$= 9$$

if $x = 5$,

$$y = 3 \times 5^2 - 4 \times 5 + 5$$

$$= 60$$

Since the value of y depends upon the value allocated to x it is called the **dependent variable**. Because we can give x any value we like it is called the **independent variable.**

Because the value of y depends upon the value allocated to x we say that y is a function of x. Instead of writing

$$y = 3x^2 - 4x + 5$$

we can write

$$f(x) = 3x^2 - 4x + 5$$

meaning that the function of x is $3x^2 - 4x + 5$.

Note carefully that $f(x)$ does not mean f multiplied by x. It is simply shorthand for the statement 'is a function of x'.

The Graph of $y = ax^2$

Example 1

Draw the graph of $y = 5x^2$, for values of x between -3 and 3.

A table may be made, as follows, which gives values of y corresponding to chosen values of x.

x	-3	-2	-1	0	1	2	3
y	45	20	5	0	5	20	45

The graph is shown in Fig. 34.1 and it is seen to just touch the x-axis at the origin. It is also symmetrical about the y-axis.

Fig. 34.1

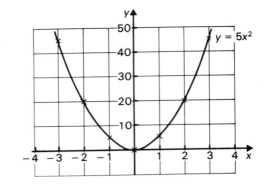

Graphs of Quadratic Functions

The function $f(x) = ax^2 + bx + c$ is called a quadratic function of x. The constants a, b and c can take any numerical value. For instance

$f(x) = x^2 - 36$
in which $a = 1$, $b = 0$ and $c = -36$

$f(x) = 5x^2 + 7x + 8$
in which $a = 5$, $b = 7$ and $c = 8$

$f(x) = 2.5x^2 - 3.1x - 2.9$
in which $a = 2.5$, $b = -3.1$ and $c = -2.9$

A quadratic function may contain only the square of the unknown quantity, as in the first of the above functions, or it may contain both the square and the first power of the unknown quantity, as in the second and third of the above functions.

When a quadratic function is plotted it always gives a smooth curve called a **parabola**.

Example 2

Plot the graph of $f(x) = 5x^2 + 13x - 6$ for values of x between -5 and 2.

A table may be made, as follows, which gives values of $f(x)$ corresponding to chosen values of x

x	-5	-4	-3	-2	-1	0	1	2
$5x^2$	125	80	45	20	5	0	5	20
$13x$	-65	-52	-39	-26	-13	0	13	26
-6	-6	-6	-6	-6	-6	-6	-6	-6
$f(x)$	54	22	0	-12	-14	-6	12	40

The graph of the function is shown in Fig. 34.2.

Fig. 34.2

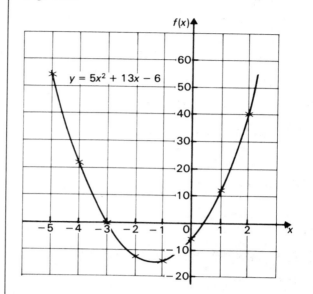

$y = 5x^2 + 13x - 6$

The Reciprocal Function

The function $f(x) = a/x$ is called the **reciprocal function**, a being a constant.

Example 3

Plot the graph of $f(x) = \dfrac{5}{x}$ for values of x between -5 and 5.

x	-5	-4	-3	-2	-1	0	1	2	3	4	5
$f(x)$	-1	-1.25	-1.67	-2.5	-5		5	2.5	1.67	1.25	1

Note that the reciprocal of a negative number is a negative number, e.g. $\dfrac{1}{-5} = -0.2$, whilst the reciprocal of a positive number is a positive number, for instance $\dfrac{1}{5} = 0.2$.

It is impossible to find the reciprocal of 0 because we cannot divide by 0. Hence the graph consists of two branches as shown in Fig. 34.3.

Fig. 34.3

f(x) graph

Curves of Natural Growth

Curves showing population growth, multiplication of bacteria, growth of invested money, etc. are called **curves of natural growth**. Expressed mathematically the exponential growth function is $f(x) = an^x$ where x is called the **exponent** and n the **base**, a being a constant.

Example 4

The population of a certain country is increasing at a rate of 4% per annum. If the population in 1975 was 2 000 000, draw a graph to represent the population growth up to the year 2000.

Since the increase is 4%,

$$n = 1 + \frac{4}{100} = 1.04.$$

Hence the graph may be constructed using the equation

$$y = 2 \times 1.04^x$$

where x is the number of years after 1975 and y is the corresponding population in millions.

To calculate the value of y which corresponds to a chosen value of x we proceed as follows:

When the year is 1980, $x = 5$ (i.e. 5 years after 1975) and

$$y = 2 \times 1.04^5$$
$$= 2.43$$

It is necessary to use a calculator, as follows:

Input	Display	
1.04	1.04	
××	1.04	
=	1.08	$(1.04)^2$
=	1.12	$(1.04)^3$
=	1.16	$(1.04)^4$
=	1.22	$(1.04)^5$
×	1.22	
2	2.	
=	2.43	(2×1.04^5)

The calculation of y for other values of x is obtained in a similar way.

Year	1975	1980	1985	1990	1995	2000
x	0	5	10	15	20	25
y	2	2.43	2.96	3.60	4.38	5.33

The graph is shown in Fig. 34.4, where it will be seen that the curve rises more steeply as the number of years increases. This is typical of graphs depicting the law of natural growth.

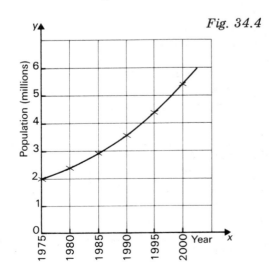

Fig. 34.4

The Gradient of a Curve

In mathematics and science we often need to know the rate of change of one variable with respect to another. For instance speed is the rate of change of distance with respect to time.

Consider the graph of $y = x^2$, part of which is shown in Fig. 34.5. As the values of x increase so do the values of y, but they do not increase at the same rate. A glance at Fig. 34.5 shows that the values of y increase faster when x is large because the gradient of the curve is increasing.

To find the rate of change of y with respect to x at a particular point we need to find the gradient of the curve at that point. If we draw a tangent to the curve at the point, the gradient of the tangent will be the same as the gradient of the curve.

Fig. 34.5

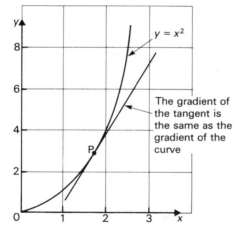

The gradient of the tangent is the same as the gradient of the curve

Example 5

Draw the graph of $y = x^2 - 3x + 7$ for values of x between -4 and $+4$. Hence find the gradient of the curve at the points where $x = -3$ and $x = 2$.

To plot the curve we draw up the following table:

x	-4	-3	-2	-1	0	1	2	3	4
y	35	25	17	11	7	5	5	7	11

The point where $x = -3$ is the point $(-3, 25)$. We draw a tangent to the curve at this point as shown in Fig. 34.6. The gradient is found by constructing a right-angled triangle (which should be as large as possible for accuracy), as shown, and measuring the length of its height and base. The gradient is negative because the tangent slopes downwards from left to right.

$$\text{Gradient at the point } (-3, 25) = \frac{\text{Height}}{\text{Base}}$$

$$= -\frac{27}{3}$$

$$= -9$$

The negative gradient indicates that the values of y are decreasing as the values of x increase.

At the point where $x = 2$ the value of y is 5. Hence by drawing a tangent and a right-angled triangle at the point $(2, 5)$,

$$\text{Gradient at the point } (2, 5) = \frac{3}{3} = 1$$

The gradient is positive because the tangent slopes upwards from left to right. A positive gradient indicates that the values of y are increasing as the values of x increase.

Fig. 34.6

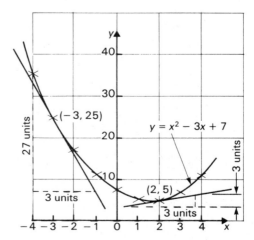

Example 6

Draw the graph of $f(x) = x^2 + 3x - 2$ for values of x between -1 and 4. Hence find the value of $f(x)$ where the gradient of the curve is 7.

To plot the graph the following table is made:

x	-1	0	1	2	3	4
y	-4	-2	2	8	16	26

The graph is shown in Fig. 34.7.

To obtain the line whose gradient is 7 we draw the right-angled triangle ABC making (for convenience) AB = 2 units to the scale on the x-axis and BC = 14 units to the scale on the y-axis.

$$\text{Gradient of AC} = \frac{14}{2}$$
$$= 7$$

Using set squares draw a tangent to the curve so that it is parallel to the line AC. From the diagram it will be seen that the tangent just touches the curve at the point (2, 8). Hence when the gradient of the curve is 7, the value of $f(x)$ is 8.

Fig. 34.7

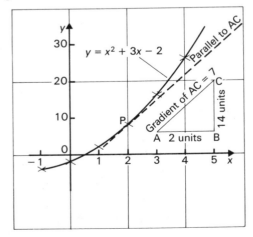

Exercise 34.1

1. Draw the graph of $f(x) = 3x^2$ for values of x between -3 and $+3$.

2. Plot the graph of $y = 2x^2 - 7x - 5$ for values of x between -4 and 12.

3. Plot the graph of $f(x) = x^2 - 4x + 4$ for values of x between -3 and $+3$.

4. Draw a graph of $f(x) = 4/x$ for values of x between -4 and $+4$.

5. The figures in the table below give corresponding values of x and y for a certain reciprocal function.

x	-4	-3	-2	-1	0	1	2	3	4
y	-2	-2.67	-4	-8		8	4	2.67	2

Draw a graph of this information using a scale of 1 large square to 1 unit on the x-axis and 1 large square to 2 units on the y-axis.

6. A bacterium multiplies according to the law $N = 2^T$ where N is the number of bacteria after a time of T hours.

   (a) Draw up a table showing the number of bacteria after a time of 1, 2, 3, 4 and 5 hours.

   (b) Draw a graph of $N = 2^T$ and use it to estimate the number of bacteria after a time of $3\frac{1}{2}$ hours. Use scales of 1 large square to 1 hour on the horizontal axis and 1 large square to 5 units on the vertical axis.

7. The following table shows how a sum of £8000 appreciates over a period of 7 years.

Year	1	2	3	4	5	6	7
Amount (£)	8800	9680	10 650	11 710	12 880	14 170	15 590

   (a) Draw a graph to depict this information using scales of 1 large square to 1 year on the horizontal axis and 1 large square to £1000 on the vertical axis.

(b) What is the amount after $4\frac{1}{2}$ years?

(c) Estimate the number of years for the amount to reach £13 500.

8. The population of a territory is increasing at the rate of 5% per annum. In 1960 the population was 2 million. Estimate the population in the year 2000, assuming that the rate of growth of the population remains the same.

9. Draw the graph of $y = 3x^2 + 7x + 3$ for values of x between -5 and $+5$. Hence find the gradient of the curve at the points where $x = -2$ and $x = 2$.

10. Draw the graph of $f(x) = 2x^2 - 5$ for values of x between -2 and $+3$. Hence find the gradient of the curve at the points where $x = -1$ and $x = 2$.

11. If $y = (1 + x)(5 - 2x)$, copy and complete the table below.

x	-2	$-1\frac{1}{2}$	-1	0	$\frac{1}{2}$	1	$1\frac{1}{2}$	2	3
y	-9		0	5		6	5	3	-4

Hence draw the graph of $y = (1 + x)(5 - 2x)$. Find the value of x at which the gradient of the curve is -2.

12. If $f(x) = x^2 - 5x + 4$, find, by plotting the curve for values of x between 4 and 12, the value of x at which the gradient of the curve is 11.

Distance–Time Graphs

In Chapter 12 it was shown that

$$\text{Average speed} = \frac{\text{Total distance travelled}}{\text{Total time taken}}$$

If the distance is measured in metres and the time in seconds, then the speed is measured in metres per second (m/s). If the distance is measured in kilometres and the time in hours, then the speed is measured in kilometres per hour (km/h).

Example 7

(a) A car travels a distance of 100 km in 2 hours. Calculate its average speed.

$$\text{Average speed} = \frac{100}{2} \text{ km/h}$$
$$= 50 \text{ km/h}$$

(b) A body travels a distance of 80 m in 4 s. Find its average speed.

$$\text{Average speed} = \frac{80}{4} \text{ m/s}$$
$$= 20 \text{ m/s}$$

Since

$$\text{Distance} = \text{Speed} \times \text{Time}$$

when the speed is constant the distance travelled is proportional to the time taken. The graph of distance against time will therefore be a straight line passing through the origin. The gradient of this graph will represent the speed.

Example 8

Fig. 34.8 shows a distance–time graph for a car. Find the average speed of the car.

Fig. 34.8

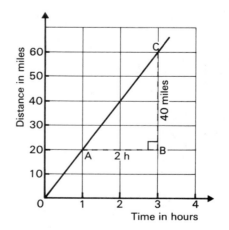

The average speed of the car is given by the gradient of the graph. Drawing the right-angled triangle ABC we see that when AB = 2 hours, BC = 40 miles. Hence

$$\text{Gradient of line} = \frac{BC}{AB}$$

$$= \frac{40}{2}$$

$$= 20$$

Hence the average speed of the car is 20 miles per hour.

Example 9

A man travels a distance of 120 km in 2 hours by car. He then cycles 20 km in $1\frac{1}{2}$ hours and finally walks a distance of 8 km in 1 hour, all at constant speed. Draw a graph to illustrate this journey and from it find the average speed for the entire journey.

Fig. 34.9

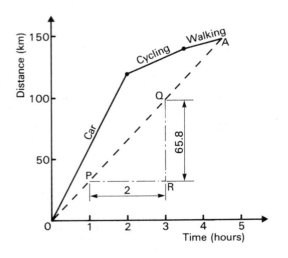

The graph is drawn in Fig. 34.9 and it consists of three straight lines. The average speed for the entire journey is found by drawing the straight line OA and finding its gradient.

To find the gradient of OA, the right-angled triangle PQR is drawn, which for accuracy should be as large as possible.

$$\text{Gradient of OA} = \frac{QR}{PR}$$

$$= \frac{65.8}{2}$$

$$= 32.9$$

Hence the average speed for the entire journey is 32.9 km/h.

Exercise 34.2

1. A vehicle travels 300 km in 5 hours. Calculate its average speed.

2. A car travels 400 km at an average speed of 80 km/h. How long did the journey take?

3. A train travels for 5 hours at an average speed of 60 km/h. How far did it travel?

4. A vehicle travels a distance of 250 km in a time of 5 hours. Draw a distance-time graph to depict the journey and from it find the average speed of the vehicle.

5. A car travels a distance of 120 km in 3 hours. It then changes speed and travels a further distance of 80 km in $2\frac{1}{2}$ hours. Assuming that the two speeds are constant, draw a distance–time graph and from it determine the average speed for the complete journey.

6. A girl cycles a distance of 20 km in 100 minutes. She rests for 20 minutes and then cycles a further 10 km, which

takes her 60 minutes. Draw a distance–time graph to represent the journey and from it find the average speed for the entire journey.

7. A man travels a distance of 90 km by car, which takes him $1\frac{1}{2}$ hours. He then cycles 18 km in a time of $1\frac{1}{2}$ hours. He rests for 15 minutes before continuing his journey on foot, during which he walks 8 km in 2 hours. By drawing a distance–time graph find his average speed for the entire journey, assuming that he travels at constant speed for each of the three parts of his journey.

Example 10

Two towns A and B are 120 km apart. A motorist starts out from A at 9.00 a.m. and drives towards B at a constant speed of 60 km/h. At 9.30 a.m. a second motorist sets out from B and drives at a constant speed of 90 km/h towards A. How far from A will the two cars meet?

The distance–time graphs for the two motorists are drawn on the same axes, as shown in Fig. 34.10. Since the speed of the first motorist is constant the graph is a straight line. If 9.00 a.m. is taken as zero time, then this line will pass through the origin.

Half an hour later the second motorist sets out from B, which is 120 km from A. As time progresses he gets nearer to A. Hence the distance–time graph for the second motorist is also a straight line, but it slopes downwards from left to right.

The vertical coordinate of the point X, where the two graphs intersect, gives the distance from A where the two motorists meet. From the graph, this distance is found to be 66 km.

Fig. 34.10

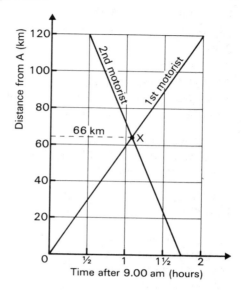

Exercise 34.3

1. A cyclist starts from a town A at 9.00 a.m. and cycles towards town B at an average speed of 15 km/h. At 10.00 a.m. a motorist starts the same journey but travels at 60 km/h. When and where will the motorist overtake the cyclist?

2. A train left London for Newcastle, a distance of 420 km, at 1500 hours. It stopped at Grantham (180 km from London) for 12 minutes. It stopped again at York (315 km from London) for 12 minutes. Except for the halts, it travelled at a steady 70 km/h. A second train left Newcastle for London at 1530 hours and it travelled non-stop at 90 km/h.

   (a) Draw a distance–time graph for the first train.

   (b) At what time did the first train reach Newcastle?

(c) On the same axes draw a distance-time graph for the second train.

(d) When and where did the two trains cross?

3. Fig. 34.11 shows an incomplete distance-time graph for a squad of soldiers walking from A to B. Copy the graph.

(a) At what speed, in kilometres per hour, was the squad walking during the first twenty minutes?

(b) Up to 2 p.m., for how long did the squad rest?

(c) The squad started off again at 2 p.m. and at the end of 3 hours and 40 minutes had walked a total distance of 22 km. On your copy of Fig. 34.11 complete the graph for the last 1 hour and 40 minutes, assuming that the squad walked at a uniform speed during this period.

(d) By drawing a suitable straight line on the graph find the average speed of the squad for the entire journey, including resting time.

Fig. 34.11

4. A woman leaves home at 9 a.m. and returns at 4.30 p.m. The graph (Fig. 34.12) shows her distance from home, in kilometres, at various times during the day.

Fig. 34.12

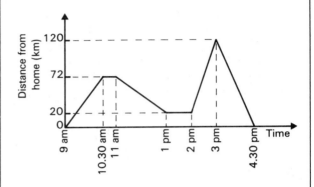

(a) Calculate her average speed between 9 a.m. and 10.30 a.m.

(b) Find the total number of kilometres she travels between 9 a.m. and 4.30 p.m.

(c) How far is she from home at 2.45 p.m.?

(d) For how many hours during the day is she moving towards home?

5. A young man sets out at noon to cycle from P to Q, a distance of 200 km. He cycles at a steady 20 km/h for $4\frac{1}{2}$ hours then stops and rests for 1 hour. He then cycles on and reaches Q at 2200 hours. A motorist using the same route sets out from Q at 1800 hours and travels towards P. He maintains a steady speed of 60 km/h.

On the same axes draw graphs of these two journeys using scales of 1 cm = 1 hour horizontally and 1 cm = 40 km vertically.

(a) At what time do the motorist and the cyclist pass each other?

(b) How far is the cyclist from Q when the motorist passes him?

(c) At what speed does the cyclist travel during the second part of his journey?

(d) At what times were the motorist and cyclist 20 km apart?

Speed–Time Graphs

If a speed–time graph is drawn (Fig. 34.13), the area under the graph gives the distance travelled. The gradient of the curve gives the acceleration, because acceleration is rate of change of speed with respect to time. If the speed is measured in metres per second then the acceleration will be measured in metres per second per second (m/s^2).

Fig. 34.13

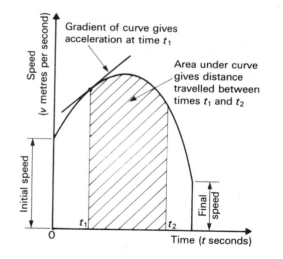

If the acceleration is uniform the speed-time graph will be a straight line (Fig. 34.14). When the speed is increasing the graph has a positive gradient (i.e. it slopes upwards to the right). If the speed is decreasing (i.e. there is deceleration, sometimes called retardation) the graph has a negative gradient and it slopes downwards to the right.

Fig. 34.14

Example 11

A car starting from rest attains a speed of 20 m/s after 5 seconds. It continues at this speed for 15 seconds and then slows down and comes to rest in a further 8 seconds. If the acceleration and retardation are constant, draw a speed–time graph and from it find

(a) the acceleration of the car

(b) the retardation of the car

(c) the distance travelled in the total time of 28 seconds

The speed–time graph is shown in Fig. 34.15.

Fig. 34.15

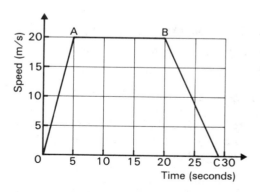

(a) The acceleration is given by the gradient of the line OA

$$\text{Acceleration} = \frac{20}{5}$$

$$= 4 \, m/s^2$$

(b) The retardation is given by the gradient of the line BC

$$\text{Retardation} = \frac{20}{8}$$

$$= 2.5 \, \text{m/s}^2$$

(c) The distance covered in the 28 seconds the car was travelling is given by the area of the trapezium OABC

$$\text{Distance travelled} = 20 \times \tfrac{1}{2} \times (15 + 28)$$

$$= 20 \times \tfrac{1}{2} \times 43$$

$$= 430 \, \text{metres}$$

Example 12

The table below gives the speed of a car v metres per second after a time t seconds.

t	0	5	10	15	20	25
v	0	2.4	5.0	7.5	9.5	10.2

t	30	35	40	45	50	55
v	9.2	5.2	2.7	2.3	2.7	3.5

Draw a smooth curve to show how v varies with t. Use the graph to find

(a) the speed of the car after 47 seconds

(b) the times when the speed is 4 m/s

(c) the acceleration after 15 seconds

(d) the retardation after 35 seconds

The graph is shown in Fig. 34.16.

(a) The speed after 47 seconds is read directly from the graph and it is found to be 2.4 m/s.

(b) The times when the speed is 4 m/s are also found directly from the graph. They are 8 seconds and 37 seconds.

(c) To find the acceleration at a time of 15 seconds draw a tangent to the curve at the point A and find its gradient by constructing the right-angled triangle ABC.

$$\text{Gradient} = \frac{AB}{BC}$$

$$= \frac{4}{10}$$

$$= 0.4$$

Hence the acceleration is 0.4 m/s².

(d) To find the retardation at a time of 35 seconds draw a tangent to the curve at the point D. Then by constructing the right-angled triangle EFG

$$\text{Gradient} = \frac{EF}{FG}$$

$$= \frac{5.6}{10}$$

$$= 0.56$$

Hence the retardation is 0.56 m/s².

Note that the car is slowing down at this time and therefore retardation occurs. This is also shown by the negative gradient of the tangent at the point D.

Fig. 34.16

Exercise 34.4

1. Fig. 34.17 shows a number of speed–time graphs. In each case state the distance travelled.

Fig. 34.17

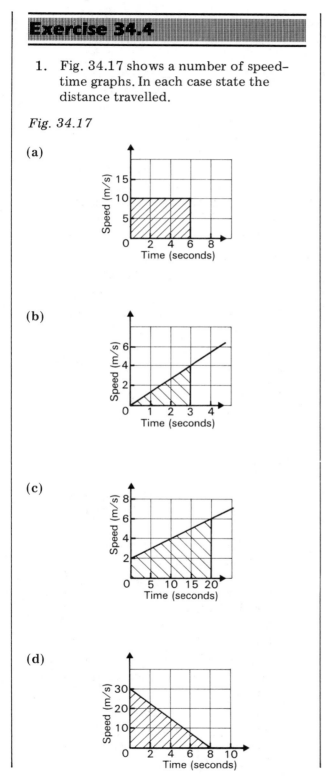

(a)

(b)

(c)

(d)

(e)

2. Fig. 34.18 shows several speed–time graphs. For each, write down the acceleration or retardation.

Fig. 34.18

(a)

(b)

(c)

(d)

3. Fig. 34.19 shows a speed–time graph for a vehicle travelling at a uniform speed. Find the total distance travelled by the vehicle in 15 seconds.

Fig. 34.19

4. The speed–time graph shown in Fig. 34.20 shows a car travelling with constant acceleration. Find

   (a) the acceleration

   (b) the initial speed

   (c) the distance travelled in 30 seconds

Fig. 34.20

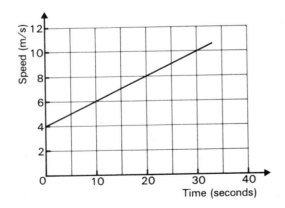

5. From the speed–time graph shown in Fig. 34.21, find

   (a) the acceleration

   (b) the retardation

   (c) the initial speed

   (d) the distance travelled in the first 10 seconds

Fig. 34.21

6. Fig. 34.22 shows a speed–time graph. Find

   (a) the acceleration after 5 seconds

   (b) the acceleration after 18 seconds

   (c) the acceleration after 40 seconds

   (d) the maximum speed reached

   (e) the total distance travelled in 50 seconds

Fig. 34.22

7. A vehicle starting from rest attains a speed of 16 m/s after it has been travelling for 8 seconds with uniform acceleration. It continues at this speed for 15 seconds and then it slows down with uniform retardation until it finally comes to rest in a further 10 seconds.

   (a) Sketch the speed–time graph.

   (b) Determine the acceleration of the vehicle.

   (c) What is the retardation?

   (d) Find the distance travelled by the vehicle in the 33 seconds depicted by the speed–time graph.

8. A car has an initial speed of 5 m/s. It then accelerates uniformly for 6 seconds at 0.5 m/s². It then proceeds at this speed for a further 25 seconds. Draw the speed–time graph and use it to find the total distance travelled in the time of 31 seconds.

9. The speed of a body, v metres per second, at various times t seconds is shown in the table below.

t	0	1	2	3	4	5	6	7	8
v	0	1	2	6	12	20	30	42	56

(a) Draw the graph showing how v varies with t. Horizontally, take 1 cm to represent 1 second and vertically, take 1 cm to represent 10 m/s.

(b) From the graph find the acceleration after times of 2 seconds and 6 seconds.

10. A body moves a distance s metres in a time t seconds so that $s = t^3 - 3t^2 + 8$. Plot a distance–time graph for this equation taking values of t between 0 and 5 seconds in half-second intervals. Hence use the graph to find the speed of the body after 3 seconds.

Miscellaneous Exercise 34

Section B

1. Part of the timetable for morning trains between four towns, Altown, Beeford, Colville and Denton is given below.

Altown	dep 0820	dep 0830	dep 0945
Beeford		arr 0857	arr 1015
		dep 0900	dep 1017
Colville			arr 1035
			dep 1037
Denton	arr 0910	arr 0930	arr 1054

(a) How long does the 0820 train from Altown take to make the journey to Denton?

(b) The 0945 train from Altown averages 66 km/h on its journey to Beeford. How far is it from Altown to Beeford?

(c) Denton is 60 km from Altown. What is the average speed, in kilometres per hour, for the journey of the 0820 train from Altown?

(d) The distance–time graph for one of the trains is sketched in Fig. 34.23, To which of the trains does it apply?

Fig. 34.23

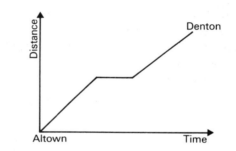

2. Copy and complete the table below, given that $y = x^2 - 6x + 8$.

x	−1	0	1	2	2.5	3	3.5	4	5
x^2	1				6.25		12.25		25
$-6x$	6				−15		−21		−30
8	8				8		8		8
y	15				−0.75		−0.75		3

Use the completed table to plot the graph of $y = x^2 - 6x + 8$, taking 2 cm to represent 1 unit on the x-axis and 1 cm to represent 1 unit on the y-axis. Hence find the gradient of the curve at the points where $x = 1$ and $x = 4$.

3. The table below gives the population (in millions) of a certain country during the years stated.

Year	1950	1955	1960	1965	1970	1975	1980
Population	3	3.48	4.03	4.67	5.42	6.28	7.28

(a) Draw a graph of this information taking the year on the horizontal axis. Suitable scales are: 1 large square to 5 years horizontally and 1 large square to 1 million of population vertically.

(b) The equation connecting year and population is $p = 3 \times 1.03^x$ where p is the population at the end of x years. Use this equation to estimate the population in the year 2000 assuming that the rate of increase of the population remains the same.

4. During the first stage of the vertical ascent of a rocket, its speed v metres per second t seconds after the start is given by the formula $v = 20t + 8t^2$.

Copy and complete the following table.

t	0	5	10	15	20	25	30
$20t$		100				500	
$8t^2$		200				5000	
v		300				5500	

(a) Use the completed table to draw a graph showing how v varies with t. Suitable scales are 1 large square to 5 seconds horizontally and 1 large square to 1000 metres per second vertically.

(b) Use your graph to determine the acceleration of the rocket 20 seconds after the start.

5. A person invests £10 000. If the interest rate is 10% per annum, draw a graph showing the appreciation of the investment over a period of 10 years. Use 1 large square to represent 2 years horizontally and 1 large square to represent £5000 vertically.

Transformational Geometry

Introduction

When we are given a geometrical shape there are many things which we can do with it.

(1) We can move the shape from one position to another by sliding it in one direction. Thus we can slide a parallelogram from position A to position A' (Fig. 35.1). We say that the parallelogram A has been **translated** to the new position A'.

Fig. 35.1

(2) A mirror reflects an image. In Fig. 35.2, the triangle B' is a **reflection** of the triangle B in the mirror line m.

Fig. 35.2

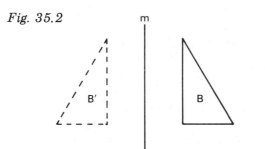

(3) We can rotate the shape from one position to another. In Fig. 35.3, the triangle C has been given a **rotation** about P so that it takes up the position C'.

Fig. 35.3

(4) A slide projector enlarges pictures. Working in a similar way we can enlarge geometric shapes. Thus in Fig. 35.4, rectangle D' is an enlargement of the rectangle D.

Fig. 35.4

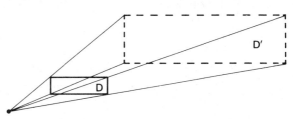

In each case we say that the figure has been **transformed**. The operation applied is called a **transformation**. The original figure is called the **pre-image** and the transformed figure is called the **image**. For example, in Fig. 35.1, A' is the image of A.

A transformation is often referred to as a **mapping**. A transformation therefore maps a pre-image on to an image.

Translation

If every point in a line or a shape moves the same distance in the same direction, the transformation is called a **translation**. Thus in Fig. 35.5, every point in the line AB has moved 2 units to the right and 3 units upwards. That is $A(2, 1)$ translates to $A'(4, 4)$ and $B(6, 3)$ translates to $B'(8, 6)$.

Fig. 35.5

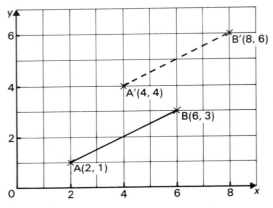

Similarly in Fig. 35.6, the triangle ABC has been translated to $A'B'C'$. Every point in ABC has been displaced by 4 units to the left and 2 units upwards. The translation maps the point $A(1, 1)$ on to $A'(-3, 3)$ and $B(3, 1)$ on to $B'(-1, 3)$ and $C(1, 4)$ on to $C'(-3, 6)$.

Fig. 35.6

Generally this transformation maps (x, y) on to $(x - 4, y + 2)$. If this translation is denoted by T_1 we may write

$$T_1 : (x, y) \rightarrow (x - 4, y + 2)$$

The translation is defined by the constants -4 and $+2$ which are added to the coordinates of the original point. We may therefore define the translation by the **column vector** $\begin{pmatrix} -4 \\ 2 \end{pmatrix}$.

Properties of Translations

If two figures are related by a translation, the corresponding segments of these figures are equal in length and they are also parallel. Thus in Fig. 19.6, $AB = A'B'$ and $AB \parallel A'B'$; $AC = A'C'$ and $AC \parallel A'C'$; $BC = B'C'$ and $BC \parallel B'C'$. When a figure undergoes a translation, the image and the pre-image are **congruent** i.e. they are the same shape and size. Thus in Fig. 35.6, the triangles ABC and $A'B'C'$ are congruent.

Example 1

The rectangle ABCD is formed by joining four points, $A(1, 5)$, $B(5, 5)$, $C(5, 3)$ and $D(1, 3)$. What is the image of ABCD under the translation $\begin{pmatrix} 3 \\ 4 \end{pmatrix}$?

The image of A is $(1 + 3, 5 + 4) = (4, 9)$

The image of B is $(5 + 3, 5 + 4) = (8, 9)$

The image of C is $(5 + 3, 3 + 4) = (8, 7)$

The image of D is $(1 + 3, 3 + 4) = (4, 7)$

The rectangle ABCD and its image under the translation are shown in Fig. 35.7.

Fig. 35.7

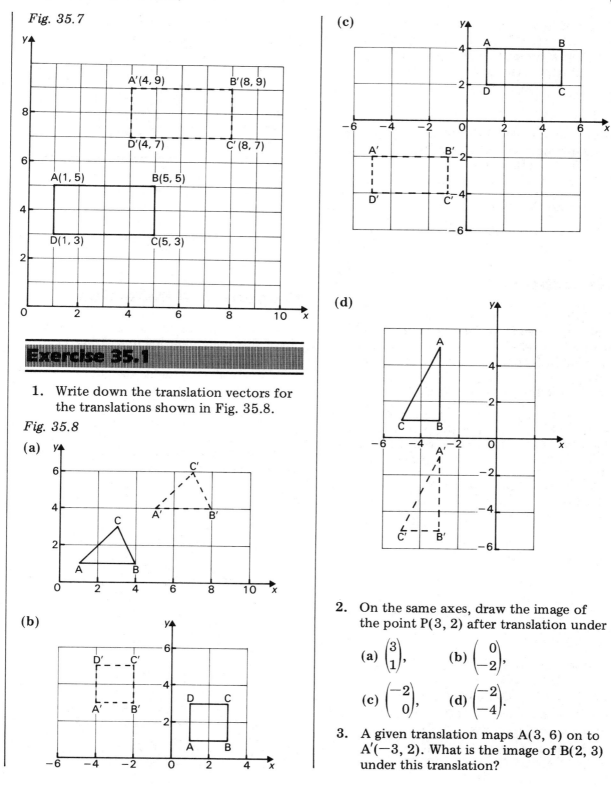

Exercise 35.1

1. Write down the translation vectors for the translations shown in Fig. 35.8.

Fig. 35.8

(a)

(b)

(c)

(d)

2. On the same axes, draw the image of the point P(3, 2) after translation under

(a) $\begin{pmatrix} 3 \\ 1 \end{pmatrix}$, (b) $\begin{pmatrix} 0 \\ -2 \end{pmatrix}$,

(c) $\begin{pmatrix} -2 \\ 0 \end{pmatrix}$, (d) $\begin{pmatrix} -2 \\ -4 \end{pmatrix}$.

3. A given translation maps A(3, 6) on to A'(−3, 2). What is the image of B(2, 3) under this translation?

4. The vertices of the square ABCD are A(0, 1), B(2, 1), C(2, 3) and D(0, 3). Draw this square on graph paper. Draw also the image of ABCD under the translation $\begin{pmatrix} 5 \\ 4 \end{pmatrix}$.

5. The triangle ABC is formed by joining the three points A(−2, −2), B(−1, −4) and C(−1, −1). Draw this triangle on graph paper and show its image A′B′C′ under the translation $\begin{pmatrix} 2 \\ -1 \end{pmatrix}$.

State the coordinates of the points A′, B′ and C′.

Reflection

The idea of a reflection is a familiar one. If you look in a mirror you see an image of yourself. A reflection is another transformation and it is very useful in the study of the properties of geometric figures and in forming wallpaper and carpet patterns.

If a point P is reflected in a mirror so that its image is P′, the mirror is the perpendicular bisector of PP′ (Fig. 35.9). We say that P′ is the **reflection** of P in the mirror line m.

Fig. 35.9

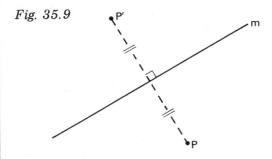

Example 2

The triangle ABC is formed by joining the points A(1, 1), B(2, 1) and C(1, 2). Draw this triangle on graph paper.

(a) Draw the image of ABC after reflection in the x-axis and mark it A′B′C′.

(b) Draw the image of ABC after reflection in the y-axis and mark it A″B″C″.

The reflections are shown in Fig. 35.10

Fig. 35.10

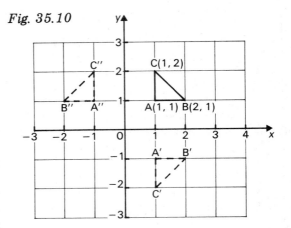

Properties of Reflection

After the reflection, the pre-image and the image are congruent figures. Thus in Fig. 35.10, triangles ABC, A′B′C′ and A″B″C″ are all congruent triangles.

Example 3

In Fig. 35.11, XY is a reflection of AB, m being the mirror line. Given that BC = 4 cm, find the length of CX.

Fig. 35.11

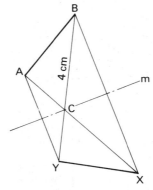

Since XY is a reflection of AB, triangles ABC and CXY are congruent. Hence

$$CX = BC$$
$$= 4 \text{ cm}$$

Look at the kite drawn in Fig. 35.12. We see that it is symmetrical about the line AC. This means that △ABC is a reflection of △ADC, AC being the mirror line (△ is shorthand notation for 'triangle'). It follows that

  (i) AC is the perpendicular bisector of BD, since the point B is a reflection of the point D,
 (ii) AC bisects the angles BAD and BCD,
(iii) Triangles ABC and ADC are congruent.

Fig. 35.12

If a figure contains its own reflection in a line m, then the figure is said to be symmetrical about m. The line m is the mirror line or the axis of symmetry.

Example 4

Complete the figure in Fig. 35.13 so that it has two axes of symmetry.

Fig. 35.13

The figure is completed by reflecting it in the line YY′ and then reflecting the double figure in the line XX′. The finished figure then has two axes of symmetry, XX′ and YY′, as shown in Fig. 35.14.

Fig. 35.14

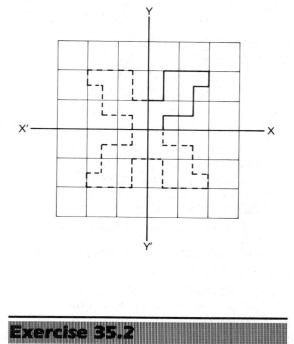

Exercise 35.2

1. Reflect the shape in Fig. 35.15 in
   (a) the *x*-axis, (b) the *y*-axis.

Fig. 35.15

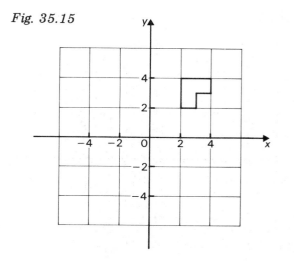

2. Copy the shapes shown in Fig. 35.16 and show all the lines of symmetry for each shape.

Fig. 35.16

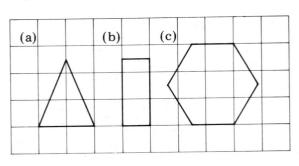

3. Plot each of the following points on graph paper and join them up in alphabetical order: A(0, 0), B(0, 2), C(2, 4), D(2, 6), E(4, 6), F(4, 4), G(6, 4), H(6, 2) and I(4, 0).

(a) Reflect this shape in the x-axis.

(b) Reflect the original shape in the y-axis.

(c) Complete the drawing so that a shape with two lines of symmetry is produced.

4. Draw the shape shown in Fig. 35.17 and then reflect it in

(a) the x-axis,   (b) the y-axis.

Fig. 35.17

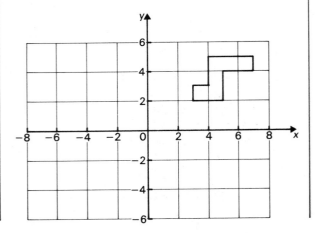

5. Complete the shape shown in Fig. 35.18 so that it has two axes of symmetry.

Fig. 35.18

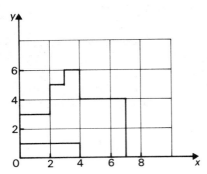

6. Draw an isosceles right-angled triangle. By drawing the axis of symmetry, obtain an angle of 45°.

7. PQRS is a square such that P is the point (2, 2), Q is the point (6, 2) and R is the point (6, 6).

(a) Write down the coordinates of the point S.

(b) Draw the square on graph paper using a scale of 1 large square to 1 unit.

(c) Draw the straight line $y = x$ on the graph paper using the same axes.

(d) Reflect PQRS in the line $y = x$.

(e) Write down the coordinates of the transformed points P′, Q′, R′ and S′.

8. The point A(2, 6) is reflected in a mirror line on to the point B(2, −2). What is the equation of the mirror line?

Rotation

A rotation is specified by the angle of rotation, its direction (assumed to be anti-clockwise if not stated) and the centre of rotation. Thus in Fig. 35.19, the shape A has been given a rotation of 60°, anticlockwise about the point P.

Fig. 35.19

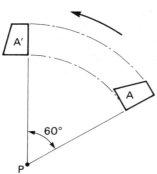

After a rotation the pre-image and the image are congruent.

Example 5

Triangle ABC has vertices A(3, 4), B(7, 4) and C(7, 6). Plot these points on graph paper and join them to form the triangle ABC. Show the image of ABC after a clockwise rotation of 90° about the origin and mark it A'B'C'. Write down the coordinates of the points A', B' and C'.

Triangle ABC is drawn in Fig. 35.20. The image A'B'C' has vertices A'(4, −3), B'(4, −7) and C'(6, −7).

Fig. 35.20

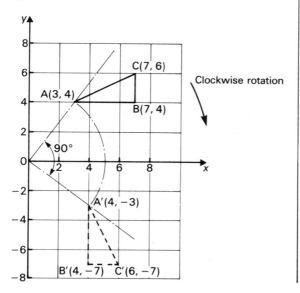

Rotational Symmetry

A figure has rotational symmetry if it coincides with its image by rotation about a point.

In Fig. 35.21, when the rectangle ABCD is rotated through 180° about O it will take up the position shown in Fig. 35.22. If the lettering is removed, there is no change in the two rectangles.

Fig. 35.21

Fig. 35.22

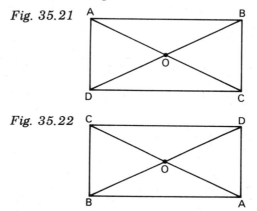

The rectangle ABCD has only two positions which are the same under rotation. We say that the rectangle has rotational symmetry of order 2.

Example 6

Draw the diagram of Fig. 35.23, then rotate it successively through 90°, 180° and 270° about O.

(a) What is the order of rotational symmetry for the original shape?

(b) What is the order of rotational symmetry for the completed figure?

Fig. 35.23

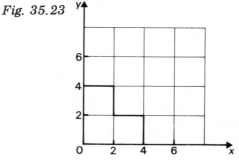

The shape after rotations of 90°, 180° and 270° is drawn in Fig. 35.24 where the images are shown by dotted lines.

Fig. 35.24

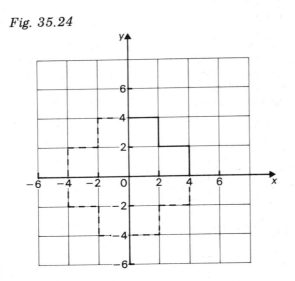

(a) The original shape has rotational symmetry of order 1 since it only gives the same shape after a rotation of 360°.

(b) The completed shape has rotational symmetry of order 4 because the shape appears unchanged after rotations of 90°, 180°, 270° and 360°.

Exercise 35.3

1. Using the origin as the centre of rotation, draw the pre-image of the point P(4, 3) and the image of P after a rotation of 30° clockwise. What are the coordinates of P′, the image of P?

2. Triangle XYZ has vertices X(3, 3), Y(6, 3) and Z(6, 5). Plot these points on graph paper and join them to form triangle XYZ. Draw the image of XYZ after a rotation of 90° about the origin and mark it X′Y′Z′. Write down the coordinates of X′, Y′ and Z′.

3. Using O as the centre of rotation, draw the image of the rectangle ABCD (Fig. 35.25) after it has been given a rotation of 45° clockwise. Write down the coordinates of the transformed points A′, B′, C′ and D′.

Fig. 35.25

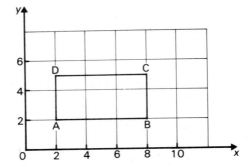

4. Draw the shape shown in Fig. 35.26. Rotate it through 90°, then 180° and finally through 270° anticlockwise.

Fig. 35.26

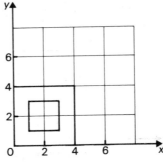

(a) Write down the order of rotational symmetry of the original shape.

(b) What is the order of rotational symmetry of the figure obtained after the rotations?

5. State the order of rotational symmetry of
   (a) a rhombus
   (b) a parallelogram
   (c) a kite
   (d) a square

Similarities

Similarities are transformations which multiply all lengths by a scale factor. If the scale factor is greater than 1 the transformation is called an **enlargement**. If the scale factor is less than 1, the transformation is called a **reduction**.

With both enlargements and reductions, angles are preserved and hence the image and the pre-image are similar figures. Thus in Fig. 35.27, the enlargement factor is $\frac{6}{2} = 3$ and ABC and A′B′C′ are similar triangles. The lengths of the sides of A′B′C′ are 3 times the lengths of the corresponding sides of ABC.

Fig. 35.27

To specify an enlargement completely we must state the scale factor and also the centre of the enlargement.

Example 7

Draw on graph paper the rectangle ABCD with vertices A(6, 4), B(16, 4), C(16, 7), D(6, 7). Draw the image of ABCD if it is enlarged by a scale factor of 2, the centre of the enlargement being the origin.

> The rectangle ABCD is shown in Fig. 35.28. To draw the image, each of the points (x, y) is mapped on to (x', y') by the mapping $(x, y) \rightarrow (2x, 2y)$.
> Thus A(6, 4) → A′(12, 8),
> B(16, 4) → B′(32, 8), etc.
> The image of ABCD is A′B′C′D′, and the transformation enlarges ABCD by a scale factor of 2.

Fig. 35.28

Example 8

Draw on graph paper the rectangle with vertices A(14, 13), B(22, 13), C(22, 25), D(14, 25). Draw A′B′C′D′ the image of ABCD if it is reduced by a scale factor of $\frac{1}{4}$, the centre of the reduction being P(−2, −3).

Fig. 35.29

The first step is to plot the points A, B, C, D and P (Fig. 35.29). Next, the radial lines PA, PB, PC and PD are drawn. Since the scale factor is $\frac{1}{4}$, PA′ is $\frac{1}{4}$ of PA, PB′ is $\frac{1}{4}$ of PB and so on. The points A′, B′, C′ and D′ can now be marked as shown in Fig. 35.29. The image of ABCD is A′B′C′D′ with A′(2, 1), B′(4, 1), C′(4, 4) and D′(2, 4). Note carefully that A′B′ is parallel to AB, that B′C′ is parallel to BC, that C′D′ is parallel to CD and that A′D′ is parallel to AD.

Example 9

In Fig. 35.30, triangle ABC is an enlargement of triangle ADE with a scale factor
of 4. If BC = 12 cm and AE = 3.5 cm,
find the lengths of DE and AC.

Fig. 35.30

Because the scale factor is 4, the corresponding sides of triangle ABC are
4 times as long as those of triangle ADE.

$$DE = \tfrac{1}{4} \text{ of } 12 \text{ cm}$$
$$= 3 \text{ cm}$$
$$AC = 4 \times 3.5 \text{ cm}$$
$$= 14 \text{ cm}$$

Exercise 35.4

1. In Fig. 35.31, (a) find the scale factor
   of the enlargement, (b) state the coordinates of the centre of the enlargement.

Fig. 35.31

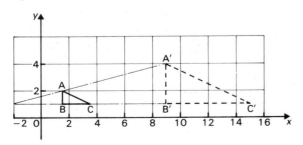

2. Copy the diagram shown in Fig. 35.32.
   Draw the lines joining A to A′, B to B′
   and C to C′. Hence find

   (a) the scale of the enlargement

   (b) the coordinates of the centre of
   the enlargement

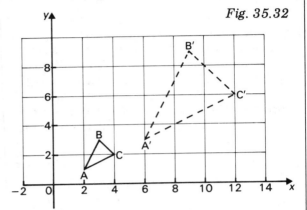

Fig. 35.32

3. Plot the points A(2, 0), B(2, 2), C(4, 3)
   and D(4, 0) on graph paper and join
   them in alphabetical order to form a
   trapezium. Draw an enlargement of
   ABCD with a scale factor of 2 and the
   origin as the centre of the enlargement.
   Label this A′B′C′D′. How many times
   bigger in area is A′B′C′D′ than ABCD?

4. For each of the diagrams of Fig. 35.33,
   write down the coordinates of the centre
   of the enlargement or reduction and
   state the scale factor. In each case, A
   has been mapped on to A′.

(a)

Fig. 35.33

(b)

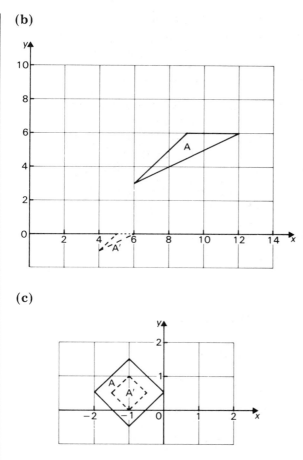

(c)

(b) Join the points A, B, C and D to form the quadrilateral ABCD.

(c) ABCD is enlarged by a scale factor of 2 with the origin as the centre of the enlargement. $A \rightarrow A'$, $B \rightarrow B'$, $C \rightarrow C'$ and $D \rightarrow D'$. Draw the quadrilateral $A'B'C'D'$ and state the coordinates of the points A', B', C' and D'.

7. In Fig. 35.34, CD is mapped on to AB by a reduction. Find the magnitude of the reduction and then find the length of AC.

Fig. 35.34

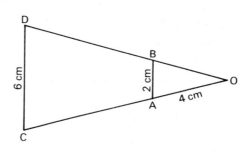

5. Draw on graph paper the quadrilateral ABCD which is formed by joining the points $A(1, 1)$, $B(1, 4)$, $C(7, 7)$ and $D(4, 1)$. ABCD is reduced by a scale factor of $\frac{1}{2}$ with $P(-6, 4)$ as the centre of the reduction. Draw $A'B'C'D'$, the image of ABCD, stating the coordinates of A', B', C' and D'.

6. Draw on graph paper the x and y-axes for values on both axes between -5 and 5, using a scale of 1 cm to 1 unit.

(a) Plot the point $A(2, 1)$ and plot points B, C and D as follows:

B is the reflection of A in the x-axis;
C is the image of A after a rotation of $180°$ about the origin;
D is the reflection of A in the line $y = x$.

8. Rectangle ABCD (Fig. 35.35) is an enlargement of rectangle EBGF with a scale factor of 4. Find the lengths of EF and CD.

Fig. 35.35

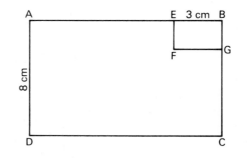

Miscellaneous Exercise 35

1. The point A(4, −2) is reflected in the line $y = x$. Find the coordinates of A′, the image of A.

2. P is the point (4, 3). Plot this point on graph paper.

    (a) The image of P when reflected in the x-axis is the point A. Mark A on your graph paper and state its coordinates.

    (b) The image of P when reflected in the y-axis is the point B. Mark B on your graph paper and state its coordinates.

3. Write down the coordinates (p, q) of the image of the point R(3, 1) under the translation $\begin{pmatrix} 3 \\ 2 \end{pmatrix}$.

4. The point P(3, −2) is given a rotation of 60° anticlockwise about the point (2, 1). Write down the coordinates of P′, the image of P, after the rotation.

5. A given translation maps P(3, 5) on to P′(7, 8). What is the image of the point Q(−2, 4) under this translation.

6. Draw on graph paper the triangle ABC formed by joining A(0, 0), B(2, 0) and C(2, 1). Draw its image under an enlargement whose centre is O (the origin) and whose scale factor is 2.

7. A trapezium is formed by joining, in alphabetical order, the points A(2, 1), B(7, 1), C(6, 3) and D(3, 3). Draw the trapezium on graph paper and indicate on it all the lines of symmetry.

8. Describe fully the single transformation which maps the shaded triangle in Fig. 35.36 on to   (a) R,   (b) T.

Fig. 35.36

Introduction to Statistics

Statistics is the name given to the science of collecting and analysing facts. Originally, statistics used only information about the state, hence the name statistics. Nowadays, however, in almost all business reports, newspapers and government publications use is made of statistical methods.

Statistical methods range from the use of tables and diagrams to the use of statistical averages, the object being to assist the reader to understand what conclusions can be made from the quantities under discussion.

Raw Data

Raw data is collected information which is not arranged in any kind of order.

Consider the marks obtained by 50 students in a test:

```
4 3 3 5 5 6 5 8 7 6 7 8 9 5 4
1 8 7 5 6 6 7 5 2 5 2 6 9 5 7
6 5 6 2 8 6 7 3 3 8 7 6 5 5 6
4 3 4 5 7
```

This is an example of raw data and we see that the information is not organized in any order.

Frequency Distributions

Frequency distributions are used to organize raw data. A tally chart is usually used to do this.

Example 1

Using a tally chart, form a frequency distribution for the data given above.

On examining the data we see that the lowest mark is 1 whilst the highest mark is 9. The marks 1 to 9 inclusive are written in column 1 of the tally chart. We now take each figure in the raw data, just as it comes and we place a tally mark in column 2 of the tally chart opposite the appropriate mark.

The fifth tally mark for each number is usually made in an oblique direction, thereby tying the tally marks into bundles of five. This makes counting easier.

When each test mark has been entered, the tally marks are counted and the numerical value found is recorded in the column headed frequency. Hence the frequency is the number of times each mark occurs in the raw data.

From the tally chart it will be seen that the mark 1 occurs once (a frequency of 1), mark 2 occurs three times (a frequency of 3), mark 6 occurs ten times (a frequency of 10) and so on.

TALLY CHART 1

Mark	Tally	Frequency
1	I	1
2	I I I	3
3	⌶⌶⌶	5
4	I I I I	4
5	⌶⌶⌶ ⌶⌶⌶ I I	12
6	⌶⌶⌶ ⌶⌶⌶	10
7	⌶⌶⌶ I I I	8
8	⌶⌶⌶	5
9	I I	2

Grouped Distributions

Grouped distributions are used when a large amount of numerical data has to be organised. The data is then grouped into classes or categories.

Example 2

The following is a record of the heights of 100 people in a random sample. The heights are given in centimetres.

```
146 194 136 157 117 151 164 131
187 166 158 140 145 176 126 146
175 194 185 126 178 129 156 152
113 187 157 136 152 158 157 149
180 117 129 180 169 127 168 182
193 164 169 153 152 177 198 137
185 122 147 173 165 108 142 148
163 164 169 159 155 147 153 142
146 130 153 166 165 151 148 153
154 151 155 158 137 151 141 153
150 145 152 154 154 140 151 148
150 157 151 125 154 140 184 148
130 160 131 134
```

Draw up a tally chart for the classes 100–109 cm, 110–119 cm, ... , 190–199 cm, and hence form a grouped frequency distribution.

TALLY CHART 2

Height (cm)	Tally	Frequency
100–109	I	1
110–119	I I I	3
120–129	ⵏⵏⵜ I I	7
130–139	ⵏⵏⵜ I I I I	9
140–149	ⵏⵏⵜ ⵏⵏⵜ ⵏⵏⵜ I I I	18
150–159	ⵏⵏⵜ ⵏⵏⵜ ⵏⵏⵜ	
	ⵏⵏⵜ ⵏⵏⵜ ⵏⵏⵜ I I	32
160–169	ⵏⵏⵜ ⵏⵏⵜ I I I	13
170–179	ⵏⵏⵜ	5
180–189	ⵏⵏⵜ I I I	8
190–199	I I I I	4

The class interval for the first class in Tally Chart 2 is 100–109 cm. The end numbers 100 and 109 are called the **class limits** for the first class.

The first number 100 is called the **lower class limit** whilst the second number 109 is called the **upper class limit.** Thus for the sixth class the lower class limit is 150 and the upper class limit is 159.

In Tally Chart 2, the heights have been recorded to the nearest centimetre. The class interval 150–159 cm theoretically includes all the heights from 149.5 cm to 159.5 cm. These numbers are called the **lower** and **upper class boundaries** respectively.

The width of a class interval is the difference between lower and upper class boundaries, i.e.

Width of class interval
$$= \text{Upper class boundary}$$
$$- \text{Lower class boundary}$$

Thus for the sixth class of Tally Chart 2

Width of class interval $= 159.5 - 149.5$

$$= 10 \text{ cm}$$

Example 3

The figures in the table below show part of a grouped frequency distribution. State the upper and lower class boundaries for the second class and hence find the width of the class interval.

Mass (kg)	Frequency
350–379	27
380–409	34
410–439	43

For the second class the upper and lower class boundaries are 409.5 kg and 379.5 kg respectively.

Width of class interval $= 409.5 \text{ kg} - 379.5 \text{ kg}$

$$= 30.0 \text{ kg}$$

Variables

A **continuous variable** is a variable which can take any value between two given values. For instance the height of an individual, which can be 159 cm, 165.2 cm or 176.831 cm, depending upon the accuracy of the measurement, is a continuous variable.

A **discrete variable** is a variable which can have only certain values. An example is the number of children in a family, which can take the value 0, 1, 2, 3, ... but not $2\frac{1}{2}$, $3\frac{1}{4}$, ... A second example is the number of pound coins in a bank, which can only be a whole number. Note that a discrete variable need not necessarily be a whole number. For instance the size of shoes can be 5, $5\frac{1}{2}$, 6, $6\frac{1}{2}$, 7, ... but not $5\frac{1}{3}$, $6\frac{1}{4}$, ... Hence the size of a shoe is a discrete variable.

Histograms

A **histogram** is a diagram which is used to represent a frequency distribution. It consist of a set of rectangles whose **areas** represent the frequencies of the various classes. If all the classes have the same width (which is usually the case) then the rectangles will all have the same width. The frequencies are then represented by the heights of the rectangles. Fig. 36.1 shows the histogram for the frequency distribution in Example 1.

Fig. 36.1

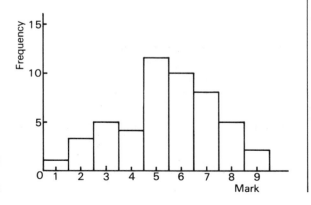

A histogram for a grouped frequency distribution may be drawn by using the mid-points of the class intervals as the centre of the rectangles. The histogram for the frequency distribution in Example 2 is shown in Fig. 36.2. Note that the extremes of each rectangle represent the lower and upper class boundaries.

Fig. 36.2

Example 4

Six coins were tossed 80 times and after each toss the number of tails was recorded as shown in the table below. Represent this information in the form of a suitable histogram.

Number of tails	Number of tosses (frequency)
0	3
1	9
2	15
3	25
4	17
5	9
6	2
Total	80

Fig. 36.3 shows a histogram for a discrete distribution. It has been drawn to represent the information given in Example 4 and takes the form of a vertical bar chart in which the bars have zero width.

Discrete data, however, are often represented by a histogram like that shown in Fig. 36.4, despite the fact that we are treating the data as though they were continuous.

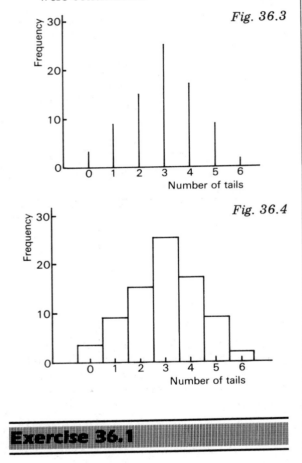

Fig. 36.3

Fig. 36.4

Exercise 36.1

1. The following marks were obtained by 50 students during a test:

Mark	1	2	3	4	5	6	7	8
Frequency	2	4	8	14	9	6	5	2

Draw a suitable histogram to represent this information.

2. The marks obtained in a recent mathematics examination were as follows:

15 37 42 19 27 42 37 27 7 33
30 12 6 31 24 25 50 29 38 35

Use the above information to obtain a grouped frequency distribution with classes 1–10, 11–20, 21–30, 31–40 and 41–50. Draw a histogram to represent this frequency distribution.

3. The distances in metres achieved by 80 competitors in a discus-throwing competition were grouped in class intervals of 10 cm. The results were as follows:

Class interval (m)	Frequency
20.16–20.25	1
20.26–20.35	4
20.36–20.45	8
20.46–20.55	12
20.56–20.65	14
20.66–20.75	15
20.76–20.85	11
20.86–20.95	7
20.96–21.05	5
21.06–21.15	3

(a) Draw a histogram to represent this information.

(b) For the third class (20.36–20.45) write down the lower and upper class boundaries and state the class width in metres.

4. A teacher asked his class of 32 pupils to write down how long a particular question took them. The following table summarises their times:

Time (min)	10	11	12	13	14	15	16	17	18	19	20
Frequency	1	1	2	7	4	5	3	3	2	2	2

Draw a suitable histogram to represent this information.

5. The speed of 120 vehicles passing a police checkpoint was measured to the nearest kilometre per hour. The results are summarised below:

Speed (km/h)	Number of vehicles
30–34	1
35–39	4
40–44	9
45–49	17
50–54	23
55–59	31
60–64	18
65–69	9
70–74	6
75–79	2

(a) For the third class (40–44) write down the lower and upper class boundaries. Hence calculate the class width.

(b) Draw a histogram to represent this information.

(c) State whether the speed is a continuous or a discrete variable.

6. Read the following passage and record the number of letters in each word to form a grouped frequency distribution with classes as follows: 1 or 2 letters, 3 or 4 letters, 5 or 6 letters, 7 or 8 letters, 9 or 10 letters, 11 or 12 letters. Hence draw a histogram to represent the frequency distribution.

'The morning was cool and cloudless. Barbara rose early and with a growing feeling of excitement hurried through her breakfast. The forecast had been good and as she slipped out of the house she gave no thought to the possibility of a change in the weather. She arrived at the quay to find an old boat-man waiting for her. The sea was calm, there was no hint of the storm to come.'

7. Classify each of the following as a continuous or a discrete variable:

(a) the diameter of motor car pistons

(b) the number of dresses sold by a shop in one day

(c) the weight of fertilizer in a sack

(d) the number of five-pound notes in a wallet

(e) the daily temperature at noon

(f) the lifetime of electric light bulbs

(g) the temperature of a furnace

(h) the number of telephone calls received by the police in one day

8. Fig. 36.5 shows the number of children in a sample classified by weight, in groups of 2 kg.

(a) Find the number of children in the survey.

(b) Is the weight a continuous or a discrete variable?

(c) Represent the information in a conventional histogram.

Fig. 36.5

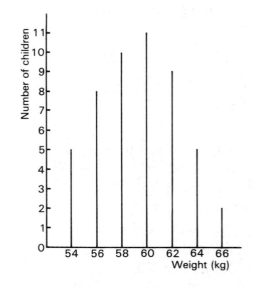

Statistical Averages

There are three types of statistical average. They are the arithmetic mean (often just called the mean), the median and the mode.

The **arithmetic mean** is found by adding all the values in a set of data and dividing by the number of items in the set. That is

$$\text{Arithmetic mean} = \frac{\text{Sum of all the values}}{\text{The number of values}}$$

Example 5

Find the arithmetic mean of 3, 5, 8, 9 and 10.

$$\text{Arithmetic mean} = \frac{3 + 5 + 8 + 9 + 10}{5}$$

$$= \frac{35}{5}$$

$$= 7$$

Example 6

The values 3, 4, 6, 8, 9 and x have a mean of 7. What is the value of x?

$$\frac{3 + 4 + 6 + 8 + 9 + x}{6} = 7$$

$$\frac{30 + x}{6} = 7$$

$$30 + x = 42$$

$$x = 12$$

When a set of values is arranged in ascending (or descending) order, the **median** is the value which lies half-way along the sequence.

Example 7

Find the median of the numbers 5, 4, 2, 8, 7, 2, 9, 7 and 3.

Arranging the numbers in ascending order we have

$$2, 2, 3, 4, 5, 7, 7, 8, 9$$

The median is 5 because there are four numbers below this value and four numbers above it.

When there is an even number of values in a set, the median is found by taking the mean of the two middle values.

Example 8

Find the median of 8, 5, 4, 7, 9, 10, 2 and 5.

Arranging the numbers in descending order we have

$$10, 9, 8, 7, 5, 5, 4, 2$$

The middle two values are 7 and 5 and therefore

$$\text{Median} = \frac{7 + 5}{2}$$

$$= 6$$

The **mode** of a set of values is that value which occurs most frequently.

Example 9

Find the mode of 11, 12, 12, 12, 13, 13, 13, 13, 13, 14, 14, 14, 15 and 15.

Looking at the set of numbers we see that 13 is the number which occurs most frequently (i.e. most times). Hence the mode is 13.

Example 10

The marks obtained in a test taken by a class of 30 children were as follows:

Mark	1	2	3	4	5
Frequency	3	7	12	6	2

Find **(a)** the mean mark, **(b)** the median mark, **(c)** the modal mark.

(a) Mean mark

$$= \frac{(1 \times 3) + (2 \times 7) + (3 \times 12) + (4 \times 6) + (5 \times 2)}{3 + 7 + 12 + 6 + 2}$$

$$= \frac{3 + 14 + 36 + 24 + 10}{30}$$

$$= \frac{87}{30}$$

$$= 2.9$$

(b) Since there are 30 children the median mark must lie between the 15th and 16th items in the distribution. We now look at the 15th and 16th items and we find that these are both 3. Hence the median mark is 3.

(c) The mode is the most frequently occurring mark. Looking at the table we see that 3 marks occurs twelve times. Hence the modal mark is 3.

Comparison of Statistical Averages

The mean: Advantages

(1) The mean is the best known average.

(2) It can be calculated exactly.

(3) It makes use of all the data.

(4) It can be used in further statistical work.

Disadvantages

(1) It is greatly affected by extreme values.

(2) When the data are discrete it can give an impossible value (e.g. 15.362 runs per innings).

(3) It cannot be obtained graphically.

The median: Advantages

(1) The median is not affected by extreme values.

(2) It can be obtained even if some of the values in the distribution are unknown.

(3) If the number of items in the distribution is odd it can represent an actual value in the distribution.

Disadvantages

(1) When only a few values are available the median may not be characteristic of the group.

(2) It cannot be used for further statistical calculations.

The mode: Advantages

(1) The mode is not affected by extreme values.

(2) It is easy to obtain from a histogram or a frequency distribution.

(3) To determine its value only values near to the modal value are needed.

Disadvantages

(1) There may be no mode or there may be more than one mode.

(2) It cannot be used in further statistical calculations.

Which Average To Use

The **mean** is the most familiar kind of average and it is extensively used in business, engineering, etc. It is easy to understand but in some cases it may be definitely misleading. For instance if the hourly rates of pay for five office workers are £4.05, £4.32, £4.81, £9.40 and £3.52, the mean rate of pay is £5.22, However, this value is greatly affected by the extreme value of £9.40. Therefore the mean gives a false impression of the wages paid in the office; it gives the impression that the wages are higher than they really are.

The **median** is not affected by extreme values and it will give a better indication of the wages paid in the office discussed above. The median wage is £4.32 which gives a better idea of the wages actually paid.

The **mode** is used when the commonest value in a distribution is required. For instance a clothing manufacturer will not be particularly interested in the mean waist size of women because almost certainly this will not be a stock size for his skirts. What he needs is a frequency distribution for his stock sizes, size 12, 14, 16, etc. From this he can obtain the modal size and organise his production accordingly. However, the average used will depend upon the particular set of circumstances.

Exercise 36.2

1. The heights of five men are 177.8 cm, 175.3 cm, 174.8 cm, 179.1 cm and 176.5 cm. What is the mean height of these five men?

2. Calculate the mean of the numbers 5, 8, 9 and 10.

3. The numbers 3, 5, 8, 10 and x have a mean of 8. Determine the value of x.

4. Find the median of the numbers 5, 3, 8, 5, 4, 2 and 8.

5. Find the median of the numbers 2, 4, 6, 5, 9, 6, 4 and 7.

6. Find the mode of 3, 5, 4, 8, 3, 6, 5, 9, 5, 4 and 7.

7. Thirteen people were asked to guess the weight of a cake to the nearest half-kilogram. Their estimates were as follows: $3\frac{1}{2}$, $2\frac{1}{2}$, 2, 1, $3\frac{1}{2}$, 2, $3\frac{1}{2}$, 3, 3, 1, $1\frac{1}{2}$, $2\frac{1}{2}$, $3\frac{1}{2}$ kg. What was
   (a) the estimated mean weight
   (b) the estimated modal weight
   (c) the estimated median weight

8. What additional number must be included in the following list to make the median of the six numbers equal to $3\frac{1}{2}$?

   3, 5, 2, 2, 8

9. Six people occupy a lift. Their masses to the nearest kilogram are 84, 67, 73, 76, 80 and 82 kg. Find
   (a) their median mass
   (b) their mean mass

10. During a census a check was made to find out how many people were living in each house in a certain street. The results are given in the table:

Number of people per house	1	2	3	4	5	6	7
Number of houses	3	9	12	9	4	2	1

   (a) How many houses were checked?
   (b) State the modal number of people per house.
   (c) Calculate the mean number of people per house.
   (d) Determine the median number of people per house.

11. The marks of 100 students were as follows:

Mark	1	2	3	4	5	6	7	8	9
Frequency	2	8	20	32	18	9	6	3	2

   (a) Calculate the mean mark.
   (b) What is the modal mark?
   (c) Find the median mark.

12. The following table shows the number of runs scored by a sample of cricketers in a limited-over competition.

Number of runs scored	18	19	20	21	22	23	24	25	26	27
Frequency	5	10	12	28	36	41	30	25	10	3

   (a) How many cricketers are in the sample?
   (b) What is the modal score?
   (c) What is the median score?
   (d) What is the mean score correct to one decimal place?

Equiprobable Events

When an unbiased die is rolled, each of the six faces has an equal chance of landing uppermost. The throwing of 1, 2, 3, 4, 5 or 6 are called **equiprobable events**.

Probability

Suppose that we want to deal an ace from a pack of 52 playing cards. Each time we deal an ace we say that there has been a **favourable outcome** or a success. Since there are four aces in a pack, the number of equiprobable events which produces a favourable outcome (or the number of successes) is 4. The probability of obtaining a favourable outcome is calculated from:

$$\text{Probability} = \frac{\substack{\text{Number of equiprobable} \\ \text{events which produces a} \\ \text{favourable outcome}}}{\substack{\text{Total number of} \\ \text{equiprobable events}}}$$

$$= \frac{\text{Number of successes}}{\text{Total number of outcomes}}$$

All probabilities have a value between 0 and 1. A probability of 0 occurs when an event can never happen. A probability of 1 occurs when an event is certain to happen (e.g. picking the winner of a one-horse race).

The **total probability** covering all possible events is 1. That is

Probability of success

+ Probability of failure = 1

Example 11
One letter is chosen from the word 'horizon'. What is the probability that it will be

(a) the letter r,   (b) the letter o,

(c) a vowel,   (d) a consonant?

(a) Since there are seven letters in the given word, there is a total of 7 equiprobable events. Only one of these is favourable because there is only one r in the word. Hence

Probability of choosing the letter r $= \dfrac{1}{7}$

(b) In this case there are two favourable outcomes because the letter o occurs twice in the given word. Hence

Probability of choosing the letter o $= \dfrac{2}{7}$

(c) The vowels are the letters a, e, i, o and u. There are three vowels in the given word, hence the number of favourable outcomes is 3.

Probability of choosing a vowel $= \dfrac{3}{7}$

(d) If a letter is not a vowel it is a consonant. Hence

Probability of choosing a vowel

+ Probability of not choosing a vowel = 1

Probability of not choosing a vowel (i.e. of

choosing a consonant) $= 1 - \dfrac{3}{7} = \dfrac{4}{7}$

Alternatively we see that the letters h, r, z and n are consonants and therefore the number of favourable outcomes is 4. Hence, as before

Probability of choosing a consonant $= \dfrac{4}{7}$

Exercise 36.3

1. A die is rolled. Calculate the probability that the result will be

   (a) a 3

   (b) a score less than 4

   (c) an even number

2. A letter is chosen from the word 'terrific'. Determine the probability that it will be

   (a) an f   (b) an r

   (c) a vowel   (d) a consonant

3. A bag contains 3 red balls, 5 blue balls and 2 green balls. A ball is chosen at random from the bag. Find the probability that it will be

   (a) green   (b) blue

   (c) red   (d) not red

4. There are seven tomatoes in a bag. Four of them are red and the remainder are green. A tomato is picked at random from the bag. What is the probability that it is red?

5. 20 discs numbered 1 to 20 are placed in a box and one disc is chosen at random. Determine the probability that the number on the disc will be

   (a) even   (b) a multiple of 3

   (c) more than 8   (d) less than 8

6. A card is drawn from a deck of 52 playing cards. Find the probability that it will be

   (a) the king of hearts

   (b) a queen

   (c) an ace, king, queen or jack

   (d) the jack of spades or the ace of diamonds

7. Two dice are thrown together and their scores added. Determine the probability that the sum will be

   (a) 7

   (b) less than 5

   (c) more than 9

8. From a shuffled pack of 52 playing cards two cards are dealt. What is the probability of

   (a) the first card being an ace

   (b) the second card also being an ace

9. The number of players in the first-team squad of a local football team is 23. Only eleven players can play in a game. What is the probability that John will *not* be chosen to play?

10. It states on the box of a certain brand of matches that the average contents is 36 matches. The contents of 9 different boxes were counted with the following results: 32, 37, 29, 35, 37, 32, 36, 40, 36. What is the probability of selecting a box from this sample which contains more than 36 matches?

Experimental Probability

Although it is possible to calculate very many probabilities in the theoretical way shown above, in a great many cases we have to rely on probabilities obtained by experiment. When an experiment is used to establish a probability it is calculated from

$$\text{Probability} = \frac{\text{Number of 'successful' trials}}{\text{Total number of trials conducted}}$$

Example 12

During a certain period of the year, the number of blooms on each of a number of rose bushes was counted and the results recorded as follows:

Number of blooms	5	6	7	8	9	10	11	12
Frequency	15	20	25	27	22	17	14	10

Calculate the probability of finding a rose bush with (a) less than 7 blooms, (b) exactly 9 blooms, (c) 10 or more blooms.

Total number of trials conducted

$$= 15 + 20 + 25 + 27 + 22 + 17$$
$$\quad + 14 + 10$$

$$= 150$$

(a) Number of rose bushes with less than 7 blooms

$$= 15 + 20 = 35$$

Probability of finding a rose bush with less than 7 blooms

$$= \frac{35}{150}$$

$$= \frac{7}{30}$$

(b) Number of rose bushes with exactly 9 blooms

$$= 22$$

Probability of finding a rose bush with exactly 9 blooms

$$= \frac{22}{150}$$

$$= \frac{11}{75}$$

(c) Number of rose bushes with 10 or more blooms

$$= 17 + 14 + 10$$

$$= 41$$

Probability of finding a rose bush with 10 or more blooms

$$= \frac{41}{150}$$

Exercise 36.4

1. Pupils in a school made a survey of the colours of front doors on a housing estate. The table below summarises the information they obtained:

Colour of door	Red	Yellow	Green
Frequency	26	9	4

Colour of door	Blue	White	Other
Frequency	15	35	11

Determine the probability of

(a) finding a house with a white door

(b) finding a house with a door that is not white

2. During a survey to find out how many people were living in each house in a certain street the following results were obtained:

Number of people per house	1	2	3	4	5	6	7
Number of houses		2	8	14	10	4	2	0

Calculate the probability that the number of people in a house selected at random will

(a) exceed 4

(b) be less than 3

(c) be exactly 6

3. The table below shows the scores obtained at ten-pin bowling:

Score	0	1	2	3	4	5	6	7	8	9	10
Frequency	1	10	12	22	25	22	16	14	12	10	6

Determine the probability that a player, chosen at random, will score

(a) more than 4

(b) less than 7

(c) exactly 9

4. The number of goals scored on a Saturday afternoon during the football season were as follows:

Number of goals	0	1	2	3	4	5
Number of teams	5	12	14	8	7	4

Calculate the probability that one team, chosen at random, scored

(a) 2 goals

(b) more than 2 goals

(c) less than 3 goals on this particular Saturday

5. The pocket money, in pence, received by 25 young children during one week of term is given below:

```
60 120   40   35 100 100   45
50   70 125   55 150   50   75
80   65   80 100   30   50   75
50   40 110   75
```

Calculate the probability that one child, chosen at random, received

(a) exactly 50 pence

(b) more than 60 pence

(c) less than 40 pence

The Addition Law of Probability

If two events could not happen at the same time the events are said to be **mutually exclusive**. For instance in a single roll of a fair die a 3 or 4 can occur but not both. Hence the events of scoring a 3 and a 4 in a single roll of a die are mutually exclusive.

Similarly it is impossible to cut a jack and a king in a single cut of a pack of cards. Hence these two events are mutually exclusive.

If $p_1, p_2, p_3, \ldots$ are the separate probabilities of a set of mutually exclusive events, then the probability of *one* of these events occurring is

$$P = p_1 + p_2 + p_3 + \ldots$$

Example 13

A die with faces numbered 1 to 6 is rolled once. What is the probability of scoring either a 3 or a 4?

$$\text{Probability of scoring a 3} = p_1$$
$$= \frac{1}{6}$$

$$\text{Probability of scoring a 4} = p_2$$
$$= \frac{1}{6}$$

$$\text{Probability of scoring a 3 or a 4} = p_1 + p_2$$
$$= \frac{1}{6} + \frac{1}{6}$$
$$= \frac{1}{3}$$

Sometimes the addition law is called the OR law because we seek the probability that one event OR another event will happen.

The Multiplication Law of Probability

An **independent** event is one which has no effect on subsequent events. For instance if a die is rolled twice, what happens on the first roll does not affect what happens on the second roll. Hence the two rolls of the die are independent events.

The events of cutting an ace from a pack of cards and tossing a head in a single toss of a fair coin are also independent events because what happens with the cut has no effect on the way in which the coin lands. If $p_1, p_2, p_3, \ldots$ are the separate probabilities of a set of independent events then the probability that *all* events will occur is

$$P = p_1 \times p_2 \times p_3 \times \ldots$$

Example 14

(a) If a coin is tossed and a card is dealt from a pack of 52 playing cards, calculate the probability that the result will be a head and an ace.

$$\text{Probability of dealing an ace} = p_1$$
$$= \frac{4}{52}$$
$$= \frac{1}{13}$$

$$\text{Probability of tossing a head} = p_2$$
$$= \frac{1}{2}$$

Probability of dealing an ace and tossing a head

$$= p_1 \times p_2$$
$$= \frac{1}{13} \times \frac{1}{2}$$
$$= \frac{1}{26}$$

(b) A coin is tossed four times. What is the probability that each toss will result in a head?

The four tosses of the coin are independent events. Therefore

Probability of tossing four heads

$$= \frac{1}{2} \times \frac{1}{2} \times \frac{1}{2} \times \frac{1}{2}$$

$$= \frac{1}{16}$$

The multiplication law is often called the AND law because we seek the probability that one event AND another event will occur.

The Probability Tree

Example 15

A coin is tossed three times. Find the probability that **(a)** a head will occur on all three tosses, **(b)** only one head will appear on three tosses.

We can work out these probabilities by using a probability tree (Fig. 36.6). Possible heads are shown in a full line and possible tails in a dotted line.

Fig. 36.6

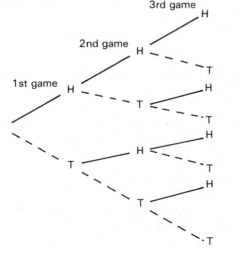

(a) The branch of the tree needed to find the probability of three heads occurring is shown in Fig. 36.7.

Probability of three heads $= \frac{1}{2} \times \frac{1}{2} \times \frac{1}{2}$

$$= \frac{1}{8}$$

Fig. 36.7

(b) The branches of the tree required to find the probability of only one head appearing are shown in Fig. 36.8.

Fig. 36.8

The diagram shows that there are three ways in which one head can occur: head, tail, tail; tail, head, tail; tail, tail, head.

Probability HTT $= \frac{1}{2} \times \frac{1}{2} \times \frac{1}{2} = \frac{1}{8}$

Probability THT $= \frac{1}{2} \times \frac{1}{2} \times \frac{1}{2} = \frac{1}{8}$

Probability TTH $= \frac{1}{2} \times \frac{1}{2} \times \frac{1}{2} = \frac{1}{8}$

Since the events HTT, THT, TTH are mutually exclusive

Probability of one head $= \frac{1}{8} + \frac{1}{8} + \frac{1}{8}$

$$= \frac{3}{8}$$

Example 16

A box contains 4 black and 6 red balls. A ball is drawn at random from the box and is not replaced. A second ball is then drawn. Find the probabilities of (a) red then red being drawn, (b) black then red being drawn, (c) red then black being drawn, (d) black then black being drawn.

The probability tree is shown in Fig. 36.9.

Fig. 36.9

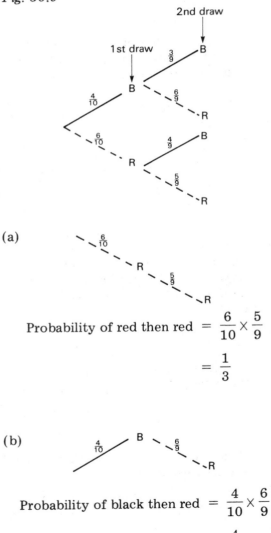

(a)

Probability of red then red $= \dfrac{6}{10} \times \dfrac{5}{9}$

$= \dfrac{1}{3}$

(b)

Probability of black then red $= \dfrac{4}{10} \times \dfrac{6}{9}$

$= \dfrac{4}{15}$

(c)

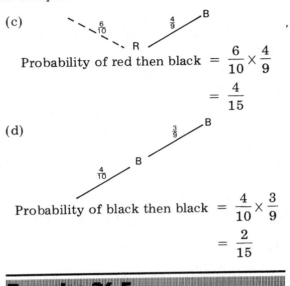

Probability of red then black $= \dfrac{6}{10} \times \dfrac{4}{9}$

$= \dfrac{4}{15}$

(d)

Probability of black then black $= \dfrac{4}{10} \times \dfrac{3}{9}$

$= \dfrac{2}{15}$

Exercise 36.5

1. A card is dealt from the top of a well-shuffled pack of playing cards. Determine the probability that it will be any ace or the king of hearts.

2. A coin is tossed and a die is rolled. Calculate the probability of

   (a) a tail and a 5 resulting

   (b) a head and an even number occurring

3. A box contains 8 red counters and 12 white ones. A counter is drawn at random from the box and then replaced. A second counter is then drawn. Determine the probabilities that

   (a) both counters will be red

   (b) both counters will be white

   (c) one counter will be white and the other red

4. A bag contains 3 red balls and 2 green ones. A ball is drawn from the bag but not replaced. A second ball is then drawn.

   (a) Draw a tree diagram showing the various probabilities.

   (b) Calculate the probability that a red ball followed by a red ball will be drawn.

(c) What is the probability of two green balls being drawn?

(d) Find the probability of a green ball being drawn followed by a red ball.

5. Six cards marked A, B, C, D, E and F are shuffled. Two cards are then drawn without replacement. What is the probability that the cards will be B and D although not necessarily in that order?

6. A box contains 3 red and 4 black balls. Draw a probability tree to show the probabilities of drawing three balls without replacement. Use the tree to answer the following

(a) Find the probability of drawing red then black then red.

(b) What is the probability of drawing red then red then black?

(c) Calculate the probability of drawing black then red then black.

(d) Determine the probability of drawing three black balls.

7. A bag contains 5 red balls and 7 blue balls. A second bag contains 6 red balls and 9 blue balls. If a ball is drawn at random from each bag, calculate the probability that each ball will be red.

Miscellaneous Exercise 36

Section A

1. The marks of 5 students in an examination were as follows: 54, 63, 49, 78 and 61. What was the mean mark?

2. The goals scored by a football team in 60 matches were:

4 1 3 2 0 1 1 1 0 2 5 0 0 4
1 1 0 1 1 5 1 2 1 1 0 2 0 1 2
3 1 1 0 1 2 0 1 0 2 0 1 3 0
0 2 0 1 2 1 0 2 1 1 2 4 0 1 2

Arrange this information in a frequency table using classes of 0, 1, 2, 3, 4 and 5 goals.

3. Draw a bar chart to depict the frequency distribution in question 2.

4. An investigation was carried out into reasons for absence or for late arrival at a school on a particular Monday morning. The result of the investigation was as follows:

Missed the bus	25%
Sickness	38%
Overslept	11%
Had a cold	21%
Bicycle had a puncture	5%

On graph or squared paper, draw a bar chart to represent these statistics. Use a scale of 2 cm to represent 5% on the vertical axis.

5. Fig. 36.10 shows the numbers of children classified by weight in groups at intervals of 2 kg.

(a) Calculate the number of children.

(b) Calculate the mean weight of the children, correct to the nearest kilogram.

Fig. 36.10

Section B

1. A letter is chosen from the word 'flagstaff'. Find the probability that it will be

   (a) l   (b) a   (c) f

2. A large transport firm had 70 lorries in different parts of the country. It wished to compile a table showing the distance travelled by each lorry in a particular period. The following information was obtained by inspectors who checked and recorded the distance, in kilometres, travelled by each lorry.

   ```
   10 12 30 56 57 36 27 32 57 20
   15 50 49 46 21 69 55 42 60 45
   35 47 22 45 44 31 67 43 17 54
   49 56 51 55 29 65 23 55 46 66
   40 63 24 33 62 58 15 57 64 37
   38 30 27 28 59 25 52 47 26 63
   48 31 31 34 65 41 16 28 39 46
   ```

   (a) Draw up a frequency distribution for the classes 0-9, 10-19, 20-29, etc.

   (b) Draw a histogram to represent this frequency distribution.

3. The information below gives the diameters of round steel bars.

Diameter (mm)	Frequency
14.96-14.98	3
14.99-15.01	8
15.02-15.04	12

   (a) Give the upper and lower class boundaries for the second class.

   (b) State the class width for each of the classes shown in the table.

   (c) What is the class interval for the first class?

4. Classify each of the following as a continuous or discrete variable:

   (a) the number of shirts sold per day

   (b) the weight of packets of chemical

   (c) the number of bunches of daffodils packed by a grower

   (d) the daily temperature at noon

   (e) the lifetime of electric light bulbs

   (f) the number of telephone calls received by a police station per day

5. In a small firm five people earn £126 per week, three earn £114 per week and two earn £132 per week. Calculate the mean wage for the ten people.

6. 50 students had their height measured with the following results:

Height (cm)	160	161	162	163	164	165	166
Frequency	1	5	10	16	10	6	2

   Calculate the mean height of the 50 students.

7. Given the set of numbers

   $$3, 5, 2, 7, 5, 8, 5, 2, 7$$

   find

   (a) the mode

   (b) the median

8. 100 pupils took a mathematics test. Their marks were as follows:

Mark	1	2	3	4	5	6	7	8	9
Frequency	2	8	20	32	18	9	6	3	2

   Determine

   (a) the median mark

   (b) the modal mark

9. A fair die is rolled. Calculate the probability of scoring

   (a) a four

   (b) a number less than four

   (c) a two or a three

   (d) an odd number

10. A loaded die shows these probabilities:

Score	1	2	3	4	5	6
Probability	0.14	0.18	0.16	0.15	0.17	0.20

(a) If the die is thrown once what is the probability of a score less than 3?

(b) If it is thrown twice, what is the probability of scoring a 4 followed by a 6?

11. A bag contains 14 yellow counters and 10 red counters. Two counters are taken out in succession and not replaced.

(a) Copy and complete the tree diagram (Fig. 36.11) by writing the correct fractions in the boxes A, B, C and D.

(b) Calculate the probability that two red counters will be taken out.

Fig. 36.11

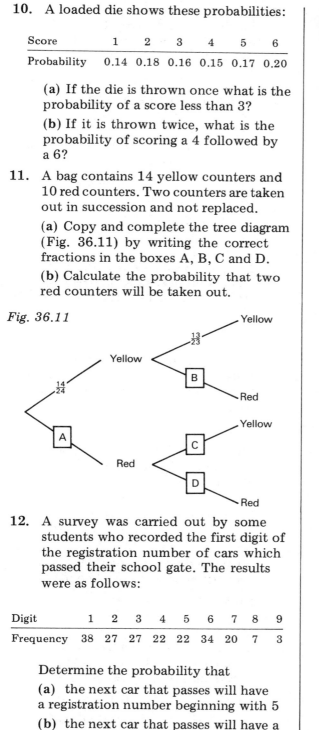

12. A survey was carried out by some students who recorded the first digit of the registration number of cars which passed their school gate. The results were as follows:

Digit	1	2	3	4	5	6	7	8	9
Frequency	38	27	27	22	22	34	20	7	3

Determine the probability that

(a) the next car that passes will have a registration number beginning with 5

(b) the next car that passes will have a registration number beginning with a digit greater than 6

Multi-Choice Questions 36

1. The tally marks below show the distribution in the subjects preferred by a sample of school children. What fraction of the sample preferred French?

English	++++
Mathematics	++++ I I
French	++++ I I I
Physics	++++ I I I
Chemistry	++++ ++++ I I

A $\frac{1}{8}$ B $\frac{7}{40}$ C $\frac{1}{5}$ D $\frac{1}{4}$

2. The mean of 12, 18, 26, 33 and 36 is

A 13   B 25   C 26   D 36

3. Ten numbers have a mean of 185. The sum of the ten numbers is

A 18.5   B 85   C 185   D 1850

4. What is the modal score for the frequency distribution which follows?

Score	5	6	7	8	9	10
Frequency	5	3	4	6	2	1

A 10   B 8   C 6   D 5

5. Find the median of the following numbers: 13, 15, 14, 11, 12, 15, 17, 13, 14, 13.

A $12\frac{1}{2}$ B 13   C $13\frac{1}{2}$ D 14

6. The marks of ten students in a test were: 17, 13, 14, 19, 18, 17, 15, 14, 17 and 15. What is the modal score?

A 19   B 17.5   C 17   D 15.9

7. The table below gives the ages of a group of students who were born in March:

Age in years	13	14	15	16	17
Number of students	1	4	2	2	1

What is the mean age of the students?

A 14   B 14.8   C 15   D 15.5

8. A box contains 5 red balls and 3 green balls. A ball is picked at random. What is the probability that it is green?

   A $\frac{1}{4}$ B $\frac{3}{8}$ C $\frac{3}{5}$ D $\frac{5}{8}$

9. A car park contains 100 cars of which 28 are blue and 34 are red. What is the probability that if a car is selected at random it is neither blue nor red?

   A $\frac{7}{25}$ B $\frac{17}{50}$ C $\frac{19}{50}$ D $\frac{31}{50}$

10. If two dice are thrown together what is the probability of scoring 5?

    A $\frac{1}{18}$ B $\frac{1}{9}$ C $\frac{5}{36}$ D $\frac{1}{6}$

Mental Test 36

1. The tally chart below shows the distribution of a sample of mixed nuts. How many of the sample were almonds and how many nuts were there in the sample altogether?

Brazils	++++
Walnuts	++++ I I
Almonds	++++ I I I
Chestnuts	++++ I I I
Cobnuts	++++ ++++ I I

2. What is the mode of the histogram shown in Fig. 36.12?

Fig. 36.12

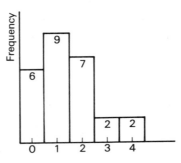

3. The probability of John being selected for the school football team is 7/25. What is the probability that he will not be selected?

4. Find the mean of £23, £42 and £25.

5. What is the median of the numbers 2, 4, 6, 7, 8, 9 and 10?

6. Find the mode of the numbers 3, 5, 2, 7, 5, 8, 2, 7, 2.

7. The following figures give the probability of an event occurring: 0, 0.6, 1 and 2. One of these figures cannot be correct. Which one is it?

8. A bag contains 7 red balls and 3 blue balls. One ball is drawn at random. What is the probability that it will be blue?

9. The table below shows the frequency distribution for the diameter of ball bearings:

Diameter (mm)	14-15	16-17	18-19
Frequency	8	11	14

   (a) Write down the lower and higher class boundaries for the second class.

   (b) What is the class width?

10. The mean of five numbers is 40. If one of the numbers is 4, what is the mean of the remaining four numbers?

11. One letter is chosen from the word 'emergency'. What is the probability that it will be an e?

Vectors

Vector Quantities

Scalar quantities are fully described by size or **magnitude**. Some examples are time, temperature and mass or weight.

Vector quantities need both direction and magnitude to describe them fully. Some examples of vector quantities are:

(1) A displacement from one point to another, e.g. 15 m due east.

(2) A velocity in a given direction, e.g. 30 km/h due north.

(3) A force of 15 newtons acting vertically downwards.

Representing Vector Quantities

If Susan walks 8 m due east, how can we represent this vector quantity in a diagram? We first choose a suitable scale to represent the magnitude. In Fig. 37.1 a scale of 1 cm to represent 1 m has been chosen. We then draw a straight line 8 cm long in an easterly direction. An arrow is placed on the line to make it clear that Susan walked to the east (and not to the west).

Fig. 37.1

Scale: 1 cm = 1 m

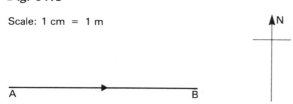

When we want to refer to a vector several times we need a shorthand way of doing this. The usual way is to name the end points of the vector. The vector described by Susan started at A and ended at B. If we now want to refer to the vector we write $\overrightarrow{AB}$ which means 'the vector from A to B'.

Draw the following vectors using a convenient scale for each. Label each vector correctly.

1. $\overrightarrow{AB}$ 5 m due north.

2. $\overrightarrow{XY}$ 8 km due south.

3. $\overrightarrow{PQ}$ 7 km due west.

4. $\overrightarrow{MN}$ 10 miles due east.

5. $\overrightarrow{RS}$ 6 m south-east.

Resultant Vectors

In Fig. 37.2 A, B and C are three points marked out in a field. A man walks from A to B (i.e. he describes $\overrightarrow{AB}$) and then walks from B to C (i.e. he describes $\overrightarrow{BC}$). Instead the man could have walked from A to C thus describing the vector $\overrightarrow{AC}$.

Fig. 37.2

Now going direct from A to C has the same effect as going from A to C via B. We therefore call $\overrightarrow{AC}$ the **resultant** of the vectors $\overrightarrow{AB}$ and $\overrightarrow{BC}$. Note carefully that the arrows on the vectors $\overrightarrow{AB}$ and $\overrightarrow{BC}$ follow nose to tail. The resultant $\overrightarrow{AC}$ is marked with a double arrow which opposes the arrows on $\overrightarrow{AB}$ and $\overrightarrow{BC}$.

Triangle Law

The resultant of any two vectors is equal to the length and direction of the line needed to complete the triangle. This is called the **triangle law.**

Example 1

Two vectors act as shown in Fig. 37.3. Find the resultant of these two vectors.

Fig. 37.3

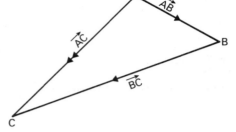

The vector triangle is drawn to scale in Fig. 37.4. The resultant is $\overrightarrow{CA}$. The length of $\overrightarrow{CA}$, written $|\overrightarrow{CA}|$, is 11.2 units. To state its direction we measure the angle ACX which is found to be 31°.

Fig. 37.4

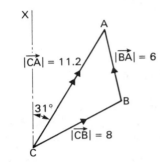

Equal Vectors

Two vectors are equal if they have the same magnitude and the same direction. Hence in Fig. 37.5, the vector $\overrightarrow{AB}$ is equal to the vector $\overrightarrow{CD}$ because the length AB is equal to that of CD and the lines AB and CD are parallel to each other.

Fig. 37.5

Inverse Vectors

In Fig. 37.6 the vector BA has the same magnitude as AB but its direction is reversed. Hence $\overrightarrow{BA} = -\overrightarrow{AB}$. $\overrightarrow{AB}$ is said to be the **inverse** of $\overrightarrow{BA}$.

Fig. 37.6

Adding Vectors

In Fig. 37.4 the arrows on the vectors $\overrightarrow{CB}$ and $\overrightarrow{BA}$ follow nose to tail. $\overrightarrow{CA}$ is the resultant of these two vectors and we write

$$\overrightarrow{CA} = \overrightarrow{BA} + \overrightarrow{CB}$$

The resultant $\overrightarrow{CA}$ is said to be the sum of the two vectors $\overrightarrow{BA}$ and $\overrightarrow{CB}$ and the sum of two vectors can always be found by making an accurate drawing.

Subtracting Vectors

Because $-\overrightarrow{CD} = +\overrightarrow{DC}$ we can treat $\overrightarrow{AB} - \overrightarrow{CD}$ as $\overrightarrow{AB} + \overrightarrow{DC}$.

Example 2

Fig. 37.7 shows two vectors v_1 and v_2. Find in magnitude and direction, the vector $v_1 - v_2$.

Fig. 37.7

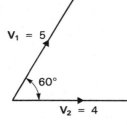

To find the vector $v_1 - v_2$ we treat $v_1 - v_2$ as $v_1 + (-v_2)$. This means that we reverse the arrow on the vector v_2 (Fig. 37.8) and then proceed in a similar way to that used in Example 1.

Fig. 37.8

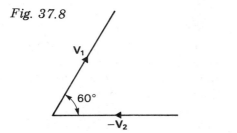

The vector triangle is drawn in Fig. 37.9. The resultant of v_1 and $-v_2$ is CA which is found to be 4.6 units. To state its direction, we measure the angle ACB which is found to be $70°$.

Fig. 37.9

Scale: 1 cm = 1 unit

Multiplying Vectors by a Scalar

If the vector $\overrightarrow{AB}$ is 5 m due west then the vector $3\overrightarrow{AB}$ is 15 m due west (Fig. 37.10). In general if k is any number, the vector $k\overrightarrow{AB}$ is a vector in the same direction as $\overrightarrow{AB}$ but k times its magnitude.

Fig. 37.10

Exercise 37.2

Find by accurate drawing, the resultant in magnitude and direction of each of the pairs of vectors shown in Fig. 37.11.

Fig. 37.11

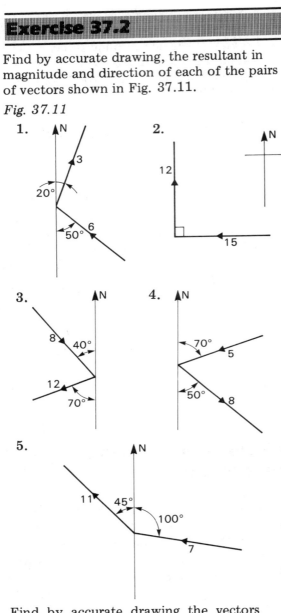

Find by accurate drawing the vectors $v_1 - v_2$ in magnitude and direction for each of the systems shown in Fig. 37.12.

Fig. 37.12

8.

9. The magnitude of a vector **v** is 8 units. Find the magnitude of the vectors

   (a) 3**v**   (b) 5**v**   (c) 9**v**

10. Draw accurately the vector **v** shown in Fig. 37.13. Now draw the vectors

    (a) 2**v**   (b) 4**v**   (c) 5**v**

Fig. 37.13

Parallelogram of Vectors

Two vectors may be added by drawing a parallelogram of vectors. This is an alternative to the triangle of vectors discussed on page 370.

Example 3

Draw the parallelogram of vectors for the two vectors shown in Fig. 37.14. Hence find their resultant in magnitude and direction.

Fig. 37.14

The two vectors are drawn to scale with the angle between them accurately made by using a protractor. The parallelogram is then completed as shown by the dotted lines in Fig. 37.15. The diagonal drawn from the point where the two original vectors meet represents the resultant in magnitude and direction. On scaling it is found to be 10.4 units acting at 25° to the 7 unit vector.

Fig. 37.15

Note that when using a parallelogram of vectors both the arrows on the original vectors must either point away from the point of intersection of the two vectors or they must point towards the point of inter-section. In Fig. 37.16(a) the arrows on the two vectors do not conform to this rule. One arrow points towards A whilst the other points away from A. However, the horizon-tal vector remains unchanged if we draw it as shown in diagram (b). Alternatively the parallelogram may be drawn as shown in diagram (c) with both arrows pointing towards A.

Note that if both arrows point away from A (diagram (b)) then the arrows on the resultant also point away from A. If both arrows point towards A then arrows on the resultant also point towards A.

Fig. 37.16

(a)

(b)

(c)

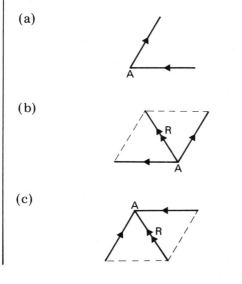

Rectangle of Vectors

The rectangle of vectors (Fig. 37.17) is a special case of the parallelogram of vectors. It is often used to solve problems in mechanics and stress analysis, where it is generally used to split a force into horizontal and vertical components.

Fig. 37.17

V_H = Horizontal component of **V**
V_V = Vertical component of **V**

Example 4

Fig. 37.18 shows a force of 20 newtons acting at 40° to the horizontal. Find its horizontal and vertical components.

Fig. 37.18

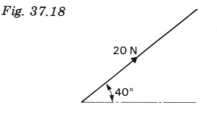

To find the two components by accurate drawing, first draw the vector to scale measuring the angle of 40° with a pro-tractor. Next construct the rectangle ABCD with the vector as diagonal. Finally measure AB = 15.3 N and AD = 12.9 N which are the horizontal and vertical components respectively (Fig. 37.19).

Fig. 37.19   Scale: 1 cm = 5 N

Exercise 37.4

By drawing a parallelogram of vectors, find the resultant vector for each of the cases shown in Fig. 37.20.

Fig. 37.20

5.

By drawing a rectangle of vectors find the vertical and horizontal components of the following vectors:

6. 5 N acting at 60° to the horizontal.

7. 30 km/h in a direction of 40° due east.

8. 80 N in a direction of 70° to the horizontal.

Coursework

Coursework tasks which are separately assessed at the examination centre may be set by the Examiner or the choice may be left to the teaching staff and the individual candidate. The content may or may not be directly related to the main core topics and the details connected with this aspect of the examination must be read through carefully in the chosen syllabus.

You will be asked to work on the various coursework assignments during the two years (or one year) before the final examinations. Coursework may include mathematical investigations as well as practical work, historical research, problem solving or perhaps a real life application of mathematics based on statistical surveys carried out locally.

A written report on each coursework study will be required to explain how the various results were obtained and in some cases the candidates may be asked for a verbal explanation to clarify some particular aspect of the work. The presentation should include all the relevant calculations and observations with diagrams, tables, graphs and constructions where necessary. The final package should contain sufficient explanation and detail for any outside reader to understand the development and conclusion. The percentage of marks allocated to coursework varies from one Examining Group to another, and an assessment will be made according to instructions received by the centre. However, the following points will certainly receive consideration:

(1) The candidate's comprehension of the task

(2) The clarity of the plan of attack

(3) Recognition and reaction to results at various stages

(4) Mathematical accuracy where appropriate

(5) Employment of equipment, graphics and suitable materials

(6) Final evaluation and presentation.

An investigation begins with some facts and ideas concerning a given situation. The way in which the clues are applied is then left to the individual who will decide which particular direction the development should take. Two students may well follow quite different lines of enquiry and could arrive at different — but equally valid — conclusions for a given set of information. Even if no final solution is obtained, the reasoning processes involved are always valuable and should receive recognition in the final presentation. Topics requiring such an approach are included under the title 'Investigations'.

The syllabus may require some form of practical work. This could involve the construction of models or a sequence of practical tasks not normally associated with everyday mathematics lessons. Coursework may include investigations of real life situations; statistical surveys could be carried out, or you may be asked to undertake some historical research involving mathematical ideas and concepts. This element of the work is covered in the 'Further Studies' section.

Mathematical problem solving requiring a logical approach or even some trial-and-error tactics in search of a finite solution, may also be included in certain coursework assignments. Examples of a variety of mathematical puzzles are offered in the 'Problems' part of this chapter.

At various times during the two-year (or one-year) GCSE preparation period you may like to attempt a selection of these different tasks. The investigations are varied and your choice will depend on the particular Examining Group's instructions that you are following. They should provide solid practice and may even stimulate a range of interesting ideas for inclusion in your own coursework.

Investigations

The first two are presented as examples and a logical sequence in pursuit of conclusions is outlined in some detail. Suggestions are made and questions are posed. In addition, you may find other useful avenues of enquiry which are not mentioned in the text.

Investigation 1

A train set contains three types of rail.

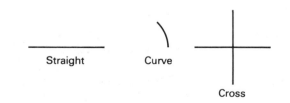

Straight   Curve   Cross

Curves, which can be used to turn left or right, are each one-eighth of a circle. This means that a single curve gives a rotation of

45°. Curves will join on to any line or on to each other.

Investigate the use of all types of rail and explore problems encountered in various track circuits.

Firstly, examine the various combinations:

Draw and explain these and other combinations which are possible with the rails.

Investigate a typical circuit which incorporates all the different types of rail.

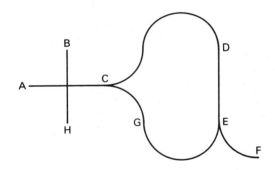

Explain why

(a) a train can travel from A to C and also from H to B but not H to C

(b) we could travel directly from A to D and A to G but not G to F

(c) we could start at A and travel back to A without stopping

(d) the train starting at F would need to stop twice if it wished to return to F.

The part of the line CGEDC is a closed circuit. Investigate the use of straights and curves in different closed circuits.

A figure-of-eight route is a particular type of closed circuit. Invent different circuits of that kind using all types of rail.

On the circuit A to F there are two junctions which present particular problems if free access to each letter is required.

Investigate modifications to each part of the track so that free travel without a stop is possible anywhere on the route.

Try to construct circuits for solving problems similar to the one presented in the next diagram.

Stations P, Q, R, S and T are to be connected by rail to the main line entering at X and leaving at Y. A continuous route is required so that any or all of the five stations may be visited on a journey from X to Y without stopping or reversing. It may, of course, be necessary to pass through some points when travelling to other stations.

Create other circuits to include engine-storage sheds and loop lines. Show two or more routes linking a pair of points. Suggest new types of rail which might usefully be included in the train set.

Investigation 2

A fly was allowed to settle on X, one vertex of a cube. It was then free to walk along the edges. The routes it creates are fly-paths.

Fly-paths are identified by the number of edges the fly uses in one journey. A fly-path does not use any edge of the cube more than once.

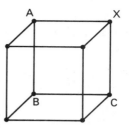

From X to A is called a 1-path.
From X to A and then to B is a 2-path.

Investigate fly-paths on the cube and other solids.

First construct a cube. One candidate used lengths of wire held together at each vertex by a blob of plasticine. Availability of materials will influence the choice of construction.

Label each vertex with a letter and record the fly-paths in a list:
1-path: XA, XC, etc.
2-path: XAB, XCB, etc.

All paths must start at X and no edge must appear twice in one path. There are 12 edges. Is there a possible 12-path? If not, what is the maximum fly-path? Is there anything particularly significant about any of the paths?

At this stage give some thought to constructing a cube using a single length of wire. It will obviously be necessary to run a double wire on one or more edges. Here is one attempt:

Does this method use the minimum possible length of wire? If the wire is allowed to follow diagonals across faces, is it possible to form a cube without duplicating any edges? Does that allow for the use of less wire? Produce diagrams to illustrate your conclusions.

Investigate these solids:

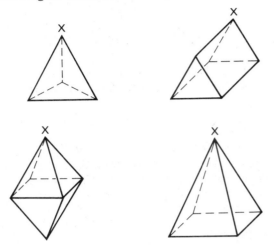

Examine fly-paths from point X. Are the numbers of faces and vertices in a given solid at all significant? Does any solid present a unique problem? Is the position of point X important?

The remainder of the investigations do not include such a comprehensive range of suggestions. Some assistance is offered, but the depth and success of each enquiry will depend upon the extent to which the given facts are questioned and examined.

Investigation 3

Spacefiller Ltd manufactures blocks having six faces, which may be squares or rectangles. There is an economy drive organised to save the paint needed to cover the surface of the blocks. Customers order blocks according to a particular volume.

Investigate the surface areas of various blocks.

One customer required a Spacefiller block

with a volume of 36 cubic centimetres. This was the first block the firm produced:

Painter's calculation:

$$2(6 \times 2) = 24 \, cm^2$$
$$2(6 \times 3) = 36 \, cm^2$$
$$2(3 \times 2) = 12 \, cm^2 \quad \text{TOTAL: } 72 \, cm^2$$

The firm's present policy is to use whole-number measurements.

Investigate the block with a volume of $36 \, cm^3$ in the light of the firm's economy drive.

Investigate blocks with other volumes such as $30 \, cm^3$, $48 \, cm^3$ and so on. If the firm decides to include half-centimetres in the measurements, will it affect the conclusions you have reached so far? Could other fractional lengths be included?

Investigation 4

Any four numbers are chosen to start this investigation.

        18   41   5   76

A second row is established from the differences between numbers next to each other and the difference between the two end numbers:

i.e. $41 - 18$ $41 - 5$ $76 - 5$ $76 - 18$

The second row therefore becomes:

        23   36   71   58

The process is continued until something significant occurs.

Investigate the sequence using any four numbers.

Does a mixture of small and very large numbers affect the final position in any way?

Does the inclusion of decimal quantities make any difference?

Investigate particular groups such as odd, even and prime numbers.

What happens with groups of three numbers and groups of five numbers?

Is there anything significant about any other particular set?

Are any patterns of behaviour worth noting?

Investigation 5

There are seven different wall tiles available at the local DIY store:

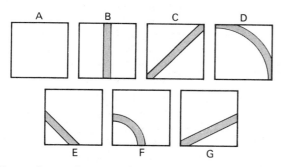

Investigate the use of the tiles:

(a) to form closed shapes

(b) to produce symmetrical patterns for bathroom or kitchen.

This is an example of a closed shape using eight tiles:

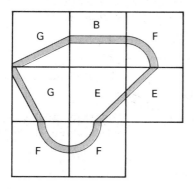

Is a closed shape possible with only three tiles?

Investigate the use of more than three tiles.

Include patterns in which tiles do not necessarily have a common edge.

Does any tile present a particular problem compared with other tiles?

Design a tile which will produce better results in the investigation.

Construct symmetrical patterns to fit a rectangle measuring 15 tiles by 12 tiles.

Investigation 6

Obtain or construct a die in the shape of a cube.

This investigation concerns the turning of the die on an edge to show a different number on top.

Is it possible to show the numbers in proper sequence as the die is turned over on an edge in any direction?

What kind of sequences are possible?

Work the die around inside this grid:

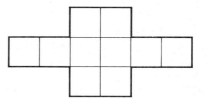

Is it possible to start and finish on the same square showing the same number on top?

Use this grid of nine squares. Always start on the centre square marked X.

H	A	B
G	X	C
F	E	D

You can leave the centre in four ways, by turning the die on its edge into squares A, C, E or G. Travel clockwise and note the sequence of numbers which appear on the top of the die until it arrives back at the centre again. Each sequence has eleven numbers.

Comment on the number patterns formed on the four separate routes.

Does the starting number affect the sequence?

How are the four routes related?

What is the effect of reversing the direction and moving anticlockwise?

Investigate 13-figure sequences produced by a figure-of-eight route such as XABCDEXAHGFEX.

Investigation 7

The firm Awkward Shapes Ltd produces plastic blocks of various shapes and sizes. Here are four of the shapes, each of which is 1 cm thick:

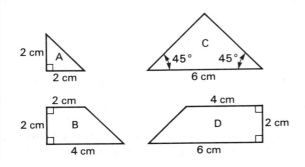

The shapes are sold in the following sets:

Set 1: 2 of A, 2 of B, 4 of C, 2 of D
Set 2: 2 of A, 2 of B, 2 of C, 2 of D
Set 3: 1 of A, 1 of B, 2 of C, 1 of D
Set 4: 3 of A, 3 of B, 2 of C, 3 of D

Investigate the various packing boxes which could be used to hold these sets of blocks.

Investigation 8

Using four lines, we can draw one square, but if five lines are used, we can produce two squares:

Using six lines, these diagrams can be constructed:

There are three squares in the first diagram. In the second diagram there are four small squares and one large square giving a total of five squares.

Investigate the use of lines to draw squares.

Is there anything significant about the maximum number of squares each time?

Make tables to include different formations and numbers of squares.

Comment on patterns shown in the tables of numbers.

Continue the investigation until you have reached patterns constructed from twelve lines.

Is it possible to predict further results?

Investigation 9

The pieces in a set of dominoes consist of rectangles made up of two squares. Each domino has two numbers from 1 to 6 and there are 28 in the set. This investigation concerns a set of trominoes. A tromino will be three squares forming a right angle. Only the numbers 1 to 4 will be used in the three squares. There are no blanks. Here are two examples:

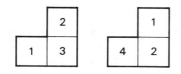

It should be noted that some formations will be mirror images of each other. These are classified as different trominoes. On the other hand, other trominoes will be simply rotated versions of one another and not actually different:

Different
(reflection)

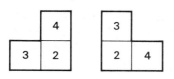

Same
(rotation)

Investigate the set of trominoes. How many are there in the set?

Invent different sets of rules which could be used to play games with the set of trominoes.

Decide on a suitable box in which the trominoes could be sold as a game.

Investigation 10

Three people start out from point A and wish to travel to point B which is 9 km away.

The three are equally athletic but can only maintain an average walking pace of 5 km/h. Each person is capable of riding a pedal cycle and a motor scooter and both machines are available at A. The maximum average speed possible on the pedal cycle is 15 km/h while the motor scooter can attain a maximum average speed of 30 km/h. Only one person can travel on either machine at any one time: no passengers are allowed. Each person will employ at least two of the methods of travel in a 9 km journey and nobody will stop longer than the time needed to make a change.

Investigate the most economic method for travelling from A to B. All three travellers start together at A and the time of travel is reckoned to be that of the last person to arrive at point B.

Construct graphs for all of the various methods of travel which you suggest they should use.

Is it possible for the three to arrive at B together?

Further Studies

Further Study 1

Time and motion studies in industry are carried out to ensure that the firm is making the best use of its workforce. To eliminate waste of time and effort, the movement of workers is studied and regulated. What kind of movement occurs at your centre when the timetable demands a change of lesson?

This is a specimen timetable for one day:

Tutorial	Room 10	Lunch	
History	Room 25	English	Room 30
Mathematics	Room 16	PE	Room 5
Break		Geography	Room 28
Practical	Room 7	Session ends	

A drawing can be constructed to show the movements through the building which a student following the timetable must make going from lesson to lesson. The drawing assumes there is a ground floor and two other floors in the building.

The scale is 1 millimetre to 1 metre for all horizontal movements on a single floor and 1 millimetre to 1 step for the use of stairs between floors.

The numbers show the positions of the rooms in the building. If a centre's rooms were all on the ground floor, the drawing could be modified to show movements left and right instead of horizontally and vertically.

Investigate parts of the timetable at your own centre. Indicate the sections of a time-table which in your opinion contain excessive movement and then make enquiries to discover whether such movement is actually necessary.

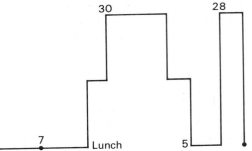

Further Study 2

As well as examining the movement of personnel for economy reasons, a firm also draws up a shopfloor plan. The plan is organised to reduce movement and make access between various machines more convenient.

As a simple exercise, examine this plan of a lounge in which certain items of furniture are to be positioned. Make a drawing of the plan twice the size of the one shown in the diagram and produce paper cut-outs of the furniture, also on twice the scale shown in the drawings.

Use the cut-outs to establish a final arrangement of furniture in the room and present a detailed report explaining the reasons for your choice of position for each item of furniture.

Now investigate one of these situations using the same approach:

(a) A kitchen contains a cooker, refrigerator, freezer, cupboards for cooking utensils, crockery and some food items, work-tops, sink, draining boards and waste disposal units. The area of the floor is a rectangle measuring about 5 metres by 3 metres. Consideration should also be given to the building of an extension to take a washing machine, drier, airing space and ironing facilities.

(b) A workshop measures 10 metres by 6 metres and sixteen students will be working at benches using wood and metal at various times. The two activities may be taking place simul-taneously on some occasions. There is a separate outside store for supplies of wood and metal, but workbenches, machines, tools and all other facilities available to the students are contained within the workshop area. An extension for light crafts is under consideration.

Further Study 3

Fashion designers sometimes use miniature models to test their creations. Draw up patterns for the design of a modern suit of clothes. This may be either a coat and trousers or a female two-piece outfit. You must not consult prepared commercial patterns.

You may simply use your imagination here or you could study the way in which various modern styles of clothing are constructed. Study the stiching to decide on the number of individual pieces of cloth that will be required. Strips of paper placed on the clothing, for instance, will help you to discover the shapes you need.

When you have obtained the necessary shapes, you then need to decide on a suitable scale according to your own choice of model. Test your patterns by attaching tabs so that the various pieces can be assembled to form the two items of clothing.

Further Study 4

A puzzle is to be constructed in which five solids will be assembled to make a cube. These notes explain how to make four of the solids, but you must work out how to produce the fifth solid yourself.

Use this net to construct four pyramids as shown:

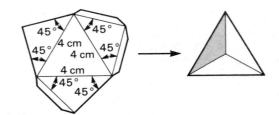

If adhesive tape is used, tabs may not be required.

On a piece of paper, draw an equilateral triangle with a side of 8 cm. Glue the four pyramids on to the triangle as shown:

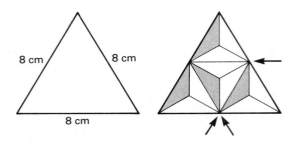

Fold the shape as indicated by the three arrows to form a cube. There will be a gap in the centre.

Decide what shape is required to fill the centre gap perfectly and construct the fifth solid according to the measurements from your cube. Finally, cut the triangular paper base to release your first four pyramids and you will then have a puzzle for constructing a cube from five other solids.

Further Study 5

Many surveys involve the use of a questionnaire. When television companies, newspapers or manufacturers of certain products wish to conduct a survey among members of the public, they produce a set of questions. The questions are either of the type which require a simple answer of 'Yes' or 'No', or a set of alternatives from which each

person taking part in the survey can select an appropriate answer.

The 'Yes' and 'No' questions would be of the following type:

Do you use this particular product?
Did you watch this particular TV programme?
Did you vote for this candidate in the election?

Alternatives can be offered as follows:

Do you rise in the morning:

(a) Before 6.00 a.m.?

(b) 6.00 to 6.30 a.m.?

(c) 6.30 to 7.00 a.m,?

(d) 7.00 to 7.30 a.m.?

(e) 7.30 to 8.00 a.m.?

(f) After 8.00 a.m.?

Questions that could have answers not appearing on the question sheet should not be asked.

Test the questionnaire on yourself or a friend before putting it into use, and make whatever improvements to the questions you consider necessary.

Analysis of the results of a survey is simple if the questions are well constructed and prepared.

Make some questionnaires for surveys on the following subjects:

(a) The belief in superstitutions among the general public

(b) Weekly purchases required for a variety of households

(c) TV viewing among young people

(d) Daily habits of people aged from 11 to 16 years

(e) The extent of active or passive sport participation

(f) People and their pets

(g) Young people and their tastes

Test one of your questionnaires on as many people as possible. Analyse and interpet the results and make a display using graphics to bring out the main conclusions.

Further Study 6

To carry out the following study, a large number of birthdays is required. These should be available from the admission rolls at your own centre. Names are not necessary.

Investigate the frequency of people born in each month of the year, each date in the month and each day of the week. Between 75 and 100 birthdays will be required.

(a) Draw up a frequency table for the months of the year. Convert the figures into any type of graph you wish.

(b) Find the frequency figures for birthdays from the first to the last day of a particular month. Comment on any conclusions you draw from certain figures.

(c) To obtain frequency figures for the days of the week, some simple mathematics is required. Using the following sequence, you can determine the day of the week on which anyone was born in this century.

You will firstly need to note the following list of figures for the twelve months of the year:

January 1 (0 in leap years); February 4 (3 in leap years); March 4; April 0; May 2; June 5; July 0; August 3; September 6; October 1; November 4; December 6

In this example, the date 13 June 1947 is used:

Last two digits of the year:	47
Divide by 4 (ignore any remainder):	11
Add the numbers:	58
Add the birth date:	13
	71
Add the month number (see list):	5
	76
Divide the total by 7:	10
	remainder 6

Look up the remainder in this list:
1: Sunday; 2: Monday; 3: Tuesday; 4: Wednesday; 5: Thursday; 6: Friday; 0: Saturday.

We obtained a remainder of 6, so 13 June 1947 was a FRIDAY.

Further Study 7

For these projects, you will need to carry out some historical research by consulting various reference books. Here are some suggestions which are worth investigating:

(a) From where did we obtain the digits 0 to 9? How did the shapes of the figures evolve and what changes occurred? Are there any other different systems presently in use elsewhere in the world?

(b) Research the history of the signs we use for addition, subtraction, multiplication and division.

(c) Where did the use of the figure 0 (zero) originate and what is the history of the decimal point?

(d) What do we know about the mathematical symbol π (pi) and its value?

(e) How and when did algebra originate?

(f) Examine the history of time measurement and the calendar.

(g) How have the units of length and weight varied in the last two thousand years?

Further Study 8

Reference books will again be required for a study involving famous mathematicians of the past.

Who was Fibonacci? Where did he live? For what is he famous?

What is the connection between Fibonacci's work in mathematics and a study of plant and animal life?

Here is a puzzle constructed on a square grid. Draw the first shape and cut it into four pieces, following the dotted lines.

What has happened to the number of squares?

How is the puzzle connected with the discoveries of Fibonacci?

Make a similar puzzle using a different number of squares.

Further Study 9

The task of a quantity surveyor is to assess the materials required for a building project after working plans have been approved. He or she decides how many bricks are needed, the lengths of various pieces of timber, the amount of sand, gravel and cement, electrical supplies, area of glass for the windows and so on.

Choose a small building at or near your centre and imagine that you have to make such a survey. Enquiries must be made about materials for mortar and cement and you will need to ask questions about how a building is put together. You are concerned with quantities and not the cost, so the question of labour is not to be considered.

Further Study 10

Construct two solids according to the instructions given in these notes and then comment on the results.

First solid:

Make eight pyramids using this net:

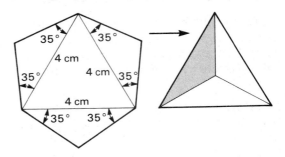

Tabs will be required if adhesive tape is not used.

On paper, draw the net of a regular octahedron and glue the eight pyramids on to its eight faces. Tabs will be necessary if liquid glue is used.

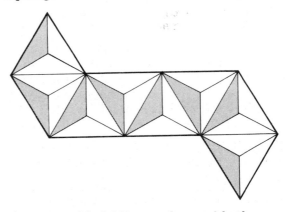

Take care with folding, and assemble the shape as one solid.

Second solid:

In a similar way, construct six pyramids using this net:

The pyramids are then glued on to the net of a cube as shown:

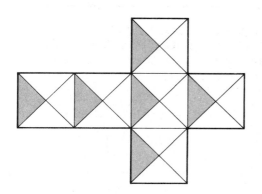

Tabs will be required throughout if adhesive tape is not used.

Fold the net to form a single solid.

Compare the two solids. Comment on the shape of each and the space left inside each one.

Problems

Problem 1

There is something interesting about various pairs of digits. For instance, let us take the four digits 1, 2, 4 and 8. We can form the two numbers 21 and 48.

By multiplication,

$$21 \times 48 = 1008$$

Reverse the digits in each number to give 12 and 84.

By multiplication,

$$12 \times 84 = 1008$$

Similar results are obtained with the digits 2, 3, 4 and 6:

$$24 \times 63 = 1512$$
$$42 \times 36 = 1512$$

Here are some clues to assist you in solving the problem:

There are other pairs you could use with the number 12.
There are pairs you could use with the digits 1 and 3.

No more clues!

Investigate the problem to find the total of fourteen sets of four digits that will produce a similar result.

Problem 2

When two squares are joined on a common side, the result is called a domino. Three squares joined so that each square is attached on at least one side to another square make a tromino. There are only two different trominoes as follows:

Four squares joined under the same rules make a tetromino. There are five different tetrominoes. Draw each one separately.

Now use this rectangle of squares:

A	B	C	D
E	F	G	H
I	J	K	L

From this pattern, we can take out the squares ABCD as a tetromino. We could also pull out BEFI to form another different tetromino.

Altogether, there are 65 different ways of constructing a tetromino from those twelve squares. How many can you find?

To prevent repetition, you will need to record all of your discoveries carefully.

Problem 3

The digits 0 to 9 are to be arranged in the form of a triangle. Before that is done, one of them is cast out and placed at the centre of the triangle so that it takes no further part in the problem. In this example, the number 0 is cast out to the centre:

$$
\begin{array}{ccccccc}
 & & & 1 & & & \\
 & & 7 & & 5 & & \\
 & 6 & & 0 & & 9 & \\
3 & & 4 & & 8 & & 2 \\
\end{array}
$$

The 0 is in the centre and can be ignored. The other nine numbers form the outer edges of the triangle so that there are four numbers arranged along each side. The total on each side of the triangle is 17.

In the next example, the number 6 has been cast out and placed in the centre and is not used in this problem.

$$
\begin{array}{ccccccc}
 & & & 0 & & & \\
 & & 7 & & 4 & & \\
 & 5 & & 6 & & 9 & \\
2 & & 3 & & 8 & & 1 \\
\end{array}
$$

The total on each side this time is 14.

Cast out each number in turn and find as many magic triangles as you can. There is more than one for each number that is placed in the centre, so there are plenty for you to discover.

Problem 4

Construct two circles which have:

(a) four common tangents

(b) only three common tangents

(c) only two common tangents

(d) only one common tangent

(e) no common tangents.

Problem 5

Wages for the staff at six stores are delivered by a security van from one bank. The stores are lettered A to F and the position of the bank is shown by X.

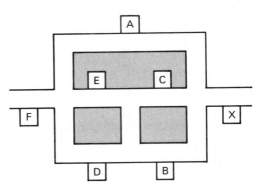

The van must proceed forward at all times and must not turn completely round in the road. The shaded portions are office blocks.

The first delivery must be to one of the stores A to E but the van must not pass in front of a store without stopping to make a delivery.

The last delivery is always to store F and each store is visited on a single journey each Thursday.

This week, the order was XBEACDF. The security firm is anxious to vary the route each week. How many different routes can you find ending at F?

Problem 6

A local jeweller had two tricky problems with chains that had suffered some damage.

(a) The first customer brought in five lengths of chain, each of which had four links. She wanted the links to join together to make a continuous loop with no break.

'I charge £1 to break a link and £1 to join it together again,' said the jeweller, 'so that will cost you £10.'

'Oh no,' replied the young lady, 'let me explain how you can do it for £8, charging the same rates.'

What was her explanation?

(b) A dowager duchess came in with a more expensive chain which had snapped into nine lengths. There were two with 5 links, three with 6 links and four separate lengths of 3 links, 4 links, 7 links and 8 links.

She also wanted all the links joined into a continuous loop. The rates of £1 for a break and £1 for a make were explained to her. Expecting no argument, the jeweller said, 'Ah, yes. I can break each link in the eight-link portion. I shall then use the eight links to join the remaining eight lengths in a continuous chain. That will cost you £16.'

'My man,' said the duchess, 'I shall expect you to work out a method which will only cost me £14.'

What was that method?

Problem 7

This problem is based on a puzzle from a mathematical magazine published in Czechoslovakia several years ago.

Four circles form two interlocking paths and at the start numbers 1 to 5 occupy the positions as shown in the first diagram. By moving the numbers, you must arrange them to appear as they are in the second drawing.

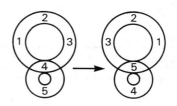

These are your rules:

(a) In the larger rings, the numbers can all move one position clockwise or anti-clockwise but must not change their order.

(b) In the smaller ring, the two numbers there at any time can change places.

(c) Moves must alternate in the large and small circles. A move in the large circle must be followed by an exchange in the small circle and vice versa. You may start in either circle.

You may take as many moves as you wish but the quickest method requires 14 moves only.

Problem 8

In the following pattern, you can only move from any letter to one of the two letters immediately below it in the next line.

A

B B

C C C

D D D D

E E E E E

F F F F

G G G

H H

I

How many different paths are there from A to I taking in one of each letter alphabetically?

Clue: Resolve your thinking into two parts, ABCDE and EFGHI.

Problem 9

In the rectangle, there are nine numbers and six mathematical signs. There are two addition signs, one multiplication sign, two equality signs and one sign which means 'less than'. Using that sign for instance we could write:

$$4 \times 2 < 16$$

All the numbers and all the signs must be used to make three correct statements. Here is one example in which all the signs are correctly used:

$$5 + 3 = 8$$
$$4 \times 4 = 16$$
$$6 + 2 < 10$$

Find at least ten other ways of producing three accurate statements.

Problem 10

The only equipment available in this problem is an ordinary fair balance that rests level when equal weights are placed on each scale-pan. Different weights on the two pans will be indicated when one scale-pan goes down.

Here is an example of how the balance may be used:

Problem: We have four weights. Three are equal, but the fourth is lighter than the others. Pick out the faulty weight using the balance twice only.

Solution: Place two weights on one side and two on the other. One pair will be lighter. Remove all the weights.

Now use only the two which were lighter, one on each pan. The faulty weight is identified as the one on the pan which goes up.

Use the same balance with these four problems.

(a) In a set of five weights, four are equal and one is lighter than the others. In only two weighings, find the faulty one.

(b) There are eight equal weights and a ninth which is lighter than the rest. Find the faulty weight in only two weighings.

(c) This is not easy. In eight weights, one is lighter than the other seven equal ones. Using the balance only twice, find the lightest weight.

(d) This is more difficult, so you are allowed to use the balance three times. In a set of twelve weights, there are eleven of equal weight and one which is lighter. How will you identify the lightest one in only three weighings?

Appendix
Important Data and Formulae

Simple Interest

$$I = \frac{PRT}{100}$$

I = simple interest

P = principal (i.e. amount invested or borrowed)

R = rate per cent

T = time in years of investment or loan

Compound Interest

$$A = P\left(1 + \frac{r}{100}\right)^n$$

A = amount after n years

r = rate per cent

P = principal

Indices

$$a^m \times a^n = a^{m+n}$$

$$a^m \div a^n = a^{m-n}$$

$$(a^m)^n = a^{mn}$$

$$a^0 = 1$$

$$a^{-m} = \frac{1}{a^m}$$

Mensuration

Circumference of a circle $= 2\pi r$ or πd

Area of a circle $= \pi r^2$

Area of a triangle $= \frac{1}{2} \times$ base $\times$ height (or altitude)

Area of a trapezium $= \frac{1}{2}(a + b)h$

Area of a rectangle $= a \times b$

Area of a square $= a^2$

Area of a parallelogram $= ah$

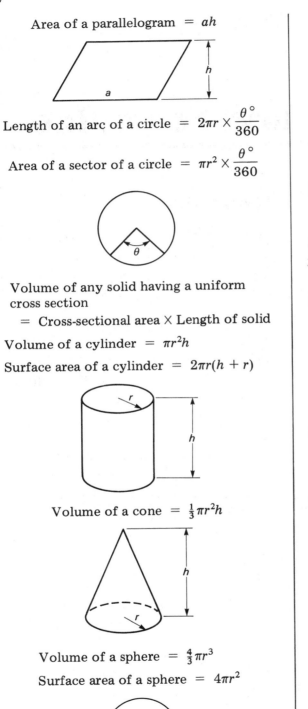

Length of an arc of a circle $= 2\pi r \times \dfrac{\theta\,°}{360}$

Area of a sector of a circle $= \pi r^2 \times \dfrac{\theta\,°}{360}$

Volume of any solid having a uniform cross section

$= $ Cross-sectional area $\times$ Length of solid

Volume of a cylinder $= \pi r^2 h$

Surface area of a cylinder $= 2\pi r(h + r)$

Volume of a cone $= \frac{1}{3}\pi r^2 h$

Volume of a sphere $= \frac{4}{3}\pi r^3$

Surface area of a sphere $= 4\pi r^2$

Volume of a pyramid $= \frac{1}{3}Ah$

$A = $ area of base

Equation of a Straight Line

$$y = mx + c$$

(m is the gradient, c is the intercept on the y-axis).

Profit and Loss

$$\text{Profit \%} = \frac{\text{Selling price} - \text{Cost price}}{\text{Cost price}} \times 100$$

$$\text{Loss \%} = \frac{\text{Cost price} - \text{Selling price}}{\text{Cost price}} \times 100$$

Pythagoras' Theorem

$$b^2 = a^2 + c^2 \quad \text{(see triangle below)}$$

Trigonometry

$$\sin A = \frac{\text{Side opposite to } \angle A}{\text{Hypotenuse}} = \frac{a}{b}$$

$$\cos A = \frac{\text{Side adjacent to } \angle A}{\text{Hypotenuse}} = \frac{c}{b}$$

$$\tan A = \frac{\text{Side opposite to } \angle A}{\text{Side adjacent to } \angle A} = \frac{a}{c}$$

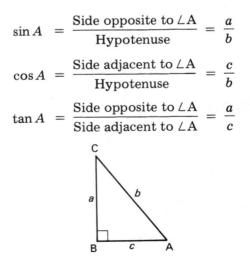

Statistics

$$\text{Arithmetic mean} = \frac{\text{Sum of the quantities in the set}}{\text{Number of quantities in the set}}$$

Probability

$$\text{Probability} = \frac{\text{Number of equiprobable events which produces a favourable outcome}}{\text{Total number of equiprobable events}}$$

Answers

Answers to Chapter 1

Exercise 1.1

1.

```
 ┼──┼──┼──┼──┼──┼──┼──┼──┼──┼──┼──┼──┼──┼──┼──┼
-8    -6    -4    -2    0       3     5     7
```

2. $+3$
3. $+2, 3, +5, 7$
4. $0, +2, 3, 5, 7$
5. $-9, -5, -4, -3, -1$
6. $+25, 18, 12, 3, -2, -6, -8, -15$
7. (a) 12 (b) -10
   (c) $+5, +7, 8, 12$
8. -5
9. $+5, 7, +11$
10. Below the freezing point of water

Exercise 1.2

1. 40
2. 60
3. 90
4. 90
5. 80
6. 700
7. 700
8. 900
9. 400
10. 900
11. 3000
12. 5000
13. 8000
14. 9000
15. 6000

Exercise 1.3

1. 16
2. 18
3. 27
4. 20
5. 30
6. 462
7. 4710
8. 58 225
9. 8 599 293
10. 140 700

Exercise 1.4

1. 13
2. -9
3. -32
4. -16
5. -4
6. -8
7. -14
8. 4
9. 3
10. -3
11. 6
12. 194
13. -227
14. -98
15. 181

Exercise 1.5

1. 4
2. 5
3. 3
4. 6
5. 5
6. 158
7. 896
8. 1429
9. 2037
10. 135 895
11. 1
12. 1
13. 8
14. 1
15. -6

Exercise 1.6

1. (a) 30 (b) 140
2. 40
3. (a) 60 (b) 280
4. 210
5. 1000
6. 3150
7. 9165
8. 17 010

9. 31 680
10. 110 880
11. -24
12. -10
13. -18
14. 8
15. -60
16. 24
17. -40
18. 60

Exercise 1.7

1. 133
2. 15
3. 27
4. 16
5. 43
6. 59
7. 18
8. 36
9. 21
10. 350
11. -4
12. -3
13. 4
14. 2
15. -2
16. 3
17. -3
18. 3
19. -2
20. 4
21. -3
22. $+2$
23. 54
24. 28
25. 1 825 740
26. 779 471
27. 7 remainder 1
28. 7 remainder 2
29. 4 remainder 3
30. 1 remainder 5
31. 6 remainder 3
32. 5 remainder 7
33. 27 remainder 5
34. 31 remainder 12
35. 268 remainder 3

Exercise 1.8

1. 19
2. 9
3. 7
4. 44
5. 33
6. 5
7. 23
8. 25
9. 7
10. 2
11. -5
12. -12
13. 56
14. 34
15. -20

Exercise 1.9

1. 768
2. 256
3. 0
4. 0
5. 18
6. 0
7. 13
8. 8
9. 0
10. 0
11. -42
12. -5

Exercise 1.10

1. 250, 1250
2. 17, 21
3. 27, 33
4. 22, 11
5. 15, 21
6. 21, 34
7. 78, 158
8. 165, 489
9. $-6, -8$
10. 3, 5
11. 6, 8
12. 21, 89
13. 82, 244
14. 245, 2189
15. 9, 1

Miscellaneous Exercise 1

Section A

1. 27
2. 18
3. 0
4. 3
5. 6, 8, 16
6. 185 min
7. 9, 5, 1
8. $8°C$

Section B

1. **(a)** $9, 5, 1$ **(b)** $16, 8, 0$
2. $900\,000$ 3. 45 4. 36
5. $36, 49$
6. **(a)** 40 **(b)** -13 **(c)** -4 **(d)** -300
7. $-5, -3, 0, 3$
8. 7866
9. $(5 + 2) \times (7 - 4) = 21$; $5 + 2(7 - 4) = 11$; 10
10. $3, 6, 96$
11. **(a)** 44 **(b)** -26
12. **(a)** $27 + (5 - 12)$ **(b)** $13 - (4 - 9)$

Multi-Choice Questions 1

1. B   2. B   3. C
4. B   5. B   6. A
7. D   8. D   9. B
10. A

Mental Test 1

1. 20 2. 20 3. 11
4. 9 5. 70 6. 90
7. 9 8. 11 9. 13
10. 12 11. -9 12. -30
13. -2 14. 3 15. -60
16. 4 17. -6 18. 54
19. 2 20. -2

Answers to Chapter 2

Exercise 2.1

1. Odd   2. Even   3. Even
4. Odd   5. Odd   6. Even
7. Even   8. Odd
9. **(a)** $17, 59, 121, 259$ **(b)** $36, 98, 136$
10. $19, 35, 89, 137$ 11. 36
12. 125
13. **(a)** 49 **(b)** 64
14. 8 15. 81
16. **(a)** 4 **(b)** 12
17. 18 18. 64 19. 90
20. 336
21. **(a)** 81 **(b)** 32 **(c)** 4096
22. 100 23. 432 24. 7
25. **(a)** 13 **(b)** 25 **(c)** 32 **(d)** 87
   (e) 125

Exercise 2.2

1. 8 2. 9 3. 9
4. 4 5. 1 6. 7
7. 9 8. 2 9. Yes
10. No   11. Yes   12. Yes
13. Yes   14. No   15. Yes
16. Yes   17. Yes   18. Yes
19. No   20. No   21. Yes
22. No   23. No   24. Yes

25. Yes   26. Yes   27. No
28. Yes   29. Yes   30. Yes
31. Yes   32. No   33. Yes
34. Yes   35. Yes   36. No
37. Yes   38. Yes   39. No
40. Yes   41. No   42. Yes
43. No   44. Yes   45. Yes

Exercise 2.3

1. **(a)** Yes   **(b)** No   **(c)** Yes   **(d)** Yes
   (e) No
2. **(a)** No   **(b)** No   **(c)** Yes   **(d)** No
   (e) Yes
3. **(a)** No   **(b)** Yes   **(c)** Yes   **(d)** No
   (e) Yes
4. **(a)** Yes   **(b)** Yes   **(c)** No   **(d)** No
   (e) Yes   **(f)** Yes   **(g)** No   **(h)** No
5. **(a)** Yes   **(b)** Yes   **(c)** Yes   **(d)** No
   (e) Yes   **(f)** No   **(g)** Yes   **(h)** Yes
   (i) No   **(j)** Yes   **(k)** No   **(l)** No
6. 7

Exercise 2.4

1. **(a)** $1, 3, 5, 15$ **(b)** $1, 2, 4, 8, 16, 32, 64$
   (c) $1, 2, 3, 4, 6, 8, 12, 16, 24, 48$
2. $23, 29$
3. **(a)** 24 **(b)** $33, 45, 61, 49$ or 27
   (c) 61 **(d)** 49
4. **(a)** 21 **(b)** 7
5. **(a)** $2^3 \times 3$ **(b)** $2^3 \times 3^2$ **(c)** $3^2 \times 5$
6. **(a)** 30 **(b)** 12 **(c)** 30 **(d)** 60
   (e) 45
7. $1, 2, 4, 5, 10, 20$; $1, 2, 3, 4, 6, 8, 12, 24$;
   $1, 2, 4$; 4
8. $6, 12, 18, 24, 30, 36, 42, 48, 54, 60$;
   $10, 20, 30, 40, 50, 60$; $30, 6$; 30
9. **(a)** 10 **(b)** 3 **(c)** 2
10. **(a)** $2, 3, 4, 6, 12$ **(b)** $6, 12, 18, 24$
11. **(a)** $1, 2, 3, 4, 6, 12$
   (b) $1, 2, 3, 4, 6, 9, 12, 18, 36$
   (c) $1, 2, 3, 4, 5, 6, 10, 12, 15, 20, 30, 60$
   (d) $1, 2, 4, 5, 10, 20, 25, 50, 100$
12. $1, 2, 3, 4, 6, 7, 12, 14, 21, 28, 42, 84$

Miscellaneous Exercise 2

Section A

1. 5 and 11
2. **(a)** $8, 16, 24, 32, 40$ **(b)** $29, 31, 37$
3. 36 4. 25 5. 28
6. 60 7. 36 8. 12

Section B

1. **(a)** $55, 60$ **(b)** $60, 63, 81$
   (c) $60, 122$
2. $2^3 \times 3^2 \times 5$

3. (a) 30   (b) 60
4. (a) 1, 2, 3, 6, 7, 14, 21, 42
   (b) 15, 20, 25
5. (a) 61; yes   (b) 5291; no
6. (a) 57   (b) 72
7. (a) 2, 8, 34   (b) 2, 3, 5, 13
   (c) 123   (d) 89
8. (a) $2^2 \times 5^2$
   (b) 6, 12, 18, 24, 30, 36, 42, 48, 54
   (c) 29, 31, 37

Multi-Choice Questions 2

1. D   2. A   3. A
4. D   5. C   6. A
7. B   8. A   9. B
10. D

Mental Test 2

1. Yes   2. No   3. 64
4. 16   5. No   6. 6
7. 40   8. 1   9. Yes
10. 1, 2, 3, 4, 6, 8, 12, 24
11. 7, 14, 21, 28, 35   12. 2×3^2
13. 10   14. 5   15. 33

Answers to Chapter 3

Exercise 3.1

1. $\frac{1}{6}$ 2. $\frac{3}{4}$ 3. $\frac{3}{8}$
4. $\frac{3}{5}$ 5. $\frac{5}{7}$

Exercise 3.2

6. 15   7. 8   8. 12
9. 12   10. 4   11. 27
12. 2   13. 3   14. 5
15. 3

Exercise 3.3

1. $\frac{1}{2}$ 2. $\frac{3}{5}$ 3. $\frac{3}{5}$
4. $\frac{1}{8}$ 5. $\frac{2}{3}$ 6. $\frac{7}{8}$
7. $\frac{2}{3}$ 8. $\frac{2}{3}$ 9. $\frac{7}{8}$
10. $\frac{5}{8}$

Exercise 3.4

1. $3\frac{1}{2}$ 2. $2\frac{3}{8}$ 3. $2\frac{4}{9}$
4. $1\frac{1}{11}$ 5. $2\frac{5}{8}$ 6. $\frac{19}{8}$
7. $\frac{51}{10}$ 8. $\frac{26}{3}$ 9. $\frac{47}{7}$
10. $\frac{38}{9}$

Exercise 3.5

1. $\frac{1}{2}, \frac{7}{12}, \frac{2}{3}, \frac{5}{6}$ 2. $\frac{3}{5}, \frac{3}{4}, \frac{7}{8}, \frac{9}{10}$
3. $\frac{17}{32}, \frac{9}{16}, \frac{5}{8}, \frac{3}{4}$ 4. $\frac{1}{5}, \frac{1}{4}, \frac{1}{3}, \frac{1}{2}$
5. $\frac{5}{18}, \frac{2}{6}, \frac{3}{8}, \frac{5}{9}$

Exercise 3.6

1. $\frac{5}{6}$ 2. $1\frac{3}{10}$ 3. $1\frac{1}{8}$ 4. $\frac{11}{20}$
5. $2\frac{1}{8}$ 6. $1\frac{5}{24}$ 7. $4\frac{15}{16}$ 8. $14\frac{4}{15}$
9. $13\frac{23}{56}$ 10. $10\frac{2}{3}$ 11. $11\frac{5}{16}$ 12. $10\frac{13}{15}$

Exercise 3.7

1. $\frac{1}{6}$ 2. $\frac{1}{24}$ 3. $\frac{2}{15}$ 4. $\frac{1}{6}$
5. $2\frac{1}{8}$ 6. $2\frac{2}{7}$ 7. $2\frac{19}{80}$ 8. $2\frac{19}{40}$
9. $\frac{39}{40}$ 10. $\frac{11}{12}$

Exercise 3.8

1. $1\frac{3}{8}$ 2. $\frac{7}{20}$ 3. $6\frac{7}{8}$ 4. $\frac{2}{3}$
5. $8\frac{13}{80}$ 6. $8\frac{9}{16}$ 7. $1\frac{3}{8}$ 8. $1\frac{1}{24}$

Exercise 3.9

1. $\frac{8}{15}$ 2. $\frac{15}{28}$ 3. $\frac{10}{27}$ 4. $1\frac{19}{36}$
5. $4\frac{9}{10}$ 6. $5\frac{5}{6}$ 7. $1\frac{32}{45}$ 8. $2\frac{53}{56}$

Exercise 3.10

1. $1\frac{1}{3}$ 2. 4   3. $\frac{7}{16}$ 4. $1\frac{1}{2}$
5. $\frac{1}{24}$ 6. 4   7. $6\frac{3}{4}$ 8. $8\frac{1}{4}$
9. 12   10. 100   11. 3   12. 2

Exercise 3.11

1. $\frac{3}{5}$ 2. 8   3. $1\frac{1}{3}$ 4. $1\frac{1}{2}$
5. $\frac{2}{3}$ 6. $\frac{25}{26}$ 7. $1\frac{1}{5}$ 8. $3\frac{5}{6}$

Exercise 3.12

1. $\frac{5}{9}$ 2. $2\frac{1}{2}$ 3. $\frac{5}{6}$ 4. $\frac{2}{3}$
5. $2\frac{1}{2}$ 6. $\frac{2}{3}$ 7. $1\frac{1}{3}$ 8. $1\frac{1}{2}$

Exercise 3.13

1. £150   2. 4
3. (a) $\frac{13}{20}$ (b) $\frac{7}{20}$
4. $262\frac{1}{2}$ litres   5. 330   6. £1.60
7. 30 min   8. £15 950

Miscellaneous Exercise 3

Section A

1. $\frac{1}{12}$ 2. $\frac{2}{9}$ 3. $2\frac{1}{2}, 2\frac{3}{4}, 3$
4. $4\frac{1}{8}$ 5. $4\frac{33}{40}$ 6. $\frac{4}{7}$
7. $\frac{2}{3}$ 8. 40 9. $1, \frac{2}{3}$
10. $\frac{27}{16}$

Section B

1. (a) $\frac{3}{4}$ (b) $\frac{3}{20}$ (c) $\frac{9}{20}$ (d) $\frac{12}{13}$
2. (a) $\frac{2}{7}$ (b) $\frac{5}{22}$ (c) $\frac{5}{24}$ (d) $4\frac{1}{12}$
3. $\frac{1}{2}$ 4. $3\frac{3}{8}, 3\frac{3}{4}, 4\frac{1}{8}$ 5. 20
6. $\frac{3}{16}$ 7. $\frac{17}{32}$ 8. $1\frac{1}{2}$
9. £2500 10. $5\frac{1}{2}$

Multi-Choice Questions 3

1. B 2. B 3. A
4. A 5. C 6. C
7. D 8. C 9. A
10. A 11. D 12. B

Mental Test 3

1. $\frac{7}{8}$ 2. $\frac{13}{20}$ 3. $\frac{1}{10}$
4. $\frac{7}{8}$ 5. $\frac{1}{10}$ 6. $\frac{2}{7}$
7. $\frac{1}{4}$ 8. 5 9. $\frac{1}{10}$
10. $\frac{1}{2}$ 11. $\frac{4}{5}$ 12. $\frac{37}{7}$

Answers to Chapter 4

Exercise 4.1

1. (a) $\frac{9}{100}$ (b) $\frac{9}{1000}$ (c) $\frac{9}{10}$ (d) 9
2. (a) 500 (b) $\frac{5}{100}$ (c) $\frac{5}{10}$ (d) $\frac{5}{100}$
3. (a) 8000 (b) $\frac{8}{1000}$ (c) $\frac{8}{10}$ (d) $\frac{8}{100}$
4. 0.396
5. 89.1
6. (a) 0.7 (b) 0.37 (c) 0.589
   (d) 0.009 (e) 0.03 (f) 0.017
7. (a) $\frac{2}{10}$ (b) $4\frac{6}{10}$ (c) $3\frac{58}{100}$
   (d) $\frac{256}{1000}$ (e) $\frac{4}{1000}$ (f) $\frac{36}{1000}$
   (g) $80\frac{29}{100}$ (h) $\frac{32}{1000}$

Exercise 4.2

1.

2. (a) 0.8 (b) 4.9 (c) 2.0
   (d) 0.3 (e) 1.2

3. $-3, -9, -29$
4. $-6, -4, 0, 3$
5. $-52.05, -5.052, 5.025. 5.205, 5.502$

Exercise 4.3

1. 3.005 2. 24.176 3. 61.75
4. 55.355 5. 8.98 6. 0.828
7. 5.4109 8. 18.69 9. 60.43
10. -79.8

Exercise 4.4

1. (a) 4.5 (b) 45 (c) 450
2. (a) 78.93 (b) 789.3 (c) 7893
3. (a) 0.58 (b) 5.8 (c) 58
4. (a) 892.346 (b) 8923.46 (c) 89 234.6
5. (a) 82.643 (b) 826.43 (c) 8264.3
6. (a) 0.63 (b) 6.3 (c) 63
7. (a) 0.0028 (b) 0.028 (c) 0.28
8. (a) 0.08 (b) 0.8 (c) 8
9. (a) 28.9 (b) 2.89 (c) 0.289
10. (a) 2.817 (b) 0.2817 (c) 0.028 17
11. (a) 82.734 (b) 8.2734 (c) 0.827 34
12. (a) 0.004 (b) 0.0004 (c) 0.000 04
13. (a) 0.0625 (b) 0.006 25 (c) 0.000 625
14. (a) 0.000 38 (b) 0.000 038 (c) 0.000 003 8
15. (a) 0.652 (b) 0.0652 (c) 0.006 52

Exercise 4.5

1. 743.0266 2. 0.949 43 3. 0.2888
4. 7.411 25 5. 0.013 76 6. 3.5
7. 13.4 8. 1.9 9. 4.25
10. 0.045 11. 48.36 12. 896.98
13. 5.32 14. 0.918 15. 0.1395

Exercise 4.6

1. (a) 19.37 (b) 19.4
2. (a) 0.007 52 (b) 0.008 (c) 0.01
3. (a) 4.970 (b) 4.97
4. 1.33 5. 0.016
6. 189.74 7. 0.1380
8. 0.000 018 9

Exercise 4.7

1. (a) 24.9358 (b) 24.94 (c) 25
2. (a) 0.008 357 (b) 0.008 36 (c) 0.0084
   (d) 0.008
3. (a) 17 359 000 (b) 17 000 000
   (c) 20 000 000
4. (a) 0.0780 (b) 0.078 (c) 0.08
5. 48.9 6. 4
7. 0.5 8. 600
9. 108 070 10. 308.9 (4 s.f.)
11. 127.0 (4 s.f.) 12. 33 (2 s.f.)
13. 47 (2 s.f.) 14. 13 (2 s.f.)

Exercise 4.8

1. $20 + 40 + 430 = 490$; 487.35
2. $80 - 40 - 10 - 30 = 0$; 1.685
3. $20 \times 0.5 = 10$; 12.54
4. $40 \times 0.25 = 10$; 10.751
5. $0.7 \times 0.1 \times 2 = 0.14$; 0.16 (correct to 2 S.F.)
6. $90 \div 30 = 3$; 2.931
7. $0.09 \div 0.03 = 3$; 2.6
8. $30 \times 30 \times 0.030 = 27$; 24 (correct to 2 S.F.)
9. $\dfrac{1.5 \times 0.01}{0.05} = 0.3$; 0.342
10. $\dfrac{30 \times 30}{10 \times 3} = 30$; 29.2

Exercise 4.9

1. 0.75
2. 0.2
3. 0.875
4. 0.8125
5. 1.625
6. 2.593 75
7. 3.234 375
8. 0.555 56
9. 0.888 89
10. 0.177 78
11. 0.454 55
12. 0.353 54
13. 212 12
14. 0.428 43
15. 0.567 16
16. $0.\dot{2}$
17. $0.1\dot{8}$
18. $0.\dot{7}$
19. $0.\dot{4}\dot{5}$
20. $0.\dot{4}$
21. $\frac{3}{10}$
22. $\frac{13}{20}$
23. $\frac{3}{8}$
24. $\frac{7}{16}$
25. $2\frac{31}{50}$
26. $1\frac{3}{4}$
27. $\frac{37}{200}$
28. $7\frac{9}{25}$
29. 0.0175
30. 1.748
31. 0.0895
32. 0.0001
33. 0.001 875

Exercise 4.10

1. 700
2. 700
3. 900
4. 3600
5. 4900
6. 50
7. 250
8. 500
9. 900
10. 17 400
11. 90
12. 380
13. 690
14. 590
15. 600

Exercise 4.11

1. 5, 198
2. -9
3. $1.57, \frac{1}{7}, -5.625, \sqrt{16}$, 6.76 and $-3\frac{1}{2}$ are all rational; $\sqrt{15}$ is irrational.
4. 0, 6
5. $-4.7, -3\frac{3}{4}, -3.4, 0, \frac{5}{8}, 2.5, 3, 7$

Exercise 4.12

1. 1.31
2. 2.93
3. 1.96
4. 3.55
5. 8.73
6. 9.51
7. 11.31
8. 18.49

Miscellaneous Exercise 4

Section A

1. 0.63
2. 0.082
3. 9
4. $0.8\dot{1}$
5. 0.87
6. 49.95
7. (a) 738.06   (b) 738.1
8. (a) 0.9375   (b) 9.16
9. 0.02
10. 56.798

Section B

1. 2.52
2. (a) 0.0495   (b) 0.049
3. (a) 15.33   (b) 16.3
4. 62
5. 0.167
6. 0.027
7. (a) $\frac{1}{3}$ (b) 0.333
8. (a) 87 400   (b) 0.0326
9. -0.68
10. $1.6, \frac{2}{3}, 0.3, -0.7, -4.2$
11. 277.155
12. (a) 0.4375   (b) 0.05
13. (a) $\frac{19}{20}$ (b) $\frac{8}{25}$
14. 9.291

Multi-Choice Questions 4

1. B
2. D
3. D
4. B
5. D
6. B
7. C
8. D
9. C
10. D

Mental Test 4

1. 2.55
2. 0.262
3. 1.12
4. 16
5. 0.06
6. $\frac{9}{100}$
7. $\frac{2}{5}$
8. 0.53
9. 4.3
10. 7.8
11. 357
12. 0.5321
13. 2.47
14. 54 000
15. 17.5
16. 0.8
17. 0.178
18. 4.75
19. 3600
20. Yes

Answers to Chapter 5

Exercise 5.1

1. 50
2. 48
3. 9
4. 2000
5. 8300
6. 28 100
7. 700
8. 1940
9. 800
10. 4800
11. 80
12. 280
13. 32.5
14. 80
15. 896.4
16. 0.54
17. 8
18. 25.634
19. 0.0543
20. 3
21. 378.64
22. 0.0935
23. 18 000
24. 6200
25. 9
26. 0.46
27. 0.058
28. 0.018
29. 0.0075
30. 0.470

Exercise 5.2

1. 4
2. 2.5
3. 3.25
4. 6.2
5. 24
6. 27.6
7. 2439
8. 118.2
9. 6
10. 4.4
11. 16.3
12. 56.15
13. 12
14. 2.9
15. 22.5
16. 403
17. 96
18. 276
19. 189.6
20. 382.8
21. 252

22. 183.6   23. 244.8   24. 1951.2
25. 3   26. 6.4   27. 5
28. 9.3

Exercise 5.3

1. 5   2. 6.98   3. 8000
4. 11 200   5. 9000   6. 39 000
7. 750   8. 45   9. 1.83
10. 0.085   11. 7000   12. 4900
13. 300   14. 9   15. 50
16. 8.57   17. 0.453   18. 0.045

Exercise 5.4

1. 4   2. 80   3. 72
4. 132.8   5. 7   6. 0.25
7. 0.75   8. 3.25   9. 448
10. 582.4   11. 13 440   12. 16 352
13. 4   14. 3.9   15. 160
16. 248   17. 6   18. 0.6
19. 4   20. 9.2

Exercise 5.5

1. 600   2. 430   3. 25
4. 50   5. 800   6. 2000
7. 300   8. 5   9. 500
10. 30   11. 5.5   12. 0.4
13. 0.08   14. 3.25   15. 0.09

Exercise 5.6

1. 3   2. 1.25   3. 2.25
4. 72   5. 50   6. 80
7. 26   8. 9   9. 320
10. 48   11. 3   12. 0.65
13. 95   14. 2   15. 25

Exercise 5.7

1. 144 lb   2. 6 fl oz   3. 20 miles
4. 40 km   5. 4.5 lb   6. 2.27 kg
7. 4.92 tons 8. 28 litres

Exercise 5.8

1. 6.315 m   2. 93 462.05 m
3. 0.2668 m   4. 2.072 kg
5. 98.05 kg   6. 68.7 cm
7. 628 mm   8. 5 cm

Exercise 5.9

1. 120.65 m   2. 17.85 m
3. 47 ; 9 cm   4. 83
5. 32 000   6. 468 complete jars
7. 48   8. 0.041

Exercise 5.10

1. £15.97   2. £2.37   3. £100.64
4. £7.13   5. £2.48   6. £190.89
7. £701.69   8. £0.78

Exercise 5.11

1. £1.44   2. £5.22   3. £217.58
4. £9.80   5. 52p   6. £2.49
7. £4.59   8. £5.72

Miscellaneous Exercise 5

Section A

1. £109.92   2. £59.85   3. £1.32
4. (a) 12 340   (b) 1.234   (c) 1.234
5. 14   6. 6.75 lb   7. 870 g
8. 40   9. 100 miles

Section B

1. (a) 2000   (b) 2056   (c) 6.72
   (d) 28.4
2. 60
3. (a) 3000   (b) 900   (c) 3.5
4. 13 ; 25 cm
5. $\frac{9}{20}$ 6. 0.8 kg   7. 17.3 kg
8. £1203.75   9. £90.10   10. £3.68

Multi-Choice Questions 5

1. B   2. B   3. C
4. C   5. D   6. D
7. B   8. C   9. C
10. C

Mental Test 5

1. 400   2. 5000   3. 2
4. 70   5. 6   6. 9200
7. 3   8. 2   9. 5280
10. 4   11. 8000   12. 7.5
13. 9200   14. 320   15. 336
16. 4   17. 100   18. 400
19. 48   20. 10   21. 3
22. 13   23. 12   24. 1950
25. 200   26. £3.75   27. £1.23
28. £5.70   29. 50p

Answers to Chapter 6

Exercise 6.1

1. (a) 800 g butter, 640 g castor sugar, 2 eggs,
       1600 g self-raising flour, 400 g currants,
       120 g mixed peel, 2 pinches of salt.
   (b) 960 g   (c) 240 g   (d) 3200 g
2. 160 kg, 320 kg
3. 40 kg

4. (a) 15 kg   (b) 20 kg
5. (a) 240 g flour, 2 eggs, 500 ml milk
   (b) 3   (c) 480 g

Exercise 6.2

1. 1:4	2. 1:3	3. 6:7
4. 5:6	5. 8:7	6. 5:4
7. 3:2	8. 5:6:7	9. 5:6:9
10. 4:5:6:7		

Exercise 6.3

1. $\frac{9}{7}$	2. $\frac{1}{2}$	3. $\frac{2}{1}$
4. $\frac{4}{5}$	5. $\frac{2}{3}$	6. $\frac{4}{5}$
7. $\frac{1}{3}$	8. $\frac{5}{6}$	9. $\frac{7}{8}$
10. $\frac{5}{9}$		

Exercise 6.4

1. 9:4	2. 4:3	3. 4:3
4. 2:5	5. 9:1	6. 7:3
7. 1:2	8. 11:6	9. 16:39
10. 48:103		

Exercise 6.5

1. 4:1	2. 1:2	3. 3:100
4. 3:1	5. 20:1	6. 1:10
7. 5:1	8. 10:3	9. 32:1
10. 100:3		

Exercise 6.6

1. £500, £300   2. 112 kg, 48 kg
3. 24 m, 36 m, 60 m
4. 168 mm, 588 mm, 924 mm
5. £280
6. 15 kg, 22.5 kg; 37.5 kg
7. £88   8. £3.60

Exercise 6.7

1. £8.48	2. £48.125	3. £1.75
4. 36 litres	5. £224	6. 7 hours
7. 90	8. 4 kg	

Exercise 6.8

1. 16 days	2. $13\frac{1}{2}$ days	3. 6
4. 6	5. $4\frac{1}{2}$ days	6. $10\frac{1}{2}$ min
7. 24		

Exercise 6.9

1. 44 km	2. 20 min	3. 80 min
4. (a) 40 miles	(b) 200 miles	
5. £1.32	6. 10 hours	7. 40 min
8. (a) £1	(b) £8	

Exercise 6.10

1. (a) 6 p per kg   (b) 72 p
2. (a) 12 litres per min   (b) 3.75 min
   (c) 84 litres
3. (a) 30 miles per gallon   (b) 150 miles
   (c) 4 gallons
4. (a) 150 km   (b) 3.5 hours
5. 13 500 kg
6. 7
7. (a) 2 m/s   (b) 175 s

Exercise 6.11

1. 57.75	2. 5500	3. 10.16
4. 177.17	5. 7658.38	6. 3.66
7. 66.80	8. 11.18	

Exercise 6.12

1. 8 m   2. 1600 m
3. (a) 2 m   (b) 6 cm
4. 25 cm
5. (a) 2.16 m   (b) 0.36 m
6. (a) 3 km   (b) 1.5 cm
7. 405 km   8. 5.76 in

Miscellaneous Exercise 6

Section A

1. £1.08	2. 31.5 km	3. 4.5 km
4. (a) 15	(b) £43.50	
5. 7:8	6. 4:1	7. $\frac{1}{8}$
8. Yes		

Section B

1. 30	2. 27.3 m	3. £178
4. 9	5. £3.48	6. 24
7. 187.50	8. 4 hours	9. 8
10. 46p	11. 56.25	12. £34
13. 15	14. 36 ℓ	15. £31.36

Multi-Choice Questions 6

1. D	2. B	3. C
4. C	5. B	6. C
7. C	8. A	9. B
10. A		

Mental Test 6

1. 10 kg	2. 3:5	3. $\frac{3}{4}$
4. 4:1	5. £80, £120	6. 18 p
7. 1 day	8. 5	9. 500
10. 4 cm		

Answers to Chapter 7

Exercise 7.1

1. 70%	**2.** 80%	**3.** 55%
4. 36%	**5.** 62%	**6.** 25%
7. 90%	**8.** 20%	**9.** 20%
10. 34%	**11.** 73%	**12.** 68%
13. 81.3%	**14.** 92.7%	**15.** 33.3%

Exercise 7.2

1. $\frac{8}{25}$	**2.** $\frac{6}{25}$	**3.** $\frac{3}{10}$
4. $\frac{9}{20}$	**5.** $\frac{3}{50}$	**6.** $\frac{3}{8}$
7. $\frac{2}{3}$	**8.** $\frac{29}{400}$	**9.** $\frac{1}{12}$
10. $\frac{5}{8}$	**11.** 0.78	**12.** 0.31
13. 0.482	**14.** 0.025	**15.** 0.0125
16. 0.0395	**17.** 0.201	**18.** 0.0196

	Fraction	Decimal	Percentage
19.	$\frac{11}{20}$	0.55	55
20.	$\frac{17}{50}$	0.34	34
21.	$\frac{7}{8}$	0.875	$87\frac{1}{2}$
22.	$\frac{19}{25}$	0.76	76
23.	$\frac{2}{25}$	0.08	8
24.	$\frac{3}{8}$	0.375	$37\frac{1}{2}$
25.	$\frac{3}{20}$	0.15	15
26.	$\frac{27}{100}$	0.27	27
27.	$\frac{9}{20}$	0.45	45
28.	$\frac{21}{400}$	0.0525	$5\frac{1}{4}$

Exercise 7.3

1. (a) 10	**(b)** 24	**(c)** 6
(d) 2.4	**(e)** 21.315	**(f)** 2.516
2. (a) $12\frac{1}{2}$	**(b)** 20	**(c)** 16
(d) 16.3	**(e)** 45.5	
3. (a) 60%	**(b)** 27	
4. 115 cm		
5. 88.667 cm		
6. (a) £7.20	**(b)** £13.20	**(c)** £187.50
7. 584 kg		
8. 39 643		

Exercise 7.4

1. £800	**2.** £20	**3.** £1000
4. 800	**5.** £1600	**6.** 2000 kg
7. £8000		

Miscellaneous Exercise 7

Section A

1. (a) $\frac{12}{25}$	**(b)** $\frac{3}{50}$
2. (a) 0.64	**(b)** 0.07

3. $a = \frac{17}{20}$; $b = 85$; $c = 0.35$; $d = 35$; $e = \frac{3}{25}$; $f = 0.12$

4. (a) £1.80	**(b)** £26.60
5. (a) £12	**(b)** £68
6. (a) 12	**(b)** 60%

Section B

1. 88%	**2.** £13	**3.** £65.25
4. £144	**5.** 1100	**6.** 28%
7. £12	**8.** £6.30	**9.** 55%; 54
10. (a) £1.20	**(b)** £120	**(c)** £110.40

Multi-Choice Questions 7

1. C	**2.** B	**3.** D
4. B	**5.** A	**6.** C
7. C	**8.** D	**9.** B

Mental Test 7

1. 30%	**2.** 40%	**3.** $\frac{7}{10}$
4. 0.24	**5.** 5	**6.** 16%
7. £20	**8.** 40 cm	**9.** 14
10. $\frac{2}{3}$		

Answers to Chapter 8

Exercise 8.1

1. £105	**2.** £5	
3. (a) £70	**(b)** £87.50	**(c)** £140
4. £3.50	**5.** £102.20	

Exercise 8.2

1. (a) £6.25	**(b)** £7.50	**(c)** £10
2. (a) £4.50	**(b)** £6.00	**(c)** £10.50
3. £15.84	**4.** £138.60	
5. (a) £7.52	**(b)** £52.64	**(c)** £184.24

Exercise 8.3

1. £19.00	**2.** £18.40	**3.** £28.50
4. £19.00	**5.** £18.50	

Exercise 8.4

1. (a) £2.00	**(b)** £6.00	**(c)** £40.00
(d) £100.00		
2. £70.00	**3.** £240.00	**4.** £97.00
5. £168.00		

Exercise 8.5

1. £443.33	**2.** £500.00	**3.** £700.00
4. £508.00	**5.** £620.00	

Exercise 8.6

1. £765
2. £350
3. £66.28
4. £720
5. £800
6. (a) £4975   (b) £414.58
7. (a) £5795   (b) £1738.50
8. £7825; £2347.50
9. (a) £7260   (b) £24 200
10. £4.62

Miscellaneous Exercise 8

Section A

1. £116
2. £750
3. £4750
4. £5
5. £3.75
6. £240
7. £90
8. 12%

Section B

1. (a) £4000   (b) £1200
2. (a) £6.00   (b) £9.00
   (c) £303.00
3. £1600
4. £19.00
5. (a) £3020   (b) £4980   (c) £415
6. £7770
7. (a) £140   (b) £5.39
   (c) £116.31
8. £138

Multi-Choice Questions 8

1. D
2. D
3. A
4. B
5. B
6. C
7. B
8. D

Mental Test 8

1. £3
2. £160
3. £3
4. £15
5. £10
6. £60
7. £450
8. £25
9. £155
10. £500

Answers to Chapter 9

Exercise 9.1

1. £25
2. £110
3. £600
4. £1600
5. £82.50
6. £180
7. £480
8. £36
9. £119
10. £945

Exercise 9.2

1. £280
2. £600
3. £200
4. £90
5. £768
6. 4 years
7. 3 years
8. 4 years
9. 7%
10. 12%
11. 10%
12. £4000
13. £200
14. £1200
15. £850
16. 8%
17. 4 years

Exercise 9.3

1. £30.75
2. £331
3. £161.25
4. £1036.02
5. £2508.80
6. £26 450
7. £1983.75
8. £10 400; £10 077.70; £322.30;
   10% p.a. simple interest
9. £2742
10. £6395
11. £750.50
12. £328.30
13. £3432
14. £8338.50

Exercise 9.4

1. £1166.40
2. £26 622.32
3. £5242.88
4. £16 875

Miscellaneous Exercise 9

Section A

1. £80
2. £20
3. (a) £24   (b) £48
4. (a) £20   (b) £42   (c) £242
5. £500

Section B

1. £80
2. £5832
3. (a) £4318   (b) £4283.02
   (c) savings bond, £34.98
4. £2178
5. £5000
6. £3973.75

Multi-Choice Questions 9

1. C
2. B
3. C
4. A
5. D
6. B
7. C
8. C

Mental Test 9

1. £10
2. £20
3. £1200
4. £146.90
5. £900

Answers to Chapter 10

Exercise 10.1

1. 25%
2. 20%
3. 50%
4. 75%
5. 20%
6. $12\frac{1}{2}\%$
7. (a) £42   (b) 25%
8. (a) £10   (b) $33\frac{1}{3}\%$

Exercise 10.2

1. £36
2. £280
3. £100
4. £15 000
5. £200
6. £17.28
7. 9p

Exercise 10.3

1. £180
2. £56
3. £77
4. £8.50
5. £0.91

Exercise 10.4

1. £276   2. £46.96   3. £184
4. £46   5. £60   6. £200
7. £20.87

Exercise 10.5

1. £216   2. £120
3. £0.90 in the £   4. 25p in the £
5. £4 644 000   6. 9.1p in the £
7. £85 000
8. (a) £87 960   (b) 55p in the £
9. £39 100   10. 2.9p in the £

Exercise 10.6

1. £37.50   2. £50.00
3. £96   4. £32
5. £5.25   6. £60.00
7. £9.54
8. (a) £6000   (b) £5241
9. £600
10. £973.38; 8.1p per km
11. 5.98p per km

Miscellaneous Exercise 10

Section A

1. (a) £80   (b) $66\frac{2}{3}$%
2. (a) £150   (b) 50%
3. £60
4. (a) £7.50   (b) £57.50
5. £500 per annum
6. £90   7. £84   8. £27 000

Section B

1. £260.40   2. £48   3. £13.50
4. £4.05   5. £17.50   6. £2700
7. (a) £835 000   (b) £7 181 000
8. £120   9. £64.84   10. £1.88

Multi-Choice Questions 10

1. B   2. D   3. B
4. C   5. C   6. D
7. D   8. C   9. B
10. B

Mental Test 10

1. £2   2. £500   3. 20%
4. 20%   5. £160   6. £7.50
7. £450   8. £5000   9. £10
10. £160

Answers to Chapter 11

Exercise 11.1

1. £25 200   2. £47.77   3. £43.24
4. £292.80   5. £260.53   6. £267.50
7. £16.67

Exercise 11.2

1. £120
2. (a) £120   (b) £12   (c) £132
   (d) £13.20
3. £24.75
4. (a) £128   (b) £512.00   (c) £61.44
   (d) £573.44   (e) £143.36
5. (a) £260   (b) £60

Exercise 11.3

1. £60   2. £144
3. (a) £93.15   (b) £11 178
4. £28.01   5. 20%

Exercise 11.4

1.

Charge per therm	Number of therms used	Cost of gas used	
		In pence	In pounds
40 p	100	4000	40.00
42 p	80	3360	33.60
48 p	135	6480	64.80
35.3 p	240	8472	84.72
37.5 p	400	15 000	150.00

2. £43.20
3. (a) 323   (b) 323
   (c) £134.72
4.
(a)

(b)

(c)

5. (a) 131 therms (b) £48.99 (c) £56.99
6. (a) 2616 (b) 3712 (c) 5264
7. (a) 4796 (b) 404
   (c) £162.63

Exercise 11.5

1.

Charge per unit	Number of units used	Cost of electricity used	
		In pence	In pounds
4.85	100	485	4.85
5.17	600	3102	31.02
6.30	450	2835	28.35
7.98	800	6384	63.84
9.80	750	7350	73.50

2. £76.44   **3.** £152.75   **4.** 116; £15.58
5. £97.20

6.

(a)

(b)

(c)

7. (a) 23 968   (b) 84 947   (c) 46 936
8. 1200

Exercise 11.6

1. £19.07   **2.** £24.79   **3.** £24.25
4. (a) £28.94 (b) £45.09 (c) £51.85
5. (a) 1252 (b) £61.35 (c) £78.64
   (d) £11.80 (e) £90.44

Miscellaneous Exercise 11

Section A

1. £6
2. (a) £45 (b) £345 (c) £39
   (d) £315
3. (a) 24 612 (b) 25 372

(c)

   (d) £51.82

4. Quarterly rental charge   £9.75
   Dialed units 600 at 4.7 p per unit   £28.20
   Total exclusive of VAT   £37.95
   VAT at 15%   £5.69
   Total payable   £43.64
5. (a) 1320 (b) 1320 (c) 307
   (d) £116.39

Section B

1. (a) £375 (b) £2875
   (c) £239.58
2. (a) £2340 (b) £53.10
   (c) £248.10
3. (a) £27 000 (b) £313.20
4. (a) £230.40 (b) £49.60
5. (a) £175 (b) £836.50
   (c) £136.50

Multi-Choice Questions 11

1. B   **2.** B   **3.** C
4. C   **5.** D   **6.** C
7. D   **8.** C

Mental Test 11

1. £40   **2.** £200   **3.** £720
4. £650 **5.** £60   **6.** £30
7. £100

Answers to Chapter 12

Exercise 12.1

1. (a) 240 min (b) 210 min
2. (a) 540 s (b) 135 s
3. 5 hours 50 min   **4.** 18 hours 7 min
5. 17 hours 15 min   **6.** 12 hours 39 min
7. 11 hours 47 min   **8.** 12 hours 6 min
9. 12 hours 39 min   **10.** 12 hours 47 min

Exercise 12.2

1. 0730; 15 min
2. 0937; 2 hours 12 min
3. 0835; 3 min
4. 1518; 1607; 2 hours 2 min
5. 1040 from Paddington; 29 min
   1202 from Stroud

Exercise 12.3

1. 2 hours	**2.** 4 hours
3. 4 s	**4.** 60 miles
5. 140 km	**6.** 480 m
7. 15 mile/h	**8.** 40 km/h
9. 16 m/s	**10.** 300 km
11. 9 km/h	**12.** 35.2 mile/h
13. 26 km/h	**14.** 75 km/h
15. 80 km/h	

Miscellaneous Exercise 12

Section A

1. (a) 2015　　(b) 30 min
2. 1815 hours
3. (a) 2 hours 30 min　　(b) 50 km/h
4. (a) 1 hour 30 min　　(b) 8 mile/h
5. (a) 2.35 p.m.　(b) 30 mile/h

Section B

1. (a) 4 hours　　(b) 64 km/h
2. (a) 1210　　(b) 40 mile/h
3. (a) 36 min　　(b) 2 hours 24 min
　　(c) 3 hours　　(d) 39 km/h
4. (a) 360 km　　(b) 4 hours 30 min
5. (a) 8 m/s　　(b) 28.8 km/h

Multi-Choice Questions 12

1. D	**2.** B	**3.** B
4. B	**5.** D	**6.** B
7. C		

Mental Test 12

1. 4 min	**2.** 240 s	**3.** 3 h
4. 300 min	**5.** 5 days	**6.** 2 h 15 min
7. 3 h 20 min	**8.** 3 h 8 min	**9.** 40 km/h
10. 50 mile/h	**11.** 150 km	**12.** 2 h
13. 1500 miles	**14.** 48 km/h	**15.** 3 h
16. 29 min	**17.** 1222	**18.** 426
19. 27 min	**20.** Yes	

Answers to Chapter 13

Exercise 13.1

1. $7a$	**2.** $5b - 7$	**3.** $5x + 3y$
4. xyz	**5.** $8mn$	**6.** $\dfrac{2p}{3q}$
7. $9r + 3s$	**8.** $\dfrac{xy}{p}$	

Exercise 13.2

1. 21	**2.** -15	**3.** 18
4. -2	**5.** 13	**6.** 1
7. 3	**8.** 2	**9.** 9

10. 0	**11.** 19	**12.** 15
13. -21	**14.** -42	**15.** -126
16. 90	**17.** 4	**18.** 2
19. 49	**20.** -71	**21.** $-\frac{7}{3}$
22. $-\frac{6}{7}$		

Exercise 13.3

1. 8	**2.** 81	**3.** 25
4. 162	**5.** 48	**6.** -48
7. 200	**8.** -750	**9.** 360
10. 13	**11.** -16	**12.** 43

Exercise 13.4

1. $8x$	**2.** $4y$	**3.** $-3p$
4. $-3q$	**5.** $-2x$	**6.** $10p$
7. $-6x$	**8.** $-2m$	
9. $13x - 9y + 15c$		**10.** $5a - 2b - 5c$

Exercise 13.5

1. $6py$	**2.** $-6mp$	**3.** $-20ab$
4. $30abc$	**5.** $12xyz$	**6.** $-14mnp$
7. $-6abc$	**8.** $-120qyz$	**9.** $-6m^2$
10. $-24p^2q^2$		

Exercise 13.6

1. $3a + 6b$	**2.** $10x - 15y$	**3.** $20p - 15q$
4. $-p + q$	**5.** $-6x + 8y$	**6.** $-20m + 15n$
7. $13a - 21b$	**8.** $-4p - 7q$	**9.** $25n$
10. $7x - 9y$		

Exercise 13.7

1. $2x$	**2.** $-\dfrac{a}{b}$	**3.** $\dfrac{x}{y}$
4. $\dfrac{2a}{b}$	**5.** $2b$	**6.** $3xy$
7. $-2ab$	**8.** $2ab$	**9.** $7ab$
10. $2xyz^2$		

Exercise 13.8

1. $3(x + 2)$	**2.** $4(y - 2)$	**3.** $2(p - 2)$
4. $5(m + 2)$	**5.** $2(x + y)$	**6.** $3(p - q)$
7. $5(x + 3y)$	**8.** $2(2x - 3y)$	**9.** $3(3x - 2y)$
10. $3(p + 3q)$		

Exercise 13.9

1. $\dfrac{2}{ab}$	**2.** $\dfrac{3y}{2x}$	**3.** $\dfrac{4qs}{r}$
4. $\dfrac{6a^2d^3}{bc}$	**5.** $\dfrac{b^2c}{a}$	**6.** $\dfrac{px}{2qy}$
7. $\dfrac{by}{ax}$	**8.** $\dfrac{pq}{mn}$	

Exercise 13.10

1. $\dfrac{9a}{20}$ 2. $\dfrac{3p - 2q}{6}$ 3. $\dfrac{11a}{20}$

4. $\dfrac{13x}{12}$ 5. $\dfrac{13}{6x}$ 6. $\dfrac{7m}{6x}$

7. $\dfrac{14x}{15y}$ 8. $\dfrac{x + 5}{3}$ 9. $\dfrac{10x - 3}{12}$

10. $\dfrac{9m + 2}{2}$ 11. $\dfrac{-9 - x}{12}$ 12. $\dfrac{9}{4x}$

13. $\dfrac{x + 6}{6}$ 14. $\dfrac{a + 12b}{6}$ 15. $\dfrac{25 - 7x}{10}$

Miscellaneous Exercise 13

Section A

1. 26   2. $\frac{1}{4}$ 3. 74
4. 30   5. $54°F$ 6. 270
7. 2

Section B

1. (a) $6a$ (b) $16a^2$ (c) 4
2. (a) $5x + 7y$ (b) $2x + 5y$ (c) $-2y$
3. 2
4. (a) $30a^2b$ (b) $5x - 6y$
5. (a) 0   (b) 0   (c) $\frac{3}{4}$
6. (a) 4   (b) -11

7. $\dfrac{x}{2}$ 8. $\dfrac{x + 8}{6}$ 9. $\dfrac{3ab^2c^2d}{2}$

10. $\dfrac{21b^2}{10ac}$

Multi-Choice Questions 13

1. C   2. D   3. D
4. D   5. A   6. D
7. D   8. B

Mental Test 13

1. $11x$ 2. $4a$ 3. $3b$
4. $4p$ 5. $6ab$ 6. p^2
7. $12m^2$ 8. $5a$ 9. $4a$
10. $4p$ 11. $3p + 3q$ 12. $-x + y$

13. $-2m - 2n$ 14. $\dfrac{ap}{bq}$ 15. $-2x$

Answers to Chapter 14

Exercise 14.1

1. 5^6 2. a^7 3. p^9
4. y^{23} 5. 2^8 6. $24a^6$
7. $15y^7$ 8. $18p^7$

Exercise 14.2

1. 3^2 2. p^2 3. d
4. q^4 5. m 6. t^5
7. m 8. $4p^3$

Exercise 14.3

1. y^8 2. 7^6 3. 5^3c^6
4. $5^5a^5b^{10}c^{15}$ 5. $\dfrac{3^4a^8}{2^4b^{12}}$ 6. m^{15}

7. 3^4b^4 8. $2^3x^6y^3$ 9. $\dfrac{3^4}{4^4}$

10. $\dfrac{1}{2^5}$

Exercise 14.4

1. 5^{-1} 2. 3^{-2} 3. a^{-3}
4. b^{-5} 5. $2x^{-3}$ 6. $5x^{-1}$
7. m^{-6} 8. $8y^{-7}$ 9. $\frac{1}{3}$

10. $\dfrac{1}{5^2}$ 11. $\dfrac{1}{a^3}$ 12. $\dfrac{1}{m^5}$

13. $\dfrac{3}{b}$ 14. $\dfrac{7}{q^3}$ 15. $\frac{1}{10}$

16. $\frac{1}{9}$ 17. $\frac{1}{16}$ 18. $\frac{1}{25}$
19. $\frac{1}{27}$ 20. $\frac{1}{1000}$ 21. 1
22. 3   23. 1   24. 1

Exercise 14.5

	10^6	10^5	10^4	10^3	10^2	10^1	10^0
1.					9	3	6
2.					7	2	5	4
3.					8	6	1	9
4.				1	8	3	4	2
5.		2	7	8	9	4	8
6.		5	6	3	7	4	0
7.	8	6	2	1	9	7	5
8.					7	3	0	0
9.			8	3	5	0	7
10.		9	0	0	0	0	7

11. $7 \times 10^2 + 5 \times 10^1 + 3 \times 10^0$
12. $3 \times 10^5 + 8 \times 10^4 + 4 \times 10^3 + 6 \times 10^2$
     $+ 8 \times 10^1 + 5 \times 10^0$

13. $1 \times 10^5 + 7 \times 10^4 + 3 \times 10^3 + 2 \times 10^2 + 7 \times 10^1 + 8 \times 10^0$
14. $2 \times 10^5 + 3 \times 10^4 + 5 \times 10^0$
15. $3 \times 10^6 + 9 \times 10^5 + 7 \times 10^3 + 6 \times 10^2$

Exercise 14.6

	10^4	10^3	10^2	10^1	10^0	10^{-1}	10^{-2}	10^{-3}	10^{-4}
1.					5	2			
2.					3	7	5		
3.					9	1	3	4	
4.	5	6	3	2	7	5	8		
5.				9	6	3			
6.			1	8	7	2	5		
7.		3	0	5	9	2	7	3	
8.					0	6	3	5	
9.		7	3	0	0	2	7	6	3
10.	8	3	5	0	7	0	0	3	8

11. (a) $5 \times 10^{-1} + 2 \times 10^{-2}$
(b) $3 \times 10^{-1} + 7 \times 10^{-2} + 5 \times 10^{-3}$
(c) $9 \times 10^{-1} + 1 \times 10^{-2} + 3 \times 10^{-3} + 4 \times 10^{-4}$
(d) $5 \times 10^4 + 6 \times 10^3 + 3 \times 10^2 + 2 \times 10^1 + 7 \times 10^0 + 5 \times 10^{-1} + 8 \times 10^{-2}$
(e) $9 \times 10^1 + 6 \times 10^0 + 3 \times 10^{-1}$
(f) $1 \times 10^2 + 8 \times 10^1 + 7 \times 10^0 + 2 \times 10^{-1} + 5 \times 10^{-2}$
(g) $3 \times 10^3 + 5 \times 10^1 + 9 \times 10^0 + 2 \times 10^{-1} + 7 \times 10^{-2} + 3 \times 10^{-3}$
(h) $6 \times 10^{-2} + 3 \times 10^{-3} + 5 \times 10^{-4}$
(i) $7 \times 10^3 + 3 \times 10^2 + 2 \times 10^{-1} + 7 \times 10^{-2} + 6 \times 10^{-3} + 3 \times 10^{-4}$
(j) $8 \times 10^4 + 3 \times 10^3 + 5 \times 10^2 + 7 \times 10^0 + 3 \times 10^{-3} + 8 \times 10^{-4}$

12. 2753
13. 84 605
14. 800.073
15. 9 076 083.26
16. 62.754

Exercise 14.7

1. 8.27×10^2
2. 1.734×10^3
3. 1.7632×10^4
4. $8.937 62 \times 10^5$
5. 8.036×10^6
6. 3200
7. 500
8. 1 870 000
9. 43 200
10. 876 200

Exercise 14.8

1. 3×10^{-1}
2. 5×10^{-2}
3. 5.6×10^{-3}
4. 7×10^{-6}
5. 6.7×10^{-5}
6. 0.002
7. 0.567
8. 0.032
9. 0.000 005
10. 0.000 41

Exercise 14.9

1. 2.1×10^3 by 1520
2. 3.95×10^4 by 30 100
3. 8.58×10^4 by 75 930
4. 2.1×10^{-2} by 0.0156
5. 1.26×10^{-1} by 0.11727
6. 1.2489×10^4
7. 1.858×10^{-1}
8. 1.04×10^1
9. 4×10^{-5}
10. 2.758×10^3
11. 8.3228×10^4
12. 8.1368×10^{-2}
13. 8.322×10^{-2}
14. 4.55×10^6
15. 1.5×10^{-2}
16. 4.22×10^{10}
17. 3.4×10^{-2}
18. 5.93×10^0

Miscellaneous Exercise 14

Section B

1. 1.2×10^6
2. 3.397×10^3
3. 0.005 01
4. 8.93×10^{-4}
5. (a) 3.2×10^3 (b) 6×10^5 (c) 2.8×10^3 (d) 1.5×10^1
6. (a) $\frac{1}{9}$ (b) 1 (c) $\frac{1}{10}$
7. 1.2×10^3 by 327
8. 10^{15}
9. $x = 4$
10. $2^6 = 64$
11. 0.08
12. (a) $n = 4$ (b) $n = 1$ (c) $n = 4$
13. 3×10^{-13}
14. 0.000 067
15. 3.57×10^{-2}
16. (a) $\frac{1}{8}$ (b) 1

Multi-Choice Questions 14

1. D
2. C
3. B
4. C
5. C
6. D
7. D
8. A
9. C
10. D

Mental Test 14

1. 8
2. $\frac{1}{3}$
3. $9a^4$
4. a^6
5. a^2
6. a^{10}
7. $\frac{a^6}{b^3}$
8. 32
9. 8×10^3
10. 3×10^{-3}

Answers to Chapter 15

Exercise 15.1

1. 2
2. 2
3. 2
4. 6
5. 6
6. 4
7. 4
8. 3
9. 12
10. 9
11. 10
12. 12
13. 8
14. 15
15. 21
16. 24
17. 10

Exercise 15.2

1. 8
2. 4
3. 2
4. -1
5. -2
6. 5
7. 6
8. 6
9. 2
10. 9
11. 2
12. -4

Exercise 15.3

1. 15
2. 21
3. 7
4. 12
5. 15
6. $\frac{6}{5}$
7. $\frac{15}{4}$
8. -15
9. $\frac{50}{47}$
10. $\frac{15}{28}$
11. $\frac{1}{2}$
12. 13

Exercise 15.4

1. 32
2. 80
3. 360
4. 144
5. 704
6. 21
7. 6

Exercise 15.5

1. 0.5
2. 5
3. 3
4. 4
5. 100
6. 8.59
7. 2.55
8. 6

Exercise 15.6

1. $V = \dfrac{c}{P}$
2. $R = \dfrac{I}{PT}$
3. $h = \dfrac{E}{mg}$
4. $a = xy$
5. $R = \dfrac{E}{I}$
6. $T = \dfrac{PV}{R}$
7. $t = \dfrac{ST}{s}$
8. $p = v - 3$
9. $q = 3 - p$
10. $a = \dfrac{v - u}{t}$
11. $r = \dfrac{n - p}{3}$
12. $x = \dfrac{b - a}{c}$
13. $m = \dfrac{y - c}{x}$
14. $D = TL + d$
15. $L = \dfrac{S(C - F)}{P}$

Exercise 15.7

1. $2 < 5$
2. $\frac{1}{2} > \frac{1}{3}$
3. $10\,\text{mm} = 1\,\text{cm}$
4. $£1 > 80\,\text{p}$
5. $9 > 1$
6. $\frac{2}{3} > \frac{4}{9}$
7. $-3 > -7$
8. $7 < 9$
9. $x^2 > x^2 - 3$
10. $2 < 4$

Exercise 15.9

1. $5, 6, 7, 8$
2. $4, 5, 6, 7, 8, 9$
3. $-2, -1, 0, 1, 2$
4. $5, 6, 7, 8$
5. $-3, -2, -1, 0, 1, 2, 3, 4$

Miscellaneous Exercise 15

Section A

1. 70
2. 28
3. 62.8
4. 78.5
5. 628
6. 11
7. (a) $5^2 - 3^2 = 4^2$
   (b) $9^2 - 3^2 > 3^2$
   (c) $\frac{5}{9} - \frac{1}{3} < \frac{2}{3}$
   (d) $\sqrt{14.4} > 1.2$
   (e) $0.5^2 > 0.5^3$

Section B

1. $x = -3$
2. (a) $x = 5$ (b) $x = 3$
3. $x = 3$
4. (a) $x = \frac{1}{2}$ (b) $x = 4$
5. (a) $x = 2$ (b) $x = 12$
6. 1080
7. 2
8. $t = \dfrac{v - u}{a}$
9. 22
10. $N = 3(T - 20)$

Multi-Choice Questions 15

1. D
2. B
3. D
4. D
5. C
6. D
7. D
8. B

Mental Test 15

1. 3
2. 8
3. 3
4. 6
5. 10
6. $N = \dfrac{P}{M}$
7. $a = xy$
8. 4

Answers to Chapter 16

Exercise 16.1

1. $x = 5, y = 3$
2. $x = 4, y = 2$
3. $x = 3, y = 4$
4. $x = 1, y = 2$
5. $x = 2, y = 1$
6. $x = 5, y = 2$
7. $x = 4, y = 3$
8. $x = 2, y = 7$

Exercise 16.2

1. $x = 2, y = 3$
2. $x = 5, y = 3$
3. $x = 3, y = 2$
4. $x = 2, y = 5$
5. $x = 1, y = 2$
6. $x = 1, y = 5$
7. $x = 5, y = -1$
8. $x = 4, y = 5$

Exercise 16.3

1. $x^2 + 4x + 3$
2. $x^2 + 6x + 8$
3. $x^2 + 6x + 5$
4. $x^2 + 9x + 18$
5. $x^2 - 3x + 2$
6. $x^2 - 7x + 12$
7. $x^2 - 8x + 12$
8. $x^2 - 9x + 20$
9. $x^2 - 2x - 3$
10. $x^2 - x - 6$
11. $x^2 - 3x - 10$
12. $x^2 + 2x - 24$

Exercise 16.4

1. $(x+3)(x+2)$ 2. $(x+4)(x+2)$
3. $(x-2)(x-5)$ 4. $(x-6)(x-5)$
5. $(x-2)(x+1)$ 6. $(x-5)(x+3)$
7. $(x+4)(x-2)$ 8. $(x+4)(x-3)$

Exercise 16.5

1. $(x+1)^2$ 2. $(x-3)^2$
3. $(x+4)^2$ 4. $(x-1)^2$
5. $(x-5)^2$ 6. $(x+1)(x-1)$
7. $(x+3)(x-3)$ 8. $(x+5)(x-5)$
9. $(x+8)(x-8)$ 10. $(x+9)(x-9)$

Exercise 16.6

1. 5 or 2   2. -3 or -6 3. 3 or -2
4. -2 or -4 5. 2 or 3   6. -5 or 2
7. -3 or 6   8. -2 9. 5
10. 3   11. -5 12. 1
13. ± 1 14. ± 4 15. ± 5
16. ± 6 17. ± 8 18. ± 7

Miscellaneous Exercise 16

Section B

1. (a) $x^2+7x+10$ (b) $x^2-9x+18$
   (c) $x^2-4x-21$
2. (a) $(x+11)(x-11)$ (b) $(x+7)^2$
3. (a) $(x-5)^2$ (b) $(x+4)(x+3)$
4. $x=\pm 10$ 5. $x=4$
6. $x=2$ or 5   7. $x=-2$ or -3
8. $x=2$ or 9   9. $x=0.5,\ y=-3$
10. $x=3,\ y=2$ 11. $x=3,\ y=-1$
12. $p=12$
13. (a) $x+y=5$ (b) $x-y=-7$
    (c) $x=-1,\ y=6$
14. $x=7,\ y=-1$

$x^2+2x-24$

Multi-Choice Questions 16

1. B   2. C   3. D   4. C
5. B   6. D   7. D

Answers to Chapter 17

Exercise 17.1

1. $240°$ 2. $135°$ 3. $72°$
4. $324°$ 5. $54°$ 6. $46.5°$
7. $57.0°$ 8. $170.3°$ 9. $36.3°$
10. $8.2°$

Exercise 17.2

1. A acute; B reflex; C acute; D obtuse;
   E reflex; F acute; G obtuse
2. (a) $60°$ (b) $22.5°$ (c) $180°$
   (d) $135°$ (e) $54°$ (f) $31.5°$

3. (a) $45°$ (b) $120°$ (c) $72°$
   (d) $60°$
4. (a) $50°$ (b) $133°$ (c) $52°$
   (d) $125°$ (e) $270°$
5. (a) $33°$ (b) $56°$ (c) $153°$
   (d) $B=82°$

Exercise 17.3

1. $a=103°$
2. $x=101°,\ y=101°$
3. $m=40°,\ n=40°,\ p=130°,\ q=310°$
4. $y=117°$
5. $p=90°,\ q=80°,\ r=60°,\ s=120°$
6. $x=72°,\ y=134°$
7. $a=85°$
8. $u=62°,\ v=72°,\ w=53°,\ x=55°,$
   $y=118°,\ z=108°$

Multi-Choice Questions 17

1. C   2. C   3. A
4. B   5. B   6. C
7. C   8. C   9. D
10. D

Mental Test 17

1. $\frac{1}{360}$ 2. $60°$ 3. $108°$
4. Acute   5. $30°$ 6. $50°$
7. (a) $x=140°$ (b) $x=130°$ (c) $x=270°$
8. $A=50°,\ B=130°$
9. $b=110°,\ c=110°,\ d=70°,\ e=70°,$
   $f=110°,\ g=70°,\ h=110°$
10. $80°$

Answers to Chapter 18

Exercise 18.1

1. (a)   (b)

   (c)   (d)

   (e)

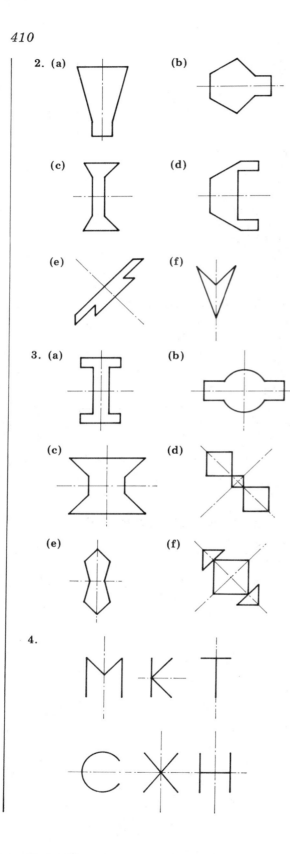

2. (a) (b) (c) (d) (e) (f)

3. (a) (b) (c) (d) (e) (f)

4.

Exercise 18.2

1. (a) 2	(b) 2	(c) Yes
2. (a) 5	(b) 5	(c) No
3. (a) 2	(b) 1	(c) Yes
4. (a) 8	(b) 8	(c) Yes
5. (a) 5	(b) 5	(c) No
6. (a) 2	(b) 4	(c) Yes
7. (a) 1	(b) 1	(c) No
8. (a) 3	(b) 3	(c) Yes

Answers to Chapter 19

Exercise 19.1

1. g, i and k 2. d, h and j
3. a, b, d, e, h and j 4. c, f and i
5. c and l

Exercise 19.2

1.

Triangle	Angles		
A	28°	67°	85°
B	37°	61°	82°
C	80°	80°	20°
D	62°	90°	28°
E	104°	13°	63°
F	90°	60°	30°

2.

Triangle	$\angle A$	$\angle B$	x
R	49°	54°	103°
S	39°	87°	126°
T	68°	94°	162°
U	107°	35°	142°
V	27°	54°	81°

3.

Triangle	$\angle A$	$\angle B$	x
G	39°	73°	112°
H	41°	67°	108°
I	65°	24°	89°
J	54°	111°	165°
K	78°	90°	168°
L	11°	86°	97°

4. (a) $x = 59°$, $y = 121°$
  (b) $x = 85°$, $y = 43°$
  (c) $x = 11°$, $y = 47°$
  (d) $x = 79°$, $y = 84°$
  (e) $x = 111°$, $y = 29°$

Exercise 19.3

1. $b = 5.9$, $c = 9.8$, $\angle C = 71°$
2. $a = 3.71$, $b = 9.44$, $\angle B = 56°$
3. $b = 6.80$, $c = 2.92$, $\angle B = 46°$
4. $b = 7$, $\angle B = 50°$, $\angle C = 80°$
5. $\angle A = 38.2°$, $\angle B = 81.7°$, $\angle C = 60.1°$
6. $\angle B = 61.3°$, $\angle C = 75.7°$, $c = 9.95$

7. $\angle B = 70.8°$, $\angle C = 47.2°$, $a = 8.42$
8. $\angle A = 56°$, $\angle C = 40°$, $c = 3.88$
9. $\angle A = 44°$, $\angle B = 36°$, $\angle C = 100°$
10. $a = 6.30$, $\angle A = 44.2°$, $\angle B = 50.8°$
11. $\angle A = 50°$, $b = 9.1$, $c = 9.8$
12. $a = 13$, $b = 8.9$, $\angle C = 30°$

Exercise 19.4

1. **(a)** AC   **(b)** BC   **(c)** AC
   (d) AB   **(e)** AC
2. **(a)** 13   **(b)** 11.1   **(c)** 10.2
   (d) 16.6   **(e)** 9.14
3. **(a)** $a = 6.40$ **(b)** $c = 13.6$ **(c)** $a = 12.7$
   (d) $b = 14.4$ **(e)** $c = 5.20$

Exercise 19.5

1. 11.9 cm   **2.** 12.3 cm   **3.** 13.1 cm
4. **(a)** $A = B = 58°$
   (b) $B = 48°$, $A = 84°$
   (c) $A = 33°$, $B = 57°$
   (d) $A = 27°$, $B = 63°$
5. **(a)** $x = 44°$, $y = 31°$
   (b) $x = 72°$, $y = 36°$
6. 10.4 cm   **7.** 10.9 cm   **8.** 7.79 cm

Miscellaneous Exercise 19

Section A

1. 9 cm   **2.** $a = 59°$, $b = 57°$ **3.** 9 cm
4. **(a)** $x = 70°$ **(b)** $y = 50°$
   (c) $\angle DAC = 40°$
5. **(a)** $52°$ **(b)** $74°$ **(c)** $52°$
   (d) $54°$ **(e)** $52°$
6. 10.8 cm   **7.** 9.17 cm

Section B

1. 13.4 cm   **2.** 53.2 cm
3. **(a)** $42°$ **(b)** $42°$ **(c)** $138°$
   (d) isosceles
4. 4 cm   **5.** 300 m
6. **(a)** 4, 4 and 4 cm
   (b) 2 cm, 5 cm and 5 cm
   (c) 3 cm, 4 cm and 5 cm
7. $x = 103°$, $y = 52°$, $z = 51°$
8. $p = 70°$, $q = 20°$, $r = 60°$, $s = 120°$
9. $m = 40°$, $n = 40°$, $p = 50°$, $q = 310°$
10. 11.3 cm

Multi-Choice Questions 19

1. B   **2.** A   **3.** B
4. A   **5.** C   **6.** B
7. B   **8.** D

Mental Test 19

1. $80°$ **2.** $80°$ **3.** $110°$
4. $70°$
5. **(a)** $60°$ **(b)** $120°$ **(c)** $60°$
   (d) $70°$
6. **(a)** 10   **(b)** 16   **(c)** 12
   (d) 39   **(e)** 24   **(f)** 75
7. **(a)** $B = 70°$, $C = 40°$, $x = 110°$
   (b) $A = 60°$, $B = 60°$, $x = 120°$
   (c) $A = 30°$, $C = 120°$, $x = 150°$
   (d) $A = 80°$, $B = 80°$, $C = 20°$
   (e) $A = 50°$, $B = 50°$, $x = 130°$
   (f) $A = 60°$, $B = 60°$, $C = 60°$
   (g) $B = 45°$, $C = 90°$, $x = 135°$
8. Equilateral   **9.** $x = 50°$, $y = 80°$
10. $30°$ **11.** $110°$
12. $60°$ **13.** Hypotenuse
14. 13 cm

Answers to Chapter 20

Exercise 20.1

1. Congruent   **2.** Congruent
3. Not congruent necessarily
4. Congruent
5. Not congruent necessarily
6. Congruent   **7.** Congruent
8. Congruent

Exercise 20.2

1. CD = 8 cm, EC = 5 cm
2. AFD, FDE, FEB, DEC, FGH, GHJ, GJD, HJE
3. TP = 8 cm, TQ = 6 cm
4. BC = 8 cm
5. ST = 8 cm, LX = 11.3 cm, RY = 11.3 cm, $\angle MLP = 40°$, $\angle SRY = 20°$
6. CD = 5 cm
7. AE = 4 cm, CD = 5 cm
8. **(a)** $\angle ADE = 80°$ **(b)** $\angle FEC = 40°$
   (c) EF = 6 cm

Answers to Chapter 21

Exercise 21.1

1. $x = 40°$, $y = 100°$, $z = 71°$
2. $x = 109°$, $y = 127°$
3. $a = 116°$, $b = 64°$
4. $\angle DAC = 65°$, $\angle BCD = 130°$, $\angle ABC = 50°$, $\angle DOC = 90°$
5. 11.7 cm   **6.** 11.3 cm
7. 10.2 cm   **8.** 10.5 m
9. **(a)** $\angle AXD = 90°$ **(b)** AX = 3 cm
   (c) trapezium
10. **(a)** trapezium
   (b) $\angle XDC = 61°$, $\angle DXB = 119°$

11. (a) $x = 42°$ (b) $y = 42°$
   (c) $\angle ACB = 96°$ (d) isosceles
12. $AO = 12$ cm, $BO = 5$ cm, $AB = 13$ cm
13. Kite and isosceles trapezium
14. Rectangle
15. $a = 35°$, $p = 65°$
16. $\angle ABE = 64°$, $\angle BED = 134°$
17. $\angle ABC = 100°$, $\angle ECD = 20°$
18. $y = 61°$

Multi-Choice Questions 21

1. B   2. C   3. C
4. D   5. C   6. A
7. D   8. C   9. A
10. C

Mental Test 21

1. $50°$ 2. $40°$
3. $\angle BEA = 90°$, $\angle ABC = 60°$
4. 5 cm   5. 4
6.

$(x^2 - 2x - 4x + 8$

7. $\angle ACD = 25°$, $\angle ACB = 80°$
8. $150°$

Answers to Chapter 22

Exercise 22.1

1. (a) $720°$ (b) $1260°$ (c) $1620°$
2. (a) $45°$ (b) $135°$
3. Each interior angle is $120°$; the hexagon has 6 axes of symmetry
4. A seven sided polygon has 7 axes of symmetry; each interior angle is $128.6°$
5. $150°$ 6. $108°$
7. 6   8. $115°$
9. (a) $20°$ (b) 18
10. $\angle AOB = 72°$, $\angle ABC = 108°$
11. 15   12. $\angle UVO = 60°$
13. $x = 45°$, $y = 45°$
14. (a) $36°$ (b) $126°$
15. $\angle AFX = 6°$ 16. 7
17. $\angle BCD = 132°$, $\angle AED = 75°$

Multi-Choice Questions 22

1. B   2. C   3. B
4. B   5. C   6. B
7. C   8. B   9. C
10. B

Mental Test 22

1. $140°$ 2. 6   3. 8
4. $36°$ 5. 6   6. Hexagon
7. 8

Answers to Chapter 23

Exercise 23.1

1. $\angle ACB = 47°$, $\angle ADB = 47°$, $\angle BDC = 43°$
2. 5 cm
3. $\angle ABO = 70°$, $\angle OAC = 20°$
4. $\angle BAC = 54°$, $\angle ACY = 27°$, $\angle YBC = 36°$
5. $\angle CBA = 90°$, $\angle ACB = 21°$, $\angle CAD = 21°$
6. $\angle AOB = 84°$, $\angle ADB = 42°$
7. $x = 63°$, $y = 48°$, $\angle BDA = 42°$
8. $\angle C = 110°$, $\angle D = 75°$, $\angle BCX = 70°$

Exercise 23.2

1. $OP = 17.9$ cm   2. $a = 78°$, $b = 12°$
3. $CB = 22.9$ cm
4. $\angle AOB = 66°$, $\angle OBD = 33°$, $\angle CBD = 57°$
5. 11.3 cm   6. $x = 59.89$ cm
7. $h = 24$ 8. $h = 57.0$ cm

Multi-Choice Questions 23

1. D   2. C   3. A
4. C   5. A   6. C
7. B   8. C   9. D
10. B

Mental Test 23

1. $60°$ 2. $80°$ 3. 10 cm
4. $40°$ 5. $80°$ 6. $10°$
7. $60°$ 8. 13 cm   9. $20°$
10. $AB = 6$ cm, $AC = 5$ cm, $BC = 5$ cm

Answers to Chapter 26

Exercise 26.1

1. 32 cm   2. 20 cm   3. 30 cm
4. 20.2 cm   5. 26.8 cm   6. 31.4 cm
7. 24.2 cm   8. 22.2 cm

Exercise 26.2

1. 35 cm^2 2. 220 mm^2 3. 280 ft^2
4. 43.16 cm^2 5. 1820 mm^2 6. 1.11 m^2
7. 0.96 m^2 8. 28.42 ft^2 9. 44 m^2
10. 174 m^2
11. (a) 52.29 m^2 (b) 36.21 m^2
    (c) 16.08 m^2
12. (a) 29 cm^2 (b) 103 m^2 (c) 64 cm^2
    (d) 52 mm^2 (e) 72 mm^2 (f) 68 cm^2
    (g) 64 cm^2 (h) 48 cm^2
13. 6 ft   14. 12 m   15. 6 cm
16. 64 cm^2 17. 144 cm^2; 12 cm

Exercise 26.3

1. 35 cm^2 2. 180 mm^2 3. 15 ft^2
4. 4 m 5. 8 in 6. 52 cm^2
7. 4.24 cm
8. (a) 81 cm^2 (b) 13.5 cm

Exercise 26.4

1. 9 cm^2 2. 28 cm^2 3. 88 in^2
4. 6.88 m^2 5. 3 cm^2
6. (a) 6 cm^2 (b) 54 cm^2 (c) 30 cm^2
   (d) 120 cm^2
7. 10 yd 8. 32 cm 9. 9.8 cm^2
10. 7.48 m^2 11. 55.4 cm^2 12. 143 cm^2
13. 30 cm^2 14. 9.61 cm^2

Exercise 26.5

1. 60 in^2 2. 40 cm^2 3. 6 cm
4. 600 in^2 5. 198 mm^2

Exercise 26.6

1. 72 m^2 2. 36 m^2 3. 266.7 mm^2
4. 3075 mm^2 5. $EF = 12 \text{ m}$ 6. 12 cm

Miscellaneous Exercise 26

Section B

1. (a) 32 cm (b) 52 cm^2
2. 25 in^2 3. 36 cm^2
4. (a) 9 cm (b) 28 cm^2
5. 5.76 cm^2 6. 3 m^2 7. 84 cm^2
8. 36 cm 9. $\frac{15}{16}$ 10. 48 cm^2

Multi-Choice Questions 26

1. D 2. A 3. D
4. D 5. D 6. D
7. A 8. B 9. B
10. B 11. B

Mental Test 26

1. 21 cm^2 2. 5 in 3. 40 cm^2
4. 8 cm 5. 16 ft^2 6. 8 cm
7. 50 cm^2 8. 8 cm^2

Answers to Chapter 27
Exercise 27.1

1. 132 cm 2. 2200 cm 3. 88 ft
4. 88 cm 5. 1760 ft 6. 270.2 cm
7. 19.9 m 8. 267.1 in 9. 134.2 mm
10. 270.7 ft 11. 201.1 m 12. 2.0 m
13. 35.0 cm 14. 263.9 m^2 15. 691.2 in

Exercise 27.2

1. 154 cm^2 2. $61\,600 \text{ mm}^2$ 3. 2464 in^2
4. 1386 m^2 5. 962.5 ft^2 6. 50.3 cm^2
7. $15\,396 \text{ in}^2$ 8. 154.0 m^2 9. 1963.8 ft^2
10. $13\,275 \text{ mm}^2$ 11. 141.4 in^2 12. 581.3 mm^2
13. 116.3 m^2 14. 3.39 in^2 15. 3.4 cm

Exercise 27.3

1. $l = 3.142 \text{ cm}$; $A = 6.284 \text{ cm}^2$
2. $l = 15.71 \text{ cm}$; $A = 78.55 \text{ cm}^2$
3. $l = 6.284 \text{ in}$; $A = 9.426 \text{ in}^2$
4. $r = 8.91 \text{ cm}$
5. $l = 7.96 \text{ m}$; $A = 31.8 \text{ m}^2$
6. $A = 9.05 \text{ yd}^2$
7. $A = 288.7 \text{ mm}^2$

Exercise 27.4

1. 14.284 cm^2 2. 76.3 cm^2 3. 292.3 cm^2
4. 5.36 cm^2 5. 13.73 cm^2 6. 23.43 cm^2
7. 57.72 cm^2 8. 42.04 cm^2

Miscellaneous Exercise 27

Section A

1. 220 cm 2. 22 yd 3. 90.28 cm^2
4. 1386 in^2 5. 33.2 cm^2

Section B

1. 28.26 cm^2 2. 14 in
3. (a) 1000 cm^2 (b) 73 cm
4. 308 cm^2 5. 28 in 6. 476.1 cm^2
7. 5.5 cm 8. 14 cm 9. 62.8 in^2
10. (a) 18.85 cm (b) 28.28 cm^2 (c) 24 cm
    (d) 99.4 cm

Multi-Choice Questions 27

1. C 2. C 3. A
4. B 5. A 6. B
7. A 8. A

Mental Test 27

1. 44 cm 2. 88 in 3. 154 cm^2
4. 4 in 5. 16 cm^2 6. $12\pi \text{ cm}^2$
7. $2\pi \text{ cm}$ 8. $12\pi \text{ cm}^2$ 9. 50 cm^2
10. 100 cm

Answers to Chapter 28
Exercise 28.1

1. 6 2. 8 3. 4
4. 6 5. 5 6. 5
7. 12 8. 8 9. 9
10. 6

Exercise 28.2

6. Slant height $= 8.73$ cm
7. (a) cylinder   (b) rectangular pyramid
   (c) rectangular pyramid

Exercise 28.3

1. $54 \, \text{m}^2$ 2. $148 \, \text{cm}^2$ 3. $3871 \, \text{cm}^2$
4. (a) $440 \, \text{cm}^2$ (b) $484 \, \text{cm}^2$
5. $176 \, \text{cm}^2$ 6. $216 \, \text{cm}^2$ 7. $130 \, \text{m}^2$
8. $204 \, \text{cm}^2$

Exercise 28.4

1. $160 \, \text{in}^3$ 2. $0.0072 \, \text{m}^3$ 3. $308 \, \text{m}^3$
4. $62\,840 \, \text{mm}^3$ 5. $60 \, \text{m}^3$ 6. $73.28 \, \text{cm}^3$
7. $13\,856 \, \text{in}^3$ 8. $300 \, \text{cm}^3$

Exercise 28.5

1. $8210 \, \text{cm}^3$ 2. $209 \, \text{in}^3$ 3. $24 \, \text{cm}^3$
4. $106.7 \, \text{ft}^3$
5. $179.6 \, \text{cm}^3$; $154.0 \, \text{cm}^2$
6. $30.2 \, \text{m}^3$ 7. $20 \, \text{m}^3$ 8. $4295 \, \text{cm}^3$

Exercise 28.6

1. $1.232 \, \ell$ 2. $1500 \, \ell$ 3. $5.75 \, \ell$
4. $42\,412 \, \ell$ 5. $38.4 \, \text{c}\ell$
6. (a) $31.8 \, \text{m}\ell$ (b) 22 glasses
7. $11.3 \, \ell$ 8. $9000 \, \ell$

Exercise 28.7

1. (a) $50 \, \text{cm}^2$ (b) $280 \, \text{cm}^3$
2. 6 in   3. $20.25 \, \text{cm}$ 4. $60.9 \, \text{in}^3$
5. $1.728 \, \text{pt}$ 6. $30 \, \text{m}$; $112 \, \text{m}^2$; $128 \, \text{m}^3$
7. $113.1 \, \text{ft}^2$; $113.1 \, \text{ft}^3$; $314.2 \, \text{ft}^2$; $523.6 \, \text{ft}^3$
8. $46.1 \, \text{cm}^2$

Miscellaneous Exercise 28

Section B

1. 6
2. (a) $48 \, \text{m}^2$ (b) $576 \, \text{m}^3$ (c) $26.9 \, \text{m}$
   (d) $323 \, \text{m}^2$
3. (a) $250 \, \text{cm}^3$ (b) $250 \, \text{cm}^2$ (c) $132 \, \text{cm}^2$
4. (a) cube   (b) none   (c) none
   (d) none   (e) tetrahedron   (f) none
   (g) none
5. (a) $29.68 \, \text{cm}^3$ (b) $237.4 \, \text{cm}^3$
   (c) 2.4 cm by 5.25 cm by 7.95 cm
6. (a) $81\,000 \, \text{cm}^3$ (b) 25 cm
7. $28.3 \, \text{m}^3$
8. (a) $5600 \, \text{cm}^2$ (b) $4400 \, \text{cm}^3$ (c) 19.6; 15.4
9. The 35-cm high packet

Multi-Choice Questions 28

1. D   2. D   3. D
4. A   5. D   6. D
7. B   8. A   9. B
10. B   11. B   12. C
13. C   14. B   15. D

Mental Test 28

1. $40 \, \text{in}^3$ 2. $100 \, \text{cm}^3$ 3. $210 \, \text{cm}^3$
4. $250 \, \text{in}^3$ 5. $200 \, \text{in}^2$ 6. $96 \, \text{cm}^3$
7. $110 \, \text{cm}^2$ 8. $100\pi \, \text{cm}^3$ 9. $28 \, \text{cm}^3$
10. $120 \, \text{cm}^3$ 11. $36\pi \, \text{cm}^2$ 12. $1:2$

Answers to Chapter 29

Exercise 29.1

1. (a) 0.707   (b) 0.866
2. (a) $18.4°$ (b) $38.7°$ (c) $79.9°$
3. (a) $43°12'$ (b) $54°42'$ (c) $43°6'$
4. (a) 0.2419   (b) 0.9563   (c) 0.4258
   (d) 0.8643   (e) 0.8111   (f) 0.9839
   (g) 0.3401   (h) 0.9713
5. (a) $66°$ (b) $79°$ (c) $47.3°$
   (d) $83.9°$ (e) $54.2°$ (f) $8.3°$

Exercise 29.2

1. (a) 11.47 cm   (b) 12.04 cm
   (c) 18.25 cm
2. (a) $36.9°$ (b) $32.2°$ (c) $53.1°$
3. 5.95 cm   4. 3.63 cm   5. 23.1 cm
6. 20.0 cm
7. $41.8°$, $41.8°$ and $96.4°$
8. $\angle B = 61.4°$, $\angle C = 28.6°$

Exercise 29.3

1. (a) 0.9511   (b) 0.2419   (c) 0.7955
   (d) 0.1461   (e) 0.5819   (f) 0.9880
2. (a) 2.74 cm   (b) 10.20 cm   (c) 5.73 cm
3. (a) $63.3°$ (b) $77.2°$ (c) $66.4°$
4. 7.47 cm
5. $47.16°$; $66.4°$; $66.4°$
6. (a) $25°$ (b) $11°$ (c) $59.3°$
   (d) $78.2°$ (e) $37.4°$ (f) $68.2°$
7. 11.1 cm
8. $\angle BAC = 60°$; $BD = 6.93$ cm

Exercise 29.4

1. (a) 0.5095   (b) 4.705   (c) 21.20
   (d) 0.2754   (e) 0.1456   (f) 1.509
2. (a) $17°$ (b) $79°$ (c) $12.3°$
   (d) $43.8°$ (e) $27.3°$ (f) $59.6°$
3. (a) 8.40 cm   (b) 9.00 cm   (c) 2.92 cm
4. (a) $33.7°$ (b) $38.7°$ (c) $60.3°$
5. 8.26 cm   6. 3.96 cm
7. 2.02 cm   8. 1.46 cm

Exercise 29.5

1. 4.82 m
2. 27.7 m
3. 42°
4. 96.2 m
5. 8.29 m
6. 23.0 m
7. 575.2 m
8. 39.8°
9. 5.87 m
10. 14.9 m

Exercise 29.6

1. 22.13 cm²
2. 31.9 cm²
3. 6 cm
4. 540.4 cm²
5. 737.5 m²
6. 8.07 m²
7. 14.0 m²
8. 3870 m²
9. 13.9 cm²
10. 173.8 m²

Exercise 29.7

1. (a) 323°   (b) 190°
2. (a) 323°   (b) 277°   (c) 350°
3. 335°   4. 248°
5. (a) 300°   (b) ∠ABC = 75°
6. 140°
7. (a) 264 km   (b) 062°
8. (a) 69.3 km   (b) 40 km

Exercise 29.8

1. (a) FH = 13 cm   (b) BH = 15.3 cm
2. (a) EF = 4 cm   (c) 71.6°
   (d) 12.6 cm
   (e) Area △VAD = 50.6 cm²
   (f) Area of pyramid = 266.4 cm²
3. 12.7 cm   4. 54.5°   5. 4.53 cm
6. EA = 14.4 cm; q = 28.7°
7. (a) 62°   (b) 1700 m²
   (c) 12 000 m³

Miscellaneous Exercise 29

1. (a) 11.5°   (b) 4.90 m
2. (a) 80°   (b) 12.86 cm
   (c) 7.66 cm
3. (a) 39.61 cm   (b) 0.3051   (c) 17.0°
4. (a) 60.8 m   (b) 37.2°
5. (a) 7.75 cm   (b) 51.6 cm²   (c) 31.8°
6. (a) 22.9°   (b) 66.4°
7. 373 m
8. (a) 60°   (b) 17.32 m
9. 027°   10. 78 km; 190°

Multi-Choice Questions 29

1. C   2. B   3. C
4. D   5. C   6. D
7. A   8. A   9. C
10. A   11. C   12. B
13. C   14. C   15. C
16. C   17. D   18. A
19. B   20. C

Answers to Chapter 30
Exercise 30.1

1. {2, 3, 5, 7, 11, 13, 17, 19}
2. {5, 10, 15, 20, 25, 30}
3. {9, 11, 13, 15}
4. {2, 4, 6, 8, 10, 12}
5. Infinite   6. Finite   7. Finite
8. Infinite   9. Null   10. Infinite
11. True   12. False   13. True
14. False   15. True   16. False
17. Horse ∉ {animals with horns}
18. Cricket ∈ {outdoor games}
19. $\frac{5}{2}$ ∈ {improper fractions}
20. 25 ∉ {multiples of 3}

Exercise 30.2

1. (a) {3, 5, 7, 9, 13, 15}
   (b) {4, 8, 10, 12, 16}
   (c) {3, 5, 7, 13}
   (d) {4, 8, 12, 16}
2. Ø, {m}, {n}, {p}, {q}, {m, n}, {m, p}, {m, q},
   {n, p}, {n, q}, {p, q}, {m, n, p}, {m, n, q},
   {m, p, q}, {n, p, q}, {m, n, p, q}; 16
3. (a) incorrect   (b) incorrect   (c) correct
   (d) incorrect
4. {9, 15, 20, 21} ⊂ {whole numbers between 3
   and 25}
   {knife, fork} ⊂ {all cutlery}
   {football, cricket, tennis} ⊂ {all outdoor games}
   {sheep, goats, cows, dogs, cats} ⊂ {all domestic
   animals}
5. 64
6.

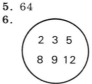

7.

Acute angle triangles

All triangles

8. M′ = {9, 11}

Shaded portion = M′

9. $A' = \{2, 7, 8, 9, 11, 12, 14\}$;
$B' = \{2, 3, 5, 6, 7, 9, 12, 14\}$

10. (a) $\{a, e, i, o, u\}$ (b) $\{u, v, w, x, y, z\}$
    (c) $\{b, c, d, f, g, h, j, k, l, m, n\}$

Exercise 30.3

1. (a) $\frac{5}{2}$ (b) $-\frac{13}{3}$ (c) $\frac{5}{1}$
    (d) $\frac{89}{100}$ (e) $\frac{6}{1}$

2.

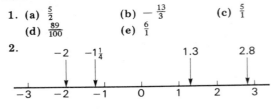

3. $A = \{3, 4, 5, 6, 7, 8, 9\}$
4. $\{x : -2 \leqslant x \leqslant 4\} = \{-2, -1, 0, 1, 2, 3, 4\}$
$\{x : -4 \leqslant x \leqslant 7\} = \{-4, -3, -2, -1, 0, 1,$
$2, 3, 4, 5, 6, 7\}$

5.

6.

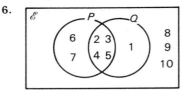

Exercise 30.4

1. (a) $\{2, 3, 4, 5, 6, 7, 8, 9\}$
    (b) $\{3, 5, 7, 9\}$ (c) $\{4, 6, 7, 9\}$
    (d) $\{2, 4, 6, 8\}$ (e) $\{2, 3, 5, 8\}$
    (f) $\{7, 9\}$ (g) $\{3, 4, 5, 6, 7, 9\}$
    (h) $\{2, 8\}$

2.

3. (a) Y (b) $A \cap B$
   (c) $(A \cup B)'$ (d) $A \cap B'$

4. (a) $\mathcal{E} = \{1, 2, 3, 4, 5, 6, 7, 8, 9, 10, 11, 12\}$
    (b) $A = \{2, 5, 7, 9\}$
    (c) $B = \{3, 6, 8, 12\}$
    (d) $A' = \{1, 3, 4, 6, 8, 10, 11, 12\}$
    (e) $B' = \{1, 2, 4, 5, 7, 9, 10, 11\}$
    (f) $A \cup B = \{2, 3, 5, 6, 7, 8, 9, 12\}$
    (g) $A \cap B = \varnothing$

5. (a)

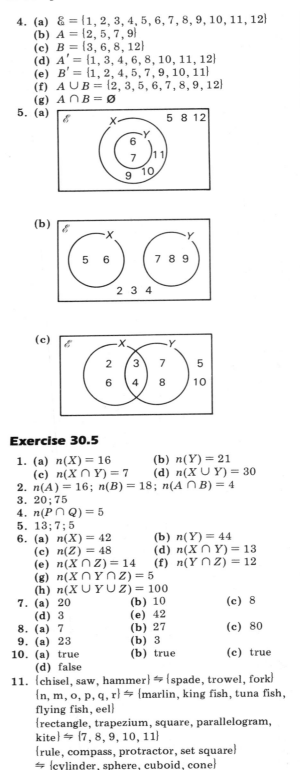

   (b)

   (c)

Exercise 30.5

1. (a) $n(X) = 16$ (b) $n(Y) = 21$
    (c) $n(X \cap Y) = 7$ (d) $n(X \cup Y) = 30$
2. $n(A) = 16$; $n(B) = 18$; $n(A \cap B) = 4$
3. $20; 75$
4. $n(P \cap Q) = 5$
5. $13; 7; 5$
6. (a) $n(X) = 42$ (b) $n(Y) = 44$
    (c) $n(Z) = 48$ (d) $n(X \cap Y) = 13$
    (e) $n(X \cap Z) = 14$ (f) $n(Y \cap Z) = 12$
    (g) $n(X \cap Y \cap Z) = 5$
    (h) $n(X \cup Y \cup Z) = 100$
7. (a) 20 (b) 10 (c) 8
    (d) 3 (e) 42
8. (a) 7 (b) 27 (c) 80
9. (a) 23 (b) 3
10. (a) true   (b) true   (c) true
    (d) false
11. {chisel, saw, hammer} ⇋ {spade, trowel, fork}
{n, m, o, p, q, r} ⇋ {marlin, king fish, tuna fish,
flying fish, eel}
{rectangle, trapezium, square, parallelogram,
kite} ⇋ {7, 8, 9, 10, 11}
{rule, compass, protractor, set square}
⇋ {cylinder, sphere, cuboid, cone}

Miscellaneous Exercise 30

Section A

1. (a)

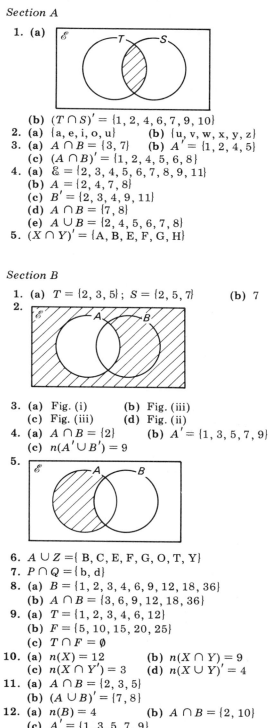

   (b) $(T \cap S)' = \{1, 2, 4, 6, 7, 9, 10\}$
2. (a) $\{a, e, i, o, u\}$ (b) $\{u, v, w, x, y, z\}$
3. (a) $A \cap B = \{3, 7\}$ (b) $A' = \{1, 2, 4, 5\}$
   (c) $(A \cap B)' = \{1, 2, 4, 5, 6, 8\}$
4. (a) $\mathscr{E} = \{2, 3, 4, 5, 6, 7, 8, 9, 11\}$
   (b) $A = \{2, 4, 7, 8\}$
   (c) $B' = \{2, 3, 4, 9, 11\}$
   (d) $A \cap B = \{7, 8\}$
   (e) $A \cup B = \{2, 4, 5, 6, 7, 8\}$
5. $(X \cap Y)' = \{A, B, E, F, G, H\}$

Section B

1. (a) $T = \{2, 3, 5\}$; $S = \{2, 5, 7\}$ (b) 7
2.
3. (a) Fig. (i)   (b) Fig. (iii)
   (c) Fig. (iii)   (d) Fig. (ii)
4. (a) $A \cap B = \{2\}$ (b) $A' = \{1, 3, 5, 7, 9\}$
   (c) $n(A' \cup B') = 9$
5.
6. $A \cup Z = \{B, C, E, F, G, O, T, Y\}$
7. $P \cap Q = \{b, d\}$
8. (a) $B = \{1, 2, 3, 4, 6, 9, 12, 18, 36\}$
   (b) $A \cap B = \{3, 6, 9, 12, 18, 36\}$
9. (a) $T = \{1, 2, 3, 4, 6, 12\}$
   (b) $F = \{5, 10, 15, 20, 25\}$
   (c) $T \cap F = \emptyset$
10. (a) $n(X) = 12$ (b) $n(X \cap Y) = 9$
    (c) $n(X \cap Y') = 3$ (d) $n(X \cup Y)' = 4$
11. (a) $A \cap B = \{2, 3, 5\}$
    (b) $(A \cup B)' = \{7, 8\}$
12. (a) $n(B) = 4$ (b) $A \cap B = \{2, 10\}$
    (c) $A' = \{1, 3, 5, 7, 9\}$
    (d) $(A \cup B)' = \{3, 7, 9\}$

Multi-Choice Questions 30

1. A	2. D	3. C
4. B	5. A	6. D
7. B	8. C	9. D
10. D	11. B	12. B

Answers to Chapter 31

Exercise 31.1

1. 1600 m   2. 25 cm
3. (a) 1 : 200   (b) 4 m by 6 m
4. (a) 1 : 50 000   (b) 5.4 km
5. 405 km
6. (a) 14.5 km   (b) 7.25 cm; 36 km

Exercise 31.2

1. (a) 15 m by 12.5 m   (b) 187.5 m^2
2. 640 m^2

3.

Name of room	Length (m)	Width (m)	Area (m^2)
Lounge	5	5	25
Dining room	5	3	15
Kitchen	5	3	15
Bathroom	4	3	12

Area of carpet required is 13 m^2.

4.

Name of room	Length (m)	Width (m)	Area (m^2)
Lounge	7	6	42
Hall	3	2	6
Bedroom	5	4	20
Kitchen	4	2	8
Bathroom	4	3	12

5. (a) 3   (b) 3   (c) 25.5 m
   (d) 13.5 m   (e) 344.25 m^2 (f) 51.64 m^3
6. $11\,875 \text{ m}^2$
7. (a) 48 m^2 (b) 1120

Exercise 31.3

1. $40\,000 \text{ m}^2$ 2. 12 m^2 3. 20 km^2
4. 10.4 km^2
5. (a) 13.4 cm^2 (b) $134\,000 \text{ m}^2$
6. 35.5 hectares

Exercise 31.4

1. $24\,000 \text{ m}^3$ 2. 1.536 m^3
3. 0.6 cm^3 4. 7812 m^3
5. 840 m^3

Exercise 31.5

1. 1 in 25   2. 1 in 40   3. 100 m
4. 30 m   5. 7.2°
6. (a) 1:4.8   (b) 20.8%
7. 15 m

Exercise 31.6

1. 62°
2. (a) 1 in 2.85   (b) 35%
3. 17.4°
4. (a) 100 m   (b) 2000 m   (c) 1 in 40
5. (a) 100 m   (b) 236 m
   (c) 1.65 km; 2.55 km   (d) 14°56'

Miscellaneous Exercise 31

1. 500 m
2. 1.75 cm × 2.25 cm × 1.25 cm
3. (a) 10 m   (b) 14 cm   (c) 20 m^2
4. (a) 80 m   (b) 5 m^2 (c) 288 000 m^3
5. (a) 50 cm   (b) 648 m^2 (c) 19.44 m^3

6.

Name of room	Length (m)	Breadth (m)	Area (m^2)
Lounge	7	6	42
Dining room	7	4	28
Bedroom 1	7	4	28
Bedroom 2	4	3	12
Hall	11	2	22
Bathroom	3	3	9
WC	2	2	4
Kitchen	5	3	15

7 windows and 8 doors

7. (a) 11.3°   (b) 1 in 5   (c) 20%

Answers to Chapter 32

Exercise 32.1

1. (a) £6.24   (b) £26.52
   (c) £215.02   (d) £174.72
   (e) £1.69   (f) £2.42
   (g) £21.37
2. (a) 28.3°C   (b) 45.9°C
   (c) 131°F   (d) 155.7°F
3. (a) 488 miles   (b) 130 miles
   (c) 63 miles
4. (a) 14.3%   (b) 1:14
5. (a) 24.13 km   (b) 30 mile/h
   (c) 93.3 km   (d) 914.12 km/h
6. (a) 37.8°C   (b) 176°F
   (c) 17.8°C   (d) 118°F
7. (a) 1.5 kg   (b) 258 kg
   (c) 4.0 lb
8. (a) 1.54 kg/cm^2 (b) 30 lb/in^2
   (c) 43 lb/in^2

Exercise 32.2

1. (a) 40   (b) 11
2.

3.

4.

5.

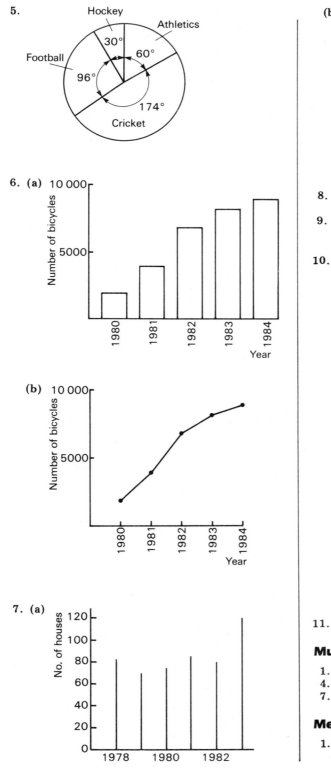

6. (a)

(b)

7. (a)

(b)

8. (a) 126°
   (b) White 8000, Brown 7000, Green 5000
9. Vegetables 190 tons, Grain 129 tons
   Potatoes 111 tons, Fruit 70 tons

10.

11. Food £22.20, Rent £12.50, Other £15.30

Multi-Choice Questions 32

1. A	2. B	3. B
4. B	5. C	6. B
7. B	8. B	9. D

Mental Test 32

1. (a) 110 miles   (b) 240 miles
   (c) 110 miles   (d) 72 miles

2. (a) $-40°F$ **(b)** $68°F$ **(c)** $38°C$
3. (a) 1600   **(b)** 880   **(c)** 460
4. 90; 150
5. (a) £60 000   **(b)** £200 000
6. (a) 1978   **(b)** 50 000
   (c) 35 000   **(d)** 200 000
7. $72°$

Answers to Chapter 33

Exercise 33.1

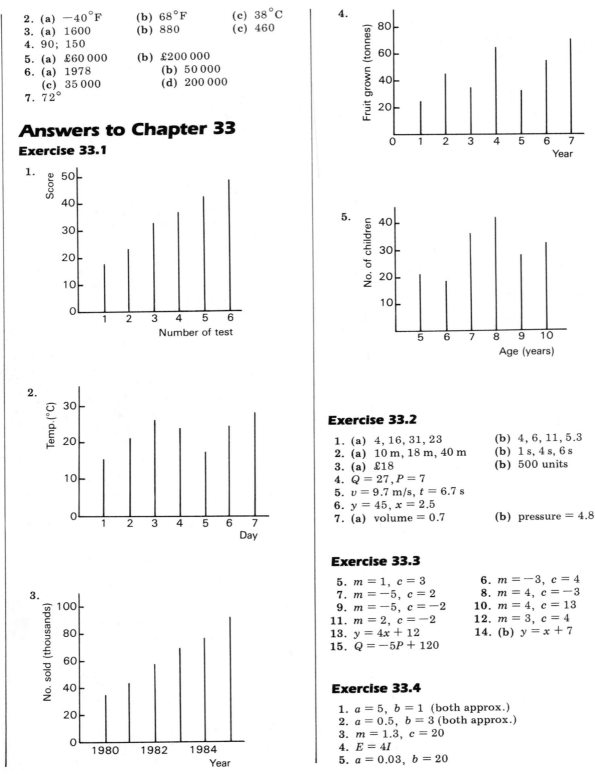

1.

2.

3.

4.

5.

Exercise 33.2

1. (a) 4, 16, 31, 23   **(b)** 4, 6, 11, 5.3
2. (a) 10 m, 18 m, 40 m   **(b)** 1 s, 4 s, 6 s
3. (a) £18   **(b)** 500 units
4. $Q = 27, P = 7$
5. $v = 9.7$ m/s, $t = 6.7$ s
6. $y = 45, x = 2.5$
7. (a) volume $= 0.7$ **(b)** pressure $= 4.8$

Exercise 33.3

5. $m = 1, c = 3$ **6.** $m = -3, c = 4$
7. $m = -5, c = 2$ **8.** $m = 4, c = -3$
9. $m = -5, c = -2$ **10.** $m = 4, c = 13$
11. $m = 2, c = -2$ **12.** $m = 3, c = 4$
13. $y = 4x + 12$ **14. (b)** $y = x + 7$
15. $Q = -5P + 120$

Exercise 33.4

1. $a = 5, \ b = 1$ (both approx.)
2. $a = 0.5, \ b = 3$ (both approx.)
3. $m = 1.3, \ c = 20$
4. $E = 4I$
5. $a = 0.03, \ b = 20$

Miscellaneous Exercise 33

Section A

1. (a) £18.50   (b) 230 units   (c) £7.50
2. (a) A$(-2, 8)$, B$(3, -4)$, C$(0, 3)$
3. (b) (i) 38.5 miles per gallon
   (ii) 15 mile/h, 78 mile/h
   (iii) 50 mile/h
4. (a) 26.4 lb   (b) 15.9 kg
5. Trapezium

Section B

1. (a) £29.50   (b) 150 units
   (c) £8   (d) 6 p
2. (b) $(0, 4)$ (c) 1
   (d) $y = x + 4$
3. (a) $y = 4x + 3$ (b) -5
4. 9 units; 18 square units
5. $c = 4$
6. (a) A$(-2, 8)$ (b) -2
   (c) $y = -2x + 4$
7. (a) A$(0, 6)$ (b) B$(-12, 0)$
   (c) 36 square units
8. (a) $c = 5$ (b) OB $= 10$

Multi-Choice Questions 33

1. B   2. A   3. D
4. A   5. C   6. B
7. C   8. A

Answers to Chapter 34
Exercise 34.1

6. (a)

T (hours)	1	2	3	4	5
N		2	4	8	16	32

   (b) 11.3
7. (b) £12 280   (c) $5\frac{1}{2}$ years
8. 14 millions
9. $-5, 19$
10. $-4, 8$
11.

x	-2	$-1\frac{1}{2}$	-1	0	$\frac{1}{2}$	1	$1\frac{1}{2}$	2	3
y	-9	-4		0	5	6	6	5	3	-4

   $x = 1.25$
12. $x = 8$

Exercise 34.2

1. 60 km/h   2. 5 h   3. 300 km
4. 50 km/h   5. 36.4 km/h   6. 10 km/h
7. 22 km/h

Exercise 34.3

1. 10.20 a.m.; 20 km from A
2. (b) 2124 hours
   (d) 1800 hours, 195 km from London
3. (a) 6 km/h   (b) 20 min
   (d) 6 km/h
4. (a) 48 km/h   (b) 344 km   (c) 95 km
   (d) 3.5 hours
5. (a) 2045 hours
   (b) 26 km   (c) 24.4 km/h
   (d) 2110 and 2020 hours

Exercise 34.4

1. (a) 60 m   (b) 6 m   (c) 80 m
   (d) 120 m   (e) 300 m
2. (a) 5 m/s^2 (b) 0.5 m/s^2 (c) -1.5 m/s^2
   (d) -2.5 m/s^2
3. 150 m
4. (a) 0.2 m/s^2 (b) 4 m/s   (c) 210 m
5. (a) 1 m/s^2 (b) 3 m/s^2 (c) 5 m/s
   (d) 100 m
6. (a) 1 m/s^2 (b) 2 m/s^2 (c) 0.5 m/s^2
   (d) 30 m/s   (e) 875 m
7. (b) 2 m/s^2 (c) 1.6 m/s^2 (d) 384 m
8. 239 m
9. (b) (i) 2.5 m/s^2 (ii) 10 m/s^2
10. 9 m/s

Miscellaneous Exercise 34

1. (a) 50 min   (b) 33 km   (c) 72 km/h
   (d) Departure time 0830
2. $-4, 2$
3. (b) 13.15 million
4.

t	0	5	10	15	20	25	30
$20t$	0	100	200	300	400	500	600
$8t^2$	0	200	800	1800	3200	5000	7200
v	0	300	1000	2100	3600	5500	7800

   (b) 340 m/s^2

Answers to Chapter 35
Exercise 35.1

1. (a) $\begin{pmatrix} 4 \\ 3 \end{pmatrix}$ (b) $\begin{pmatrix} -5 \\ 2 \end{pmatrix}$ (c) $\begin{pmatrix} -6 \\ -6 \end{pmatrix}$
   (d) $\begin{pmatrix} 0 \\ -6 \end{pmatrix}$

2. (a) $\begin{pmatrix} 6 \\ 3 \end{pmatrix}$ (b) $\begin{pmatrix} 3 \\ 0 \end{pmatrix}$ (c) $\begin{pmatrix} 1 \\ 2 \end{pmatrix}$
   (d) $\begin{pmatrix} 1 \\ -2 \end{pmatrix}$

3. $\begin{pmatrix} -4 \\ -1 \end{pmatrix}$

4. A′(5, 5), B′(7, 5), C′(7, 7), D′(5, 7)
5. A′(0, −3), B′(1, −5), C′(1, −2)

Exercise 35.2

1.

4.

2.

5.

3.

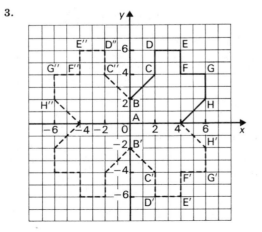

6.

45°

7. **(a)** S(2, 6)
   (e) P'(2, 2), Q'(2, 6), R'(6, 6), S'(6, 2)

8. $y = 2$

Exercise 35.3

1. P'(5, 0.6)
2. X'(−3, 3), Y'(−3, 6), Z'(−5, 6)
3. A'(2.8, 0), B'(7.1, −5.0), C'(9.2, −2.1), D'(4.9, 2.1)
4.

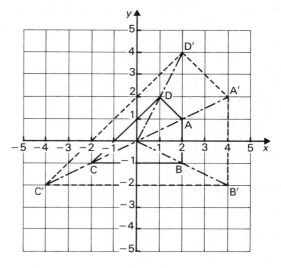

    (a) 4   **(b)** 4
5. **(a)** 2   **(b)** 2   **(c)** 1
   (d) 4

Exercise 35.4

1. 3; (−2, 1)
2. **(a)** 3   **(b)** (0, 0)
3. A'(4, 0), B'(4, 4), C'(8, 6), D'(8, 0); 4 times as large
4. **(a)** 2; (−2, −3)   **(b)** 3; (3, −3)
   (c) 2; (−1, 1)
5. A'(−2.5, 2.5), B'(−2.5, 4), C'(0.5, 5.5), D'(−1, 2.5)

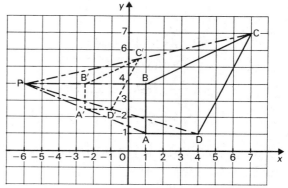

6. A'(4, 2), B'(4, −2), C'(−4, −2), D'(2, 4)

7. $\frac{1}{3}$; 8 cm
8. EF = 2 cm; CD = 12 cm

Miscellaneous Exercise 35

1. $A'(-2, 4)$
2. (a) $A(4, -3)$ (b) $B(-4, 3)$
3. $(6, 3)$
4. $P'(5, 0.5)$
5. $(2, 7)$
6. $A'(0, 0)$, $B'(4, 0)$, $C'(4, 2)$

7.

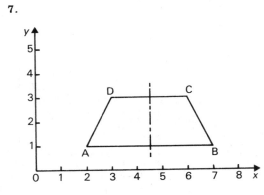

8. (a) R is the reflection of $\triangle ABC$ in the y-axis
   (b) T is the translation of $\triangle ABC$ under the

   vector $\begin{pmatrix} -5 \\ 3 \end{pmatrix}$

Answers to Chapter 36

Exercise 36.1

2.

Mark	1–10	11–20	21–30	31–40	41–50
Frequency	2	3	6	6	3

3. (b) $20.355, 20.455, 0.1$
5. (a) $39.5, 44.5, 5$ (c) continuous
6.

No. of letters	1–2	3–4	5–6	7–8	9–10	11–12
Frequency	15	35	5	14	2	1

7. (a) cont.   (b) disc.   (c) cont.
   (d) disc.   (e) cont.   (f) cont.
   (g) cont.   (h) disc.
8. (a) 50   (b) continuous

Exercise 36.2

1. 176.7 cm   2. 8 3. 14
4. 5 5. 5.5 6. 5
7. (a) $2\frac{1}{2}$ kg   (b) $3\frac{1}{2}$ kg   (c) $2\frac{1}{2}$ kg
8. 4
9. (a) 78 kg   (b) 77 kg
10. (a) 40 (b) 3 (c) 3.3
    (d) 4
11. (a) 4.34 (b) 4 (c) 8
12. (a) 200 (b) 23 (c) 23
    (d) 22.6

Exercise 36.3

1. (a) $\frac{1}{6}$ (b) $\frac{1}{2}$ (c) $\frac{1}{2}$
2. (a) $\frac{1}{8}$ (b) $\frac{1}{4}$ (c) $\frac{3}{8}$ (d) $\frac{5}{8}$
3. (a) $\frac{1}{5}$ (b) $\frac{1}{2}$ (c) $\frac{3}{10}$ (d) $\frac{7}{10}$
4. $\frac{4}{7}$
5. (a) $\frac{1}{2}$ (b) $\frac{3}{10}$ (c) $\frac{3}{5}$ (d) $\frac{7}{20}$
6. (a) $\frac{1}{52}$ (b) $\frac{1}{13}$ (c) $\frac{4}{13}$ (d) $\frac{1}{26}$
7. (a) $\frac{1}{6}$ (b) $\frac{1}{6}$ (c) $\frac{1}{6}$
8. (a) $\frac{1}{13}$ (b) $\frac{1}{17}$
9. $\frac{12}{23}$
10. $\frac{1}{3}$

Exercise 36.4

1. (a) $\frac{7}{20}$ (b) $\frac{13}{20}$
2. (a) $\frac{3}{20}$ (b) $\frac{1}{4}$ (c) $\frac{1}{20}$
3. (a) $\frac{8}{15}$ (b) $\frac{18}{25}$ (c) $\frac{1}{15}$
4. (a) $\frac{7}{25}$ (b) $\frac{19}{50}$ (c) $\frac{31}{50}$
5. (a) $\frac{4}{25}$ (b) $\frac{14}{25}$ (c) $\frac{2}{25}$

Exercise 36.5

1. $\frac{5}{52}$
2. (a) $\frac{1}{12}$ (b) $\frac{1}{4}$
3. (a) $\frac{4}{25}$ (b) $\frac{9}{25}$ (c) $\frac{6}{25}$
4. (b) $\frac{3}{10}$ (c) $\frac{1}{10}$ (d) $\frac{3}{10}$
5. $\frac{1}{30}$
6. (a) $\frac{4}{35}$ (b) $\frac{4}{35}$ (c) $\frac{6}{35}$ (d) $\frac{4}{35}$
7. $\frac{1}{6}$

Miscellaneous Exercise 36

Section A

1. 61
2.

Number of goals	0	1	2	3	4	5
Frequency	16	24	12	3	3	2

5. (a) 50 (b) 59 kg

Section B

1. (a) $\frac{1}{9}$ (b) $\frac{2}{9}$ (c) $\frac{1}{3}$
2.

Distance	0–9	10–19	20–29	30–39
Frequency	0	6	12	13

Distance	40–49	50–59	60–69
Frequency	15	14	10

3. (a) 14.985 mm; 15.015 mm
   (b) 0.03 mm   (c) 14.96–14.98 mm

4. (a) disc. (b) cont.
   (c) disc. (d) cont.
   (e) cont. (f) disc.
5. £123.60
6. 163.1 cm
7. (a) 5 (b) 5
8. (a) 4 (b) 4
9. (a) $\frac{1}{6}$ (b) $\frac{1}{2}$ (c) $\frac{1}{3}$ (d) $\frac{1}{2}$
10. (a) 0.32 (b) 0.03
11. (a) A $= \frac{10}{24}$ (or $\frac{5}{12}$); B $= \frac{10}{23}$; C $= \frac{14}{23}$; D $= \frac{9}{23}$;
    (b) $\frac{15}{92}$
12. (a) 0.11 (b) 0.15

Multi-Choice Questions 36

1. C 2. B 3. D
4. B 5. C 6. C
7. B 8. B 9. C
10. B

Mental Test 36

1. 8, 40 2. 1 3. $\frac{18}{25}$
4. £30 5. 7 6. 2
7. 2 8. $\frac{3}{10}$
9. (a) 15.5, 17.5, 2 mm
10. 38 11. $\frac{1}{3}$

Answers to Chapter 37

Exercise 37.1

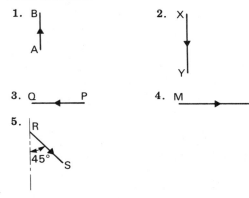

Exercise 37.2

1. 7.5 at $28°$ to N 2. 19.2 at $51°$ to N
3. 11.9 at $31°$ to N 4. 7.0 at $11°$ to N
5. 17.4 at $58°$ to N
6. 8.2 units at $67°$ to the horizontal
7. 11.1 units at $19°$ to the horizontal
8. 7 units at $22°$ to the horizontal
9. (a) 24 units (b) 40 units (c) 72 units
10. (a) 6 units (b) 12 units (c) 15 units
    all at $50°$ to the horizontal

Exercise 37.3

1. 7.7 at $34°$ to 4 vector
2. 12.2 at $22°$ to 6 vector
3. 20.0 at $37°$ to 16 vector
4. 7.8 at $73°$ to 5 vector
5. 20.7 at $18°$ to 12 vector
6. 4.3 N and 2.5 N
7. 19.3 km/h and 23.0 km/h
8. 75.2 N and 27.4 N

Answers to Chapter 38

Investigations

2. Maximum 9-path including a return to X. Tetrahedron 5-path, Triangular prism 7-path, Octahedron 12-path, Pyramid 6-path
4. Four numbers: Line of 0's; Three numbers: 101, 110 and 011 will repeat; Line of 0's always possible using 2^x numbers
5. Tile G has limited use. Its reflection would be a useful inclusion.
6. In any single sequence, two numbers are always missing. In all sequences one number never appears.
8. Diagrams produce this table:

Lines	Squares				
4	1				
5	2				
6	3	5			
7	4	8			
8	5	11	14		
9	6	14	20		
10	7	17	26	30	
11	8	20	32	40	
12	9	23	38	50	55

10. Minimum time 42 minutes when they all arrive together. Drop points: 4 km and $7\frac{1}{2}$ km. One cycles for 30 min, then walks for 12 min. One rides the scooter for 8 min, walks for 28 min and cycles for 6 min. The other walks for 32 min and then rides the scooter for 10 min.

Problems

1. 12×84, 12×63, 12×42, 13×62, 13×93, 14×82, 23×64, 23×96, 24×84, 24×63, 26×93, 34×86, 36×84, 46×96

2.
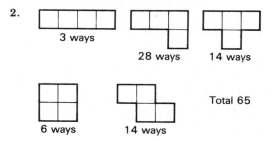

3 ways 28 ways 14 ways

6 ways 14 ways Total 65

3. Two examples for each number:

```
      2              9              3              2              9
    1  7          5  3          7  8          7  5          1  0
  9  0  6       1  0  4       5  1  2       3  1  6       7  2  6
8  3  4  5    8  6  2  7    4  0  9  6    9  4  0  8    3  8  4  5

      0              0              2              5              3
    7  6          9  7          5  4          6  7          5  2
  5  2  9       5  3  6       7  3  9       2  4  0       7  4  9
4  8  3  1    1  4  8  2    1  6  8  0    8  1  3  9    0  6  8  1

      8              4              0              0              5
    3  9          7  2          9  4          7  4          4  0
  1  5  0       3  5  9       1  6  8       5  6  9       3  7  6
7  4  6  2    1  6  8  0    7  2  3  5    2  3  8  1    8  1  2  9

      3              5              0              1              8
    6  2          6  9          9  4          6  5          4  6
  4  7  9       1  8  0       1  8  7       4  9  7       0  9  1
1  8  5  0    4  3  7  2    6  2  3  5    2  3  8  0    5  3  7  2
```

4. (a) **(b)** **(c)**

(d) **(e)**

5. XBCAEDF, XBCADEF, XBDECAF,
XBDACEF, XBEACDF, XBEDCAF,
XCDABEF, XCDEBAF, XCBAEDF,
XCBADEF, XCEABDF, XCEDBAF,
XAECBDF, XAEBCDF, XADCBEF,
XADBCEF

6. (a) Break 4 links in one length and use the four
to join the other four lengths in a loop.
Cost: £8.

(b) Break each link in the 3 and 4 lengths. Use
the seven separate links to join the remaining seven pieces in a loop. Cost: £14.

7. Using C for clockwise, A for anticlockwise and
X for exchange, there are two solutions:

XCXAXCXCXCXAXA
CXAXCXCXCXAXAX

8. Paths by route: 1, 16, 36, 16, 1. Total: 70.

9. Fifteen solutions supplied by a group of
students:

$4 \times 4 = 16$	$4 \times 2 = 8$
$6 + 2 = 8$	$6 + 10 = 16$
$5 + 3 < 10$	$5 < 4 + 3$
$6 + 4 = 10$	$5 \times 2 = 10$
$2 \times 4 = 8$	$4 + 4 = 8$
$5 + 3 < 16$	$6 + 3 < 16$

$5 + 3 = 8$	$6 + 4 = 10$
$4 \times 4 = 16$	$3 + 2 = 5$
$6 + 2 < 10$	$16 < 4 \times 8$
$8 \times 2 = 16$	$4 + 4 = 8$
$6 + 4 = 10$	$6 + 10 = 16$
$5 < 4 + 3$	$2 < 5 \times 3$
$4 + 6 = 10$	$6 + 10 = 16$
$5 + 3 = 8$	$2 + 3 = 5$
$4 \times 2 < 16$	$8 < 4 \times 4$
$6 + 10 = 16$	$4 + 4 = 8$
$5 + 3 = 8$	$3 \times 2 = 6$
$4 < 4 \times 2$	$5 < 10 + 16$
$5 + 3 = 8$	$8 + 2 = 10$
$10 + 6 = 16$	$4 \times 4 = 16$
$2 < 4 \times 4$	$6 < 3 + 5$
$4 + 4 = 8$	
$5 \times 2 = 10$	
$6 < 3 + 16$	

There are 18 basic arrangements using the
equality signs. The 'less than' sign then gives
a total of 54 solutions.

10. (a) Weigh two against two. If they are equal,
the odd one is the lightest. If not, weigh
the lighter pair against each other.

(b) Split into three sets of three. Weigh any
two sets against each other to identify the
light set. Any two from that set weighed
against each other will locate the light one.

(c) Any three are weighed against any three
others. If equal, weigh the pair against each
other. If not, proceed as in the last problem.

(d) Weigh six against six. Then weigh three
against three. Finish by weighing any one
against any other.

Index